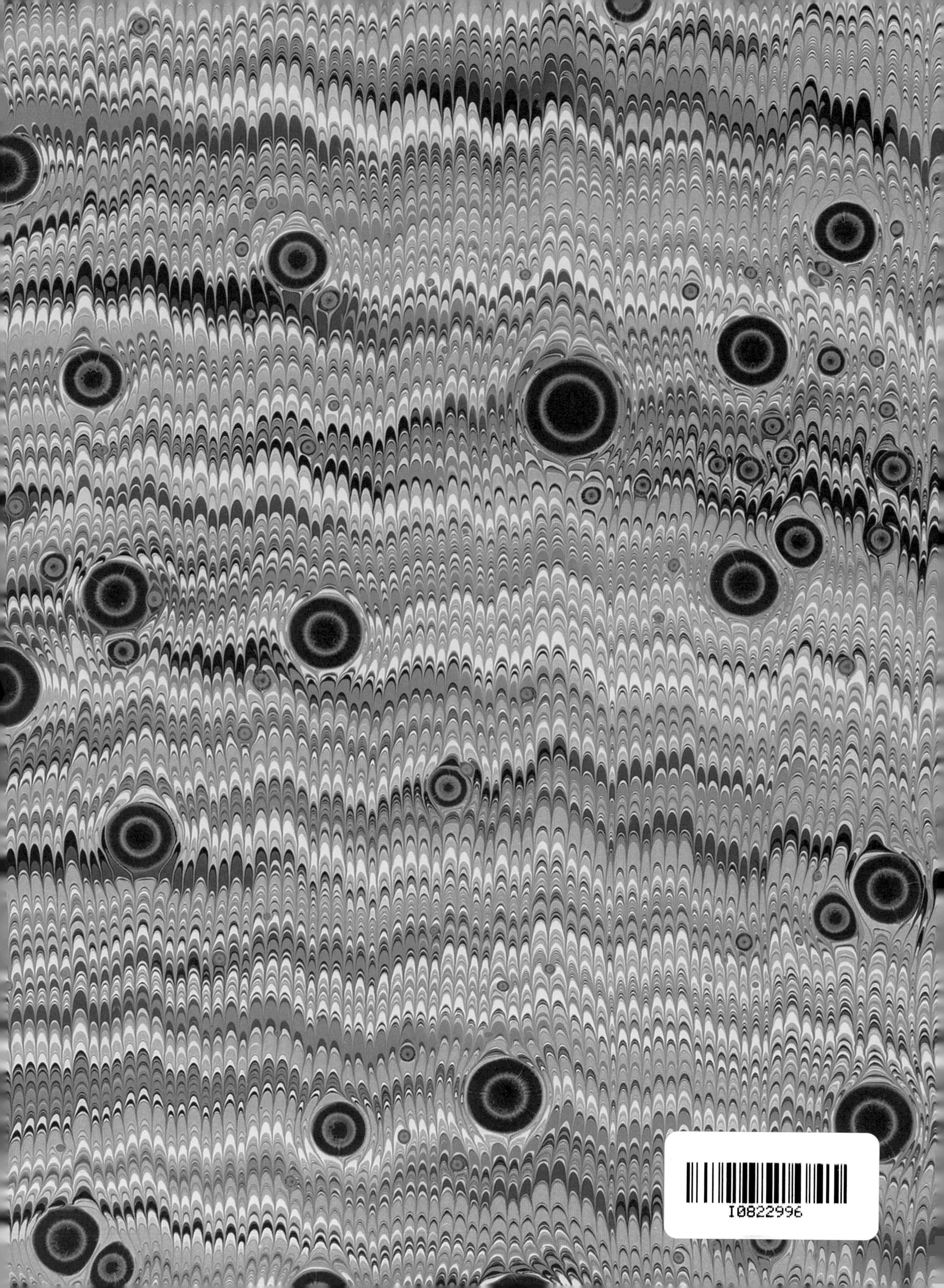
I0822996

CONJURING THE VOID

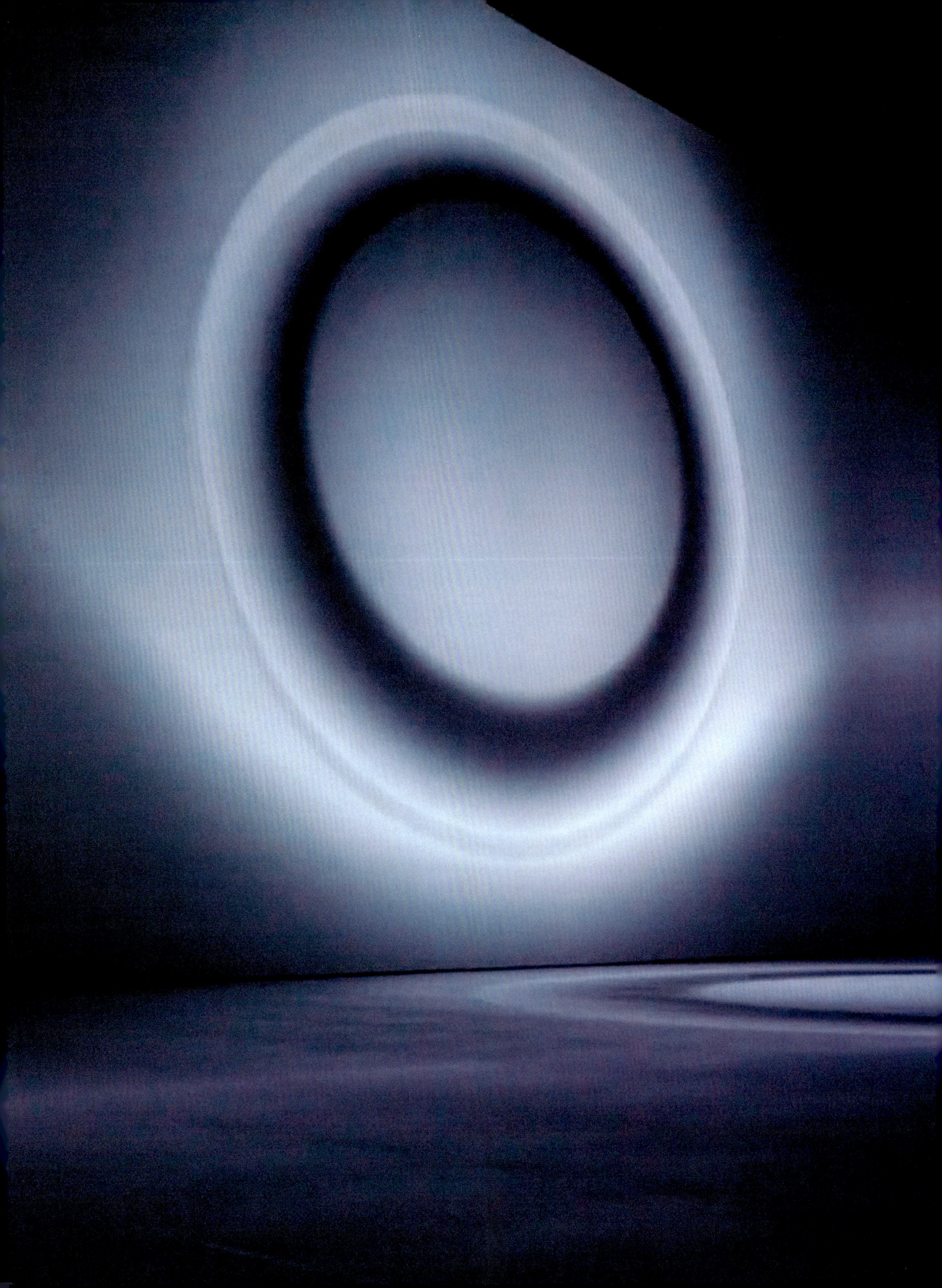

CONJURING THE VOID

THE ART OF BLACK HOLES

Lynn Gamwell

Foreword by Neil deGrasse Tyson

The MIT Press
Cambridge, Massachusetts
London, England

For Shep Doeleman, founding director of the Event Horizon Telescope, and
the more than two hundred scientists in twenty countries
who made the first image of a black hole.

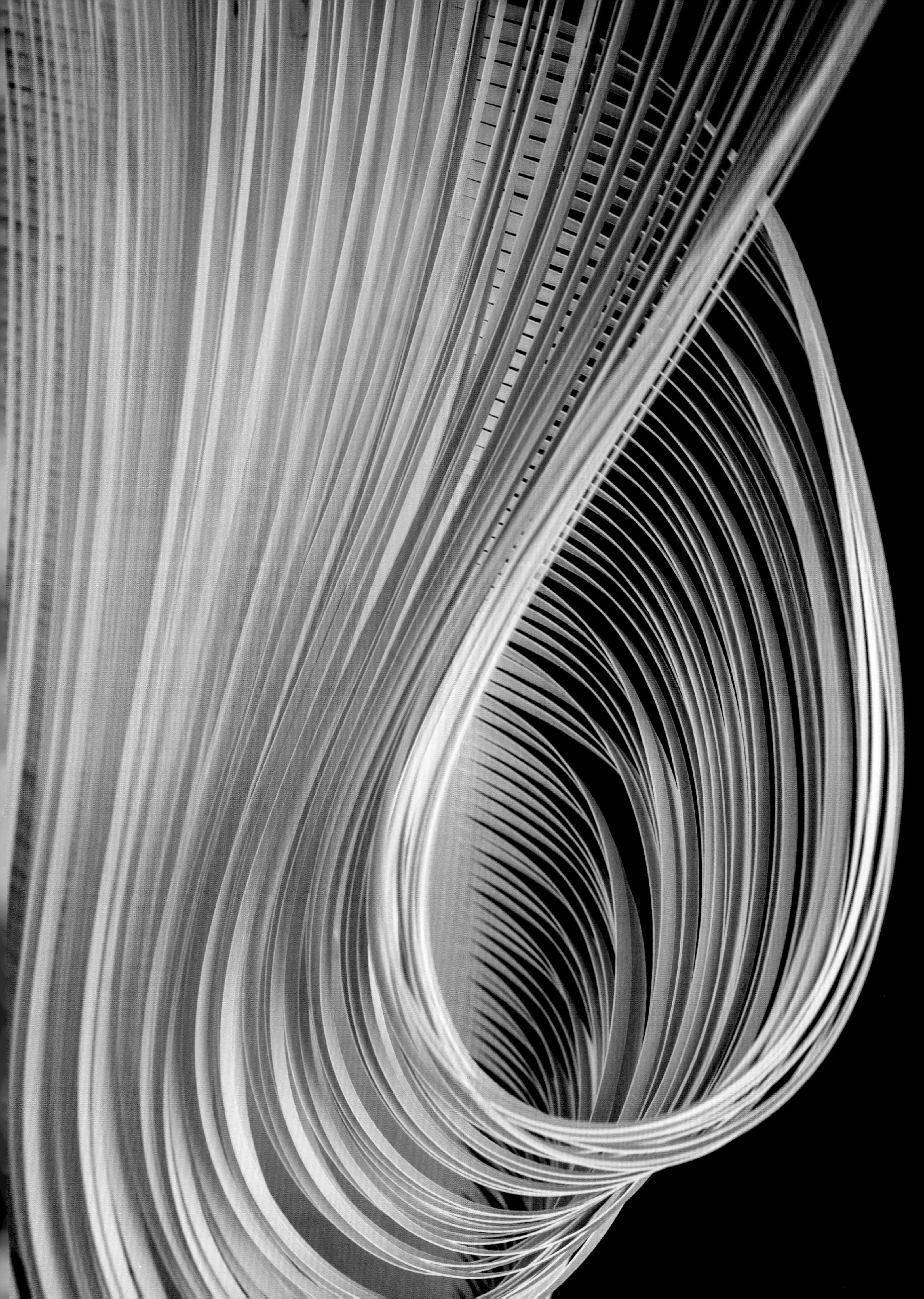

CONTENTS

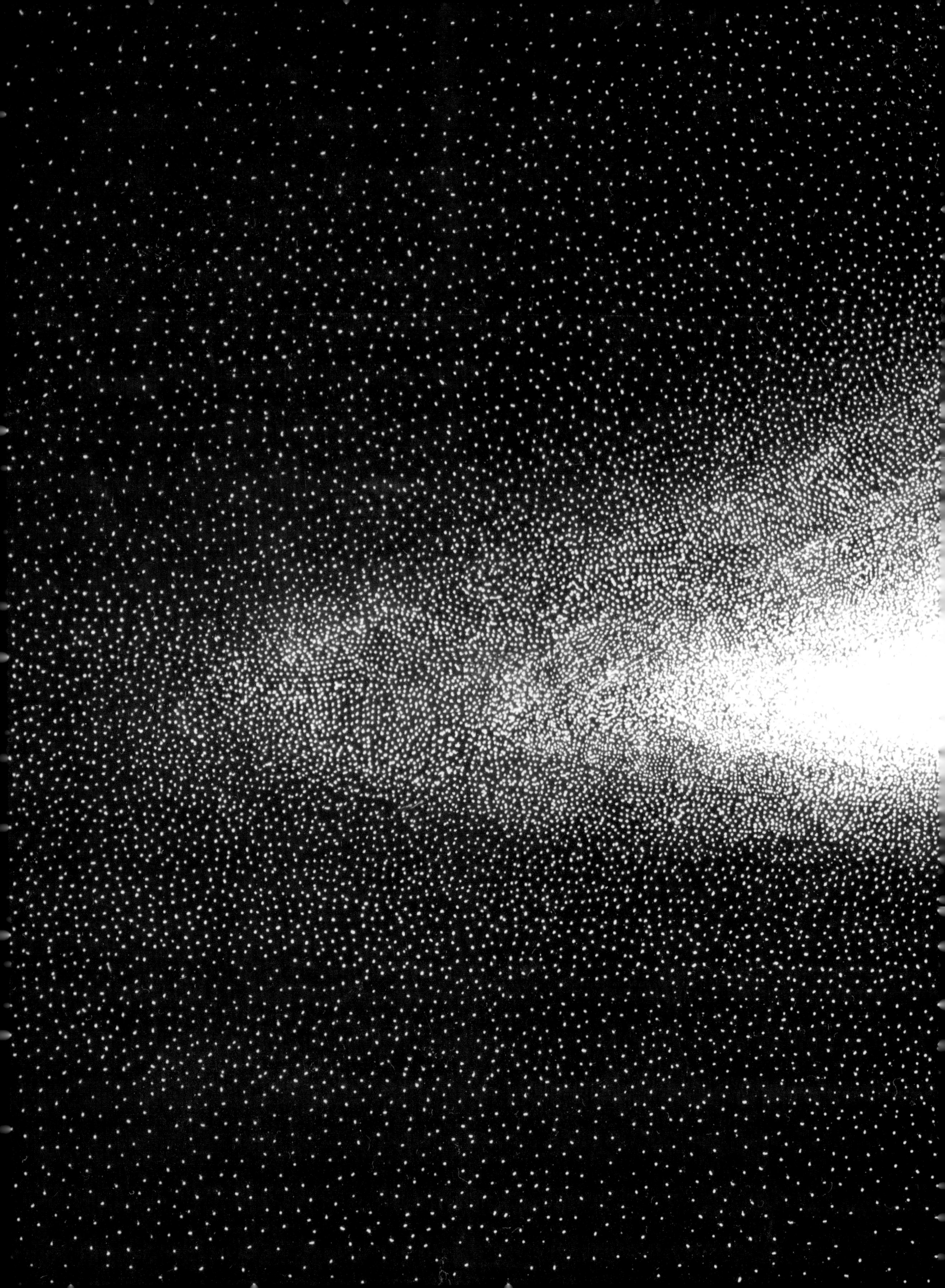

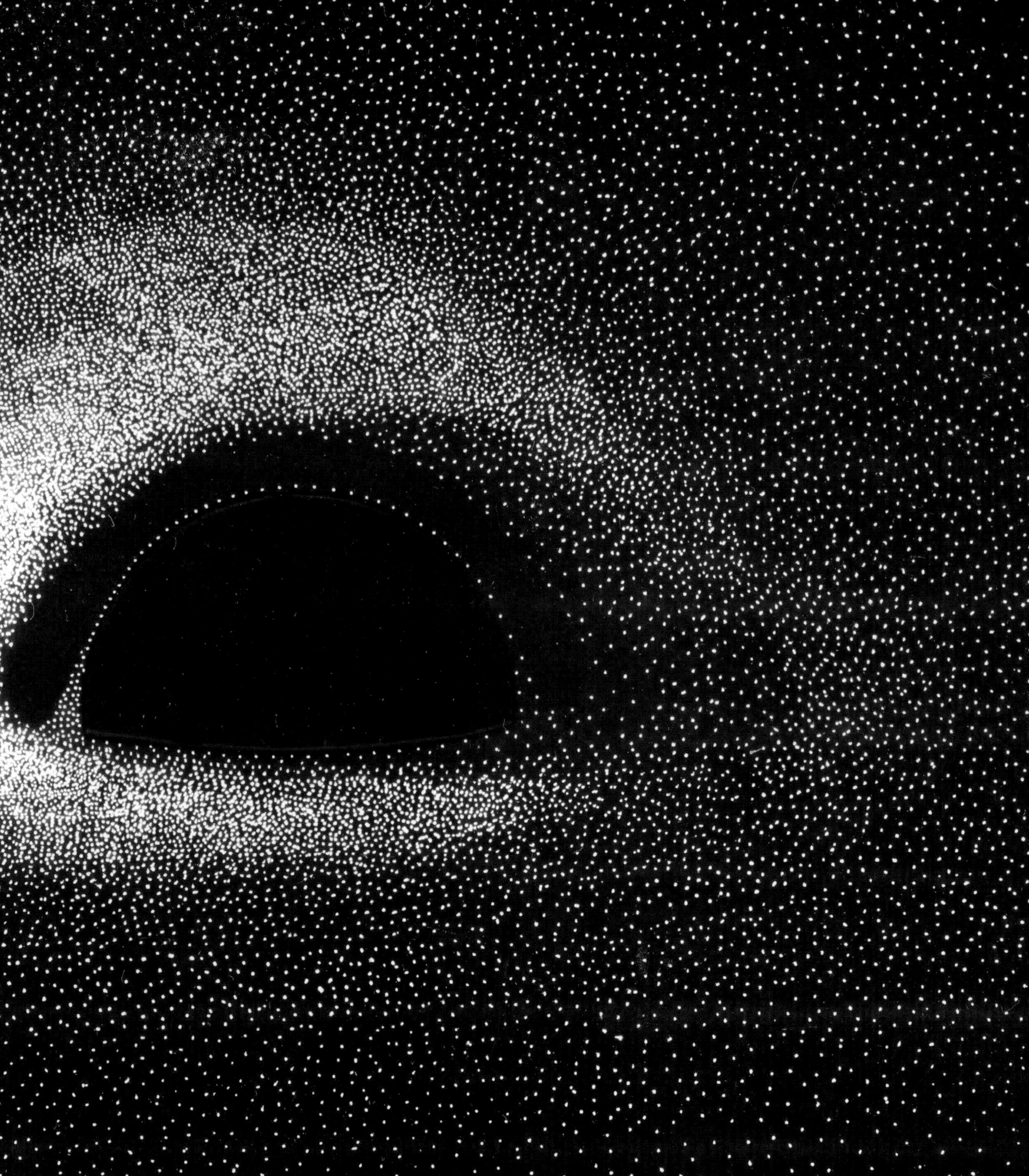

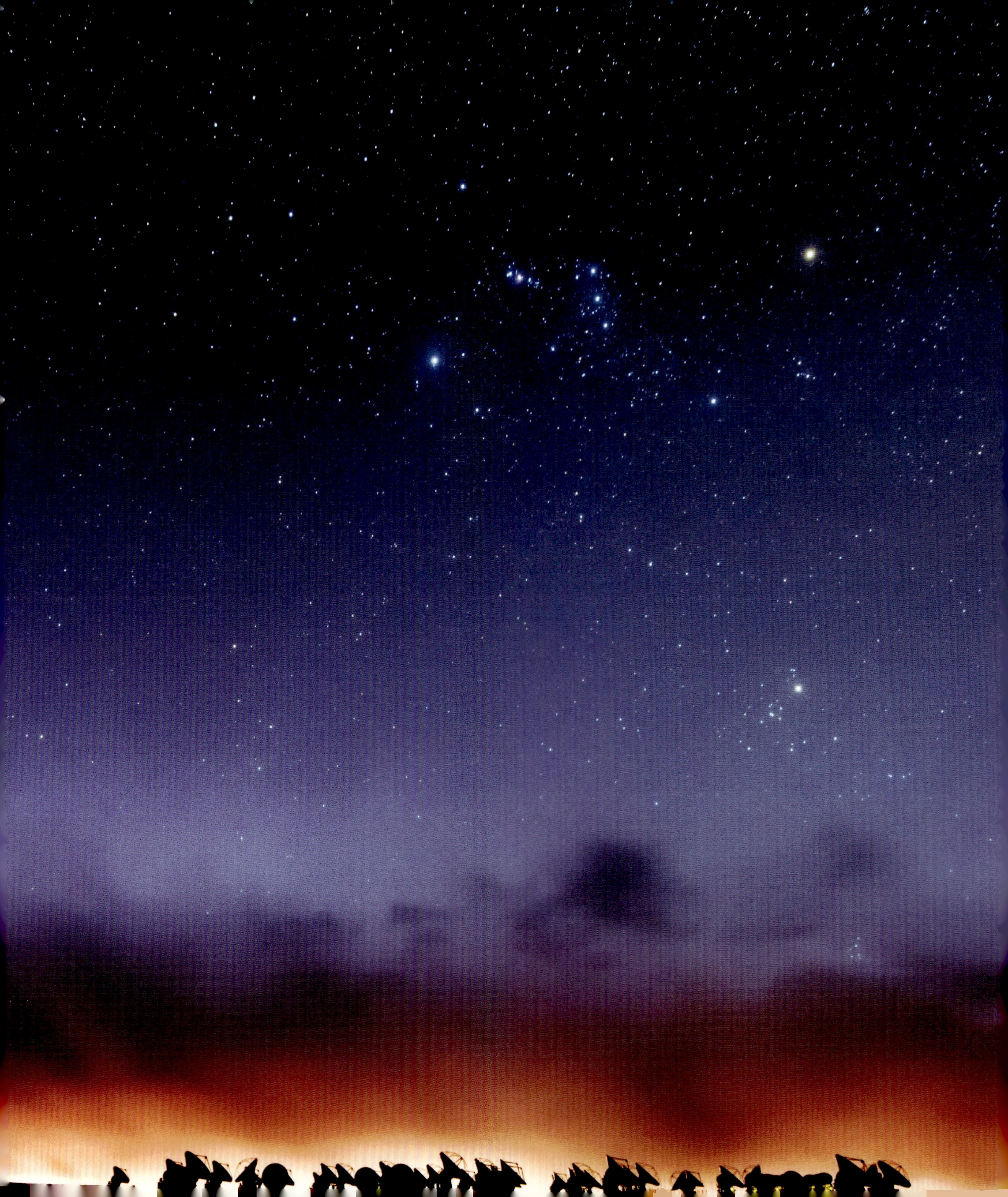

FOREWORD

Neil deGrasse Tyson

The role of the artist as the scientist's companion, and that of a scientist as the artist's muse, has evolved over the centuries, with the biggest shift occurring after photography was invented in the nineteenth century. For the first time, scientists no longer needed artists to capture objective reality. And gone were the risks of perception errors when portraying actual objects as art. Instead of thwarting artistic creativity, photography would, of course, free artists to create their own impressions of things—on purpose, and not as a shortcoming of their ability to capture reality.

The artist still offered value to the scientist. Not as before, but as a conduit to all that fell beyond the reach of the scientist's capacity to measure. Planetariums, until recent decades, employed artists to convey in space shows all that was yet to be photographed. Before we ever landed on the moon, what might those terrains have looked like if you were there? What would Jupiter look like when viewed from Europa, one of its icy moons? How might Saturn appear when seen through the thick atmosphere of its largest moon, Titan? Or if you were standing on Mars, what might a solar eclipse look like with either of its two tiny moons, Phobos or Deimos? Informed by science, these visualizations helped to propel our imaginations forward. In the presence of this visionary art, we were compelled to feel the vistas rather than just describe them from their physical properties.

Alas, images from space probes and planetary landers have further diminished the role of artists. And by way of modern detectors, we can "see" light that would otherwise be invisible to the unaided eye, coming to us from every other band of the electromagnetic spectrum—radio waves, microwaves, infrared, ultraviolet, X-rays, and gamma rays. What can the artist now do for science? Not to worry. There's plenty left in the cosmos to

PREVIOUS PAGES

0.1 Jean-Pierre Luminet, *Black Hole* (detail), 1979. Ink on paper, reversed photographically, in "Image of a Spherical Black Hole with Thin Accretion Disk," *Astronomy and Astrophysics* 75, nos. 1–2 (1979): 231, fig. 11.

OPPOSITE

0.2 The Atacama Large Millimeter Array (ALMA), Chile. ESO/B. Tafreshi.

ALMA consists of sixty-six radio telescopes and is the largest astronomical telescope project in the world. ALMA is situated more than three miles (five kilometers) above sea level in the Atacama Desert in Cerro Chajnantor, Chile, where dry air and thin atmosphere allow a spectacular view. ALMA was part of the Event Horizon Telescope, which made the first image of a black hole (M87*), released in April 2019 (see figure 3.138).

be visualized. How about the Big Bang itself? Or the multiverse? Or how about dark matter, dark energy, and black holes, none of which emit photographable light, yet all greatly influence their environments, with consequences to the evolution of the universe itself.

In particular, the maw of a black hole is legendary. So are the invisible distortions of spacetime that surround it. Both are in desperate need of scientifically informed art to represent them. The inevitable accretion disk that spirals down the black hole while flaying nearby stars is also something that we've never witnessed close up. Nor can we see what's causing the jets of material that spew forth from them. Our telescopic resolution is insufficient to see these details. All we measure is the blended light of the spiraling material, just before it descends out of view. The basic laws of thermodynamics, when combined with the orbital mechanics of gas flows, tell us what should be happening there, as our artists provide visual conduits to this front-row seat in the universe.

Meanwhile, as our computer models and attendant computer graphics of cosmic phenomena get better and better, the scientist becomes the artist. But the artist, still inspired by the universe, continues undaunted. As with the advent of photography, artists are now free—not to illustrate black holes and spacetime, but to react to them. This gift to us all is not art imitating life, but art extending life into new dimensions on the cosmic continuum of creativity.

In compiling this collection of images, Lynn Gamwell combed the world, if not the cosmos itself, to display what extraordinary efforts artists and scientists alike have invested in portraying black holes. Enjoy the art. And along the way, learn some astrophysics, too.

Neil deGrasse Tyson

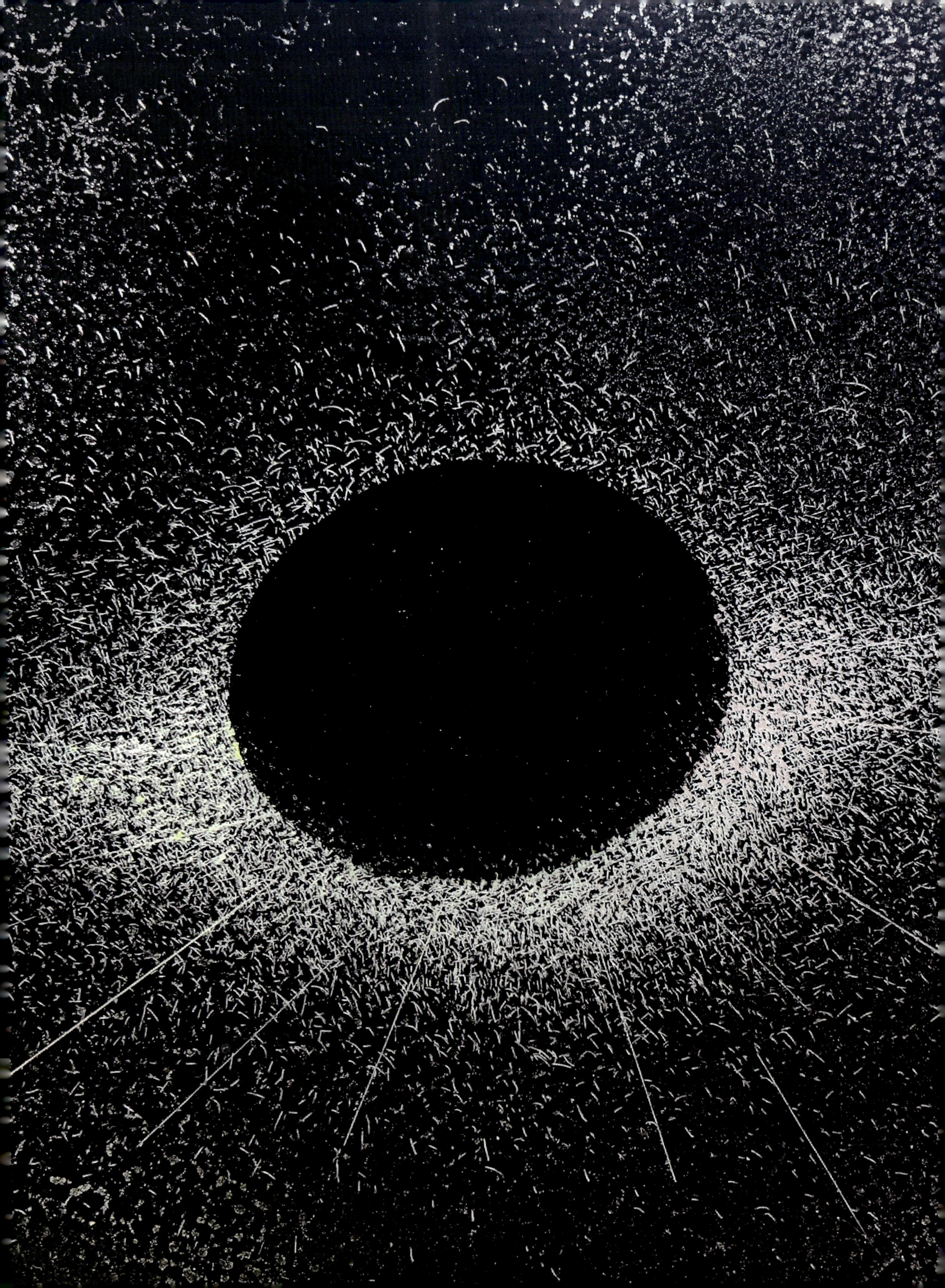

INTRODUCTION

A black hole is a region of spacetime where gravity is so strong that nothing, not even light, can escape. Any mass that crosses the black hole's event horizon—the point of no return—disappears into a void. But, paradoxically, a black hole is the opposite of a void because any mass inside the black hole is pulled by gravity to its center, where mass is compressed to infinite density and becomes a singularity.

Soon after Albert Einstein published the general theory of relativity in 1915, the German physicist Karl Schwarzschild declared that the mathematics of Einstein's theory predicts what today we call a black hole. Most scientists, including Einstein, thought black holes and singularities were mathematical oddities that didn't exist in nature. Indeed, Einstein thought a singularity was evidence that his theory of general relativity was incomplete because he believed a complete theory would be singularity-free. Then, in the 1960s, astronomers discovered powerful X-rays in the constellation Cygnus. When they looked at the source with a telescope that observed in the visible range, they saw nothing. What could it be? In the early 1970s astronomers confirmed that the invisible source of X-rays is a black hole, prompting decades of research. Today, scientists know that black holes are woven into the fabric of spacetime. They are found in the early universe soon after the Big Bang, after stars collapse in a supernova, and at the center of all large galaxies. Indeed, 13.8 billion years ago the universe itself exploded into existence from a point of infinite density.

Inescapable and mysterious, black holes have kindled public curiosity and fired the imaginations of artists. From Beijing to New York, artists have created works that embody concepts associated with black holes—gripping terms like *event horizon*, *singularity*, and *wormhole* that are drawn from the

0.3 Lucas J. Rougeux (American, born 1995), Untitled (black hole; detail), 2021. Spray paint and acrylic on mirror, 8 × 10 in. (20.3 × 25.4 cm). Courtesy of the artist.

scientist's own vocabulary. You can count on one hand the number of artworks about neutron stars. By contrast, there are countless works of art about black holes in large part because Eastern philosophies and Western modern art share themes that resonate with the science of black holes. Throughout history people have conceived of primordial energy, nothingness, and silence in their attempt to make sense of the world, and in modern times the black hole's paradoxical relation of nothing and everything has served as a cultural symbol for the enigmas of life.

This book traces our growing scientific knowledge of black holes and related phenomena while freely mixing in artworks inspired by the science described. Chapter 1 recounts how scientists predicted and eventually discovered black holes. Chapter 2 gives the ancient and modern background for imaging black holes. And Chapter 3 describes how artists and scientists have created images of these invisible objects during the last half-century.

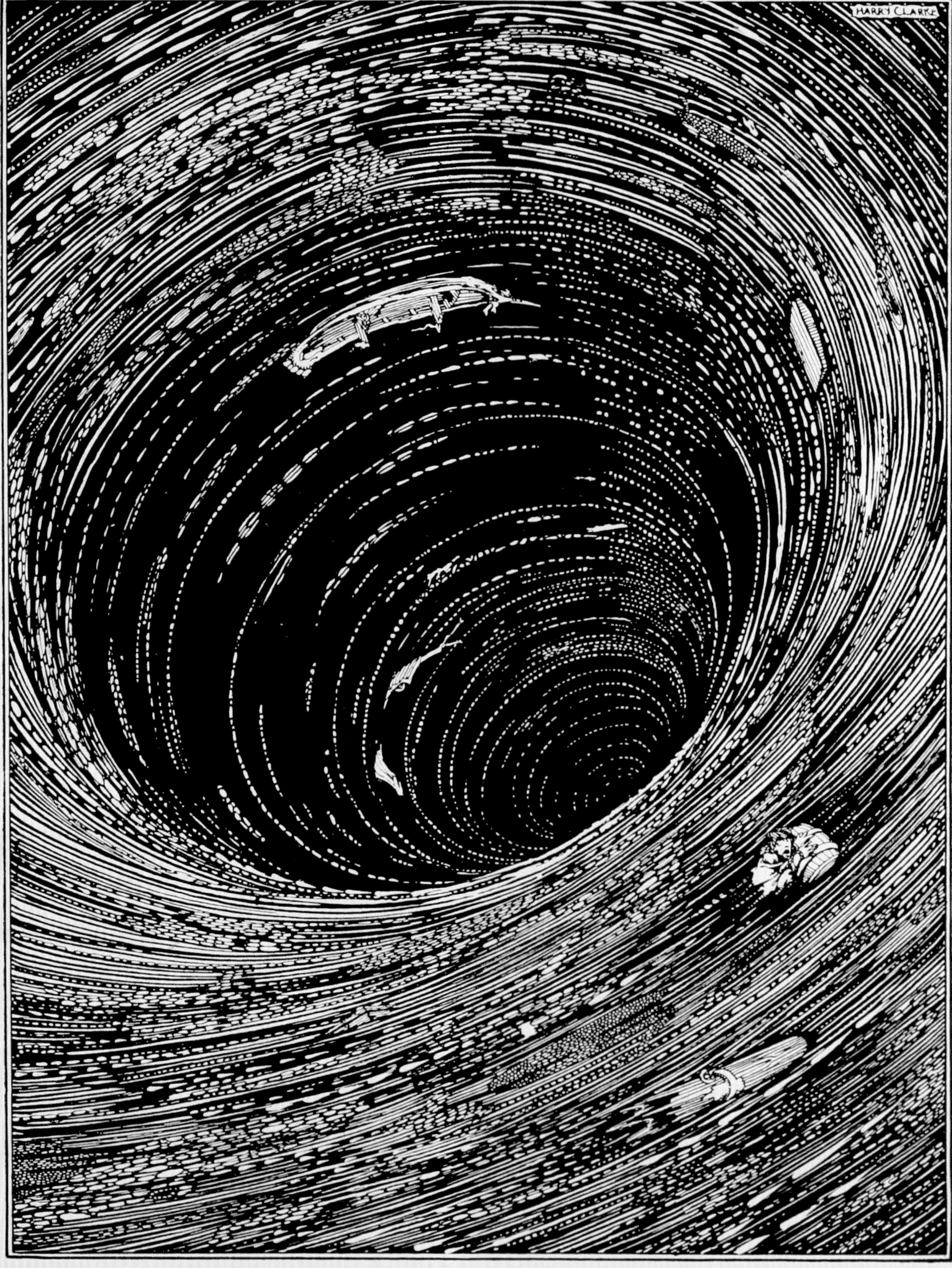
HARRY CLARKE

1

BLACK HOLE BASICS

What one imagines a black hole to be depends on one's understanding of light and gravity. In the seventeenth and eighteenth centuries, classical mechanics included Isaac Newton's definitions of light and gravity. In the early twentieth century, Albert Einstein created a revolution when he gave new definitions of light in quantum mechanics and gravity in his general theory of relativity.

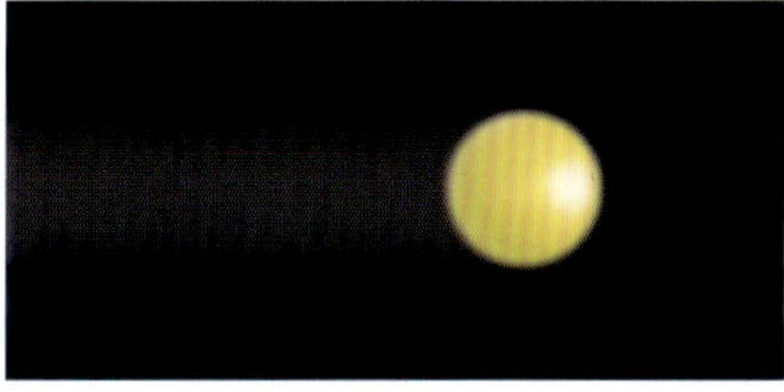

1.1 Light as a particle.

LIGHT AND GRAVITY IN CLASSICAL MECHANICS

Newton declared in *Opticks* (1704) that light was made up of particles (or "corpuscles") traveling in a straight line (figure 1.1). In *Philosophiæ Naturalis Principia Mathematica* (1687), he imagined the universe as a vast, immobile volume of space in which every massive body is attracted to every other mass by the force of gravity. Within Newton's framework, black holes are possible: If the mass of an object becomes very dense and, hence, its gravity very strong, then light cannot escape from it. The eighteenth-century British philosopher John Michell called these invisible celestial objects "dark stars" (figure 1.3)—what we now call black holes.[1] In the English language of the era, "dark" had the connotation of "concealed with a sinister purpose" as Shakespeare wrote in *King Lear*: "We will express our darker purposes.... Know that we have divided in three our kingdom" (*King Lear* 1.37). Although dark stars are invisible, Michell said they would be detectable by the gravitational effect they have on nearby visible objects. The eighteenth-century French scientist Pierre-Simon Laplace mused: "It is therefore possible that the largest luminous bodies in the universe, by the very fact that they are largest, are invisible."[2]

OPPOSITE
1.2 Edgar Allan Poe, "A Descent into the Maelström" (1841) in *Tales of Mystery and Imagination*, ill. Harry Clarke (London: Harrap, 1919), 96.

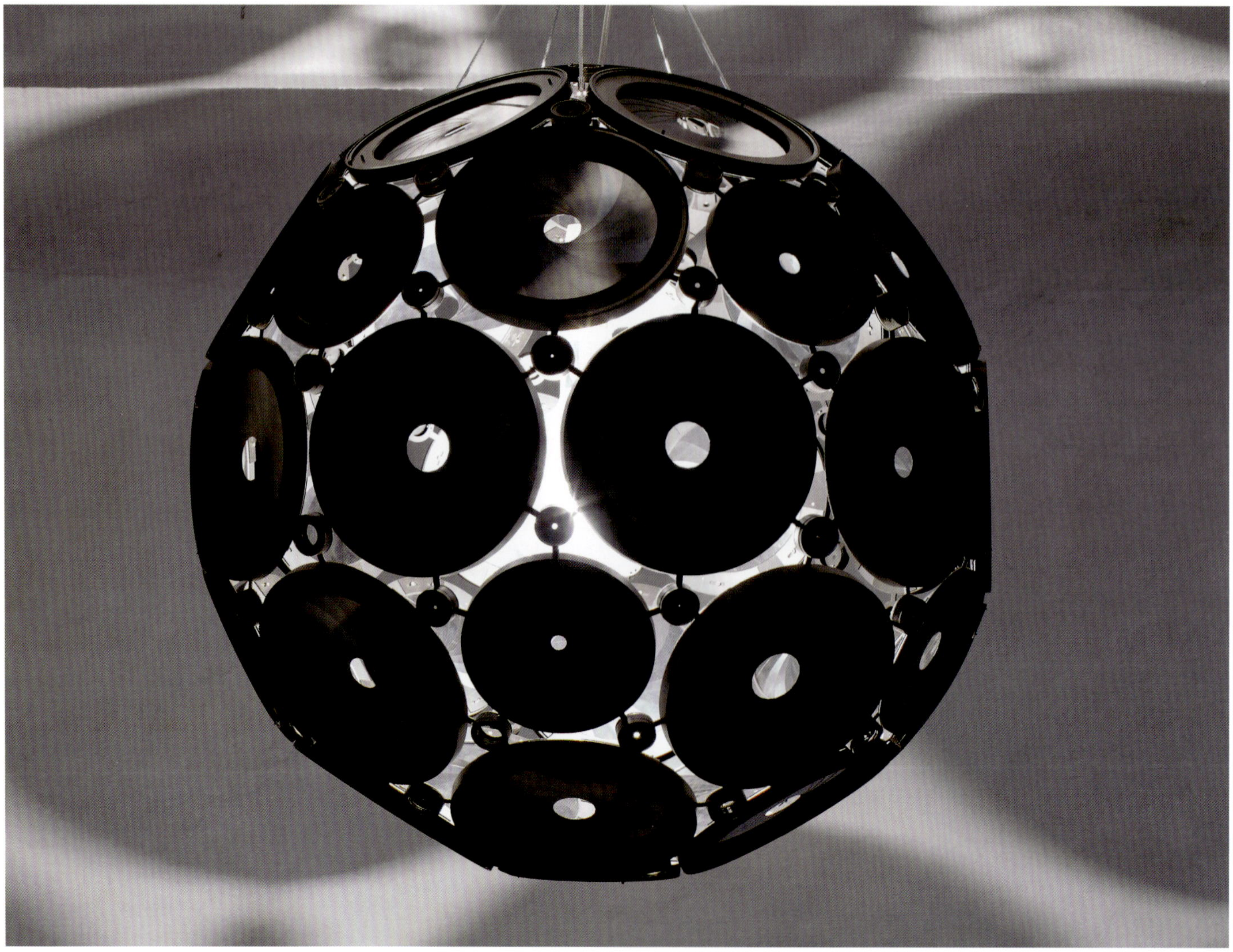

ABOVE

1.3 Olafur Eliasson (Danish, born Iceland 1967), *Dark star*, 2006. Stainless steel, aluminum, lightbulb, and iris diaphragms, 35½ in. (90 cm) diameter. Installation view at neugerriemschneider, Berlin, 2006.

Eliasson used iris diaphragms to make this sculpture that allows viewers to adjust the light it emits. Alternatively, a viewer can turn off the lightbulb, making the work a sphere that produces no light—a *Dark star*, as the artist titled his sculpture.

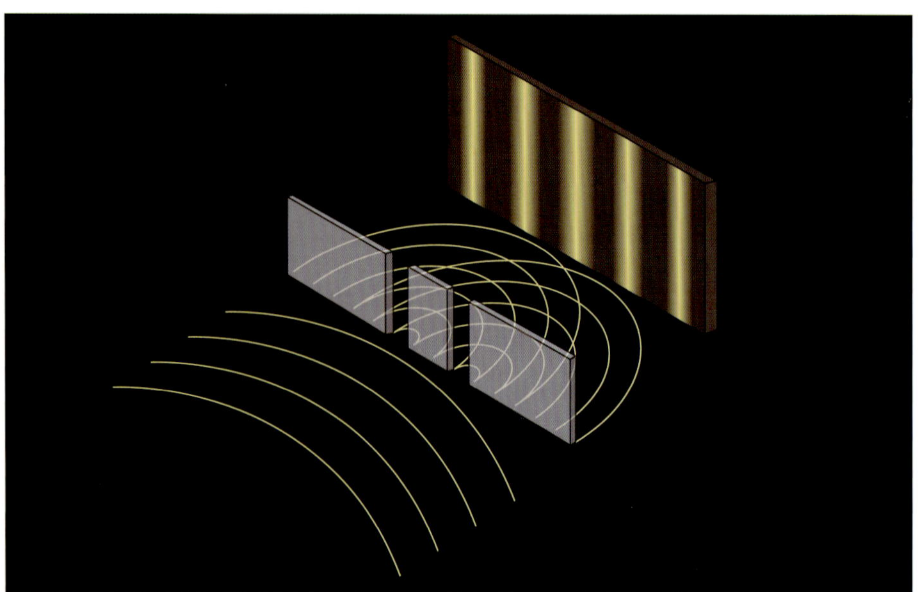

TOP RIGHT

1.4 Double-slit experiment.

When a wave passes through two slits, new waves are formed; sometimes the waves reinforce each other, and sometimes they cancel each other out, resulting in an interference pattern. Thomas Young passed light through two slits, and it formed an inference pattern, proving that light is a wave.

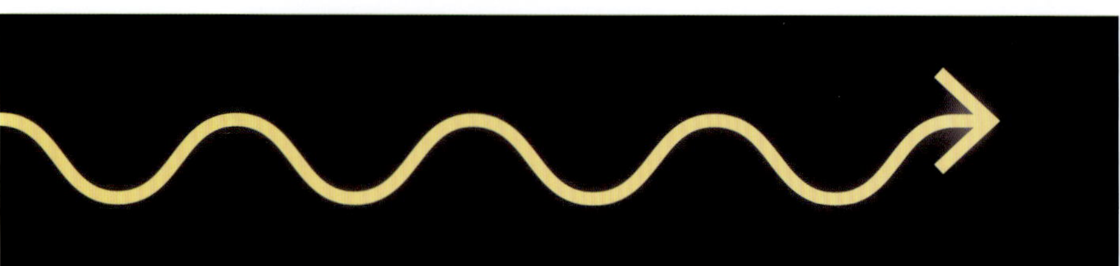

BOTTOM RIGHT

1.5 Light as a wave.

1.6 Edgar Allan Poe, "A Descent into the Maelström" (detail), 1841.

Then, around 1800, the English natural philosopher Thomas Young did a double-slit experiment that proved light consisted of waves (figures 1.4 and 1.5). Since waves of energy have no mass, they are not affected by gravity. Consequently, nineteenth-century scientists forgot about Michell's dark stars—though the writer Edgar Allan Poe did not. He declared: "We know that there exist *non-luminous suns*—that is to say, suns whose existence we determine through the movements of others."[3] Poe wrote the short story "A Descent into the Maelström" (1841) about a whirlpool that, like a dark star, is a force of nature so powerful that nothing can escape from it. Three brothers go on a fishing trip off the coast of Norway, where their boat gets caught up in a dark, dangerous vortex. Two terrified brothers drown as their ship is sucked into the abyss (figures 1.2 and 1.7), while the third saved himself by holding onto a barrel (figure 1.6).

1.7 The Moskstraumen, engraving in Vincenzo Coronelli, *Atlante Veneto* (Venice, 1693), vol. 1, fig. 41. David Rumsey Map Collection, David Rumsey Map Center, Stanford Libraries.

Edgar Allan Poe was inspired by the Moskstraumen, a natural whirlpool in the ocean between the Lofoten Islands and the Norwegian mainland. Long known to sailors, the whirlpool is created by tides, and thus Poe set his story "A Descent into the Maelström" at the full moon, when the tide is strongest.

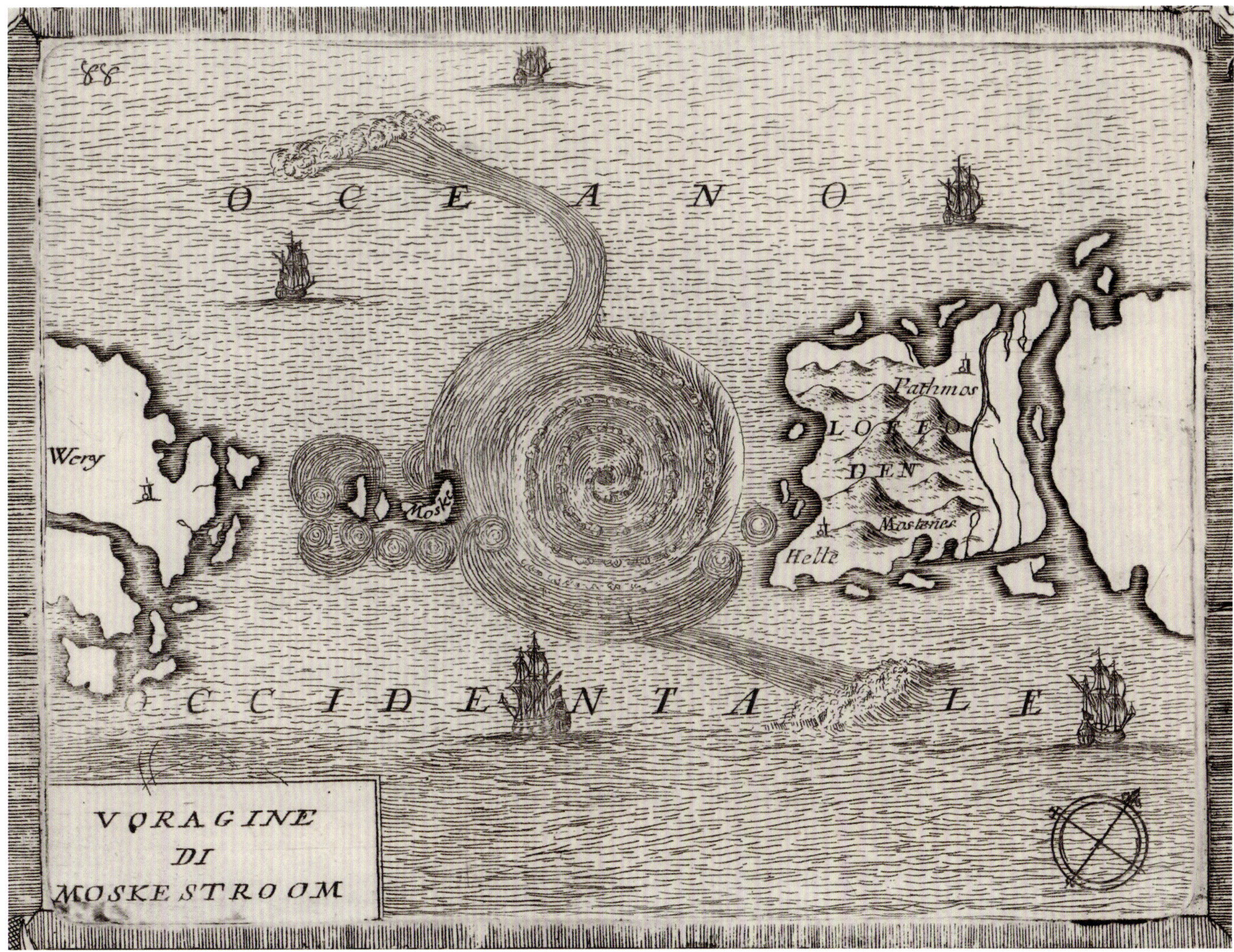

LIGHT IN QUANTUM MECHANICS AND GRAVITY IN THE GENERAL THEORY OF RELATIVITY

In the early twentieth century, Einstein redefined light and gravity, culminating in his photon theory of light (1905) and his concept of gravity as the warping of spacetime in his general theory of relativity (1915). This revolution in physics and astronomy began in the nineteenth century, when scientists thought that electricity and magnetism were two forces until the British scientist Michael Faraday determined they are one, which he named *electromagnetism*. Faraday believed that light was a form of electromagnetism, but, being untutored in mathematics, he lacked the skills to prove his hypothesis. His well-educated friend James Clerk Maxwell proved Faraday correct: light is an electromagnetic wave (figure 1.8).

In 1899 the German physicist Max Planck declared that electromagnetism comes in units of energy with a precise quantity (*quantum* in Latin); as the energy of light increases or decreases, so does the frequency of its electromagnetic wave, with the two maintaining the same proportion to each other (Planck's constant). As light comes in units of a fixed size, Einstein realized that a light wave—a form of energy with no mass—also acts as a particle (a *photon*) with a quantum of energy related to the frequency of its electromagnetic wave. In contrast to Newton, who thought light was made up of particles of matter, and Thomas Young, who described light as a wave of energy, Einstein declared that light acts both as a particle *and* a wave. Einstein's theory of light is a foundation of *quantum mechanics*, which describes events that happen at small (atomic and subatomic) scales.

In Newton's universe, space and time are independent of each other: space is a three-dimensional volume, and time is measured by points on a line. In Einstein's universe, time is always taken into account when giving location. Einstein joined space and time into one concept—*spacetime*—and described events using a four-dimensional geometry: three dimensions of space plus time.

In November 1915, during World War I, Einstein published the general theory of relativity.[4] The centerpiece of general relativity is a set of equations that describe gravity as the deformation of spacetime caused by massive bodies such as the sun. Where Newton understood gravity as a force between massive bodies, Einstein saw no gravitational force. Instead, he proposed that there is deformed spacetime that produces the effects we attribute to gravity. Light follows the path of deformation in spacetime (figure 1.9). A month after Einstein published his theory, Karl Schwarzschild (serving at the time in the German army at the Russian front) wrote a letter to Einstein announcing that he had found a solution to the equations for a spherical nonrotating body (today known as a black

BOTTOM LEFT

1.8 Light as an electromagnetic wave.

Light is a self-propagating composite of electric and magnetic waves that are *polarized*, meaning that the waves move in only one plane. Furthermore, these planes are perpendicular to each other. A "self-propagating composite" means that the moving electric wave produces a moving magnetic wave, which in turn produces a moving electric wave, and so on. In an 1861–1862 paper "On Physical Lines of Force," James Clerk Maxwell declared that this pulsing, self-propagating phenomenon happens at only one speed—the speed of light.

BOTTOM RIGHT

1.9 Spacetime curvature.

A light wave moves on a curved path through a two-dimensional plane in the gravitational field of spacetime that is distorted by the presence of massive bodies.

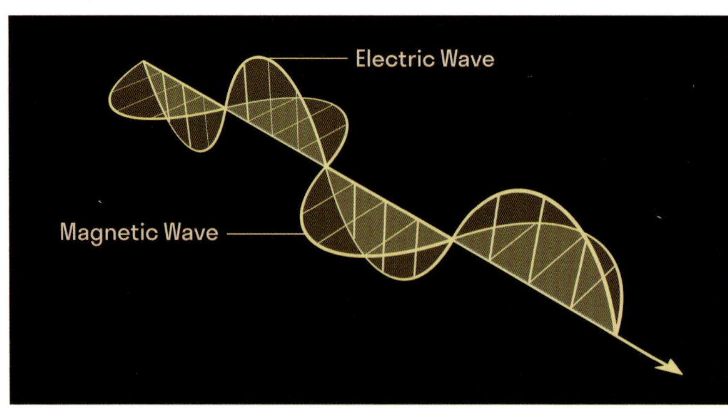

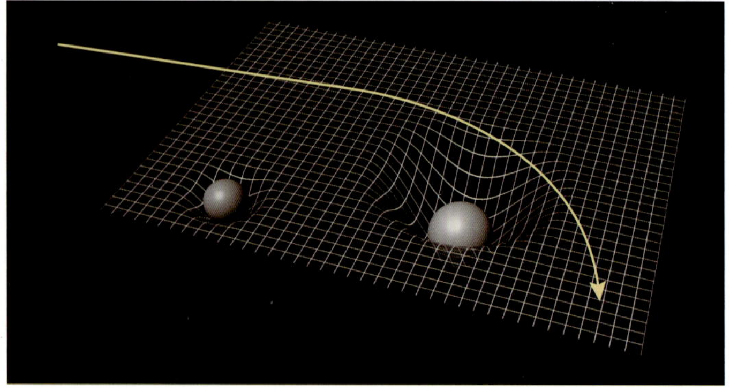

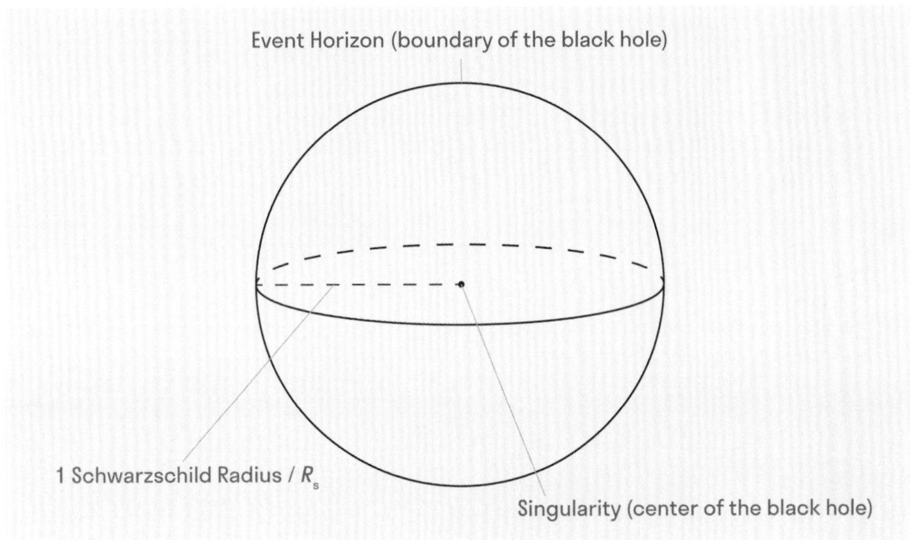

1.10 Schwarzschild black hole.

Karl Schwarzschild described a black hole that is spherical and stationary. Anything that crosses the boundary of the black hole (the event horizon) is pulled by gravity to its center (the singularity). The Schwarzschild radius is a unit of measurement—written R_S—that calculates the distance from the center of the black hole to its outer boundary (its "surface"). The Schwarzschild radius of an object is proportional to its mass—it is the size an object would have to be compressed to in order to become a black hole. The Schwarzschild radius of Earth is about 0.35 in. (9 mm). In other words, if the mass of Earth were compressed to the size of a pea, it would be a small black hole.

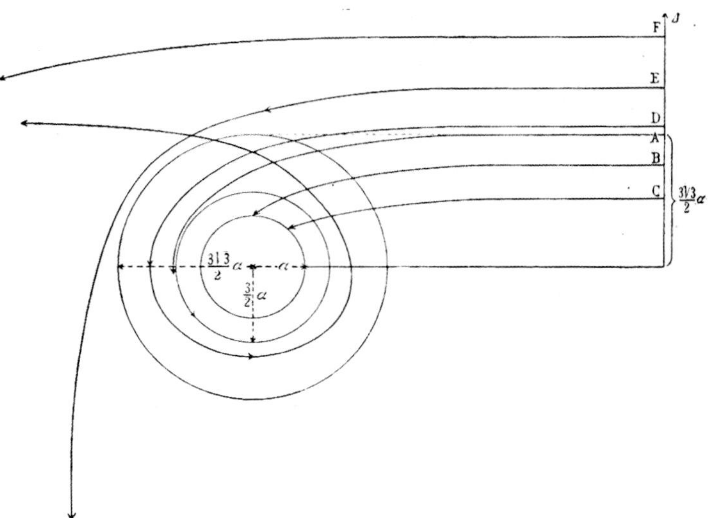

1.11 David Hilbert's calculation of the path of light near a black hole in Max von Laue, *Die Relativitätstheorie* (Braunschweig: Vieweg, 1921), 2:226, fig. 23.

In a 1916 lecture, Hilbert calculated the curved path of light near a black hole. A few years later the calculation was published in Laue's book.

hole). Schwarzschild, an astrophysicist by training, was director of the observatory at the University of Göttingen from 1901 until 1909, after which he was director of the Astrophysical Observatory in Potsdam. When war broke out in August 1914, he was exempt from military service because of his age (he was forty) and his position (he was in charge of Germany's largest refractor telescope), but he volunteered and, because he was educated in mathematics, was given the task of calculating missile trajectories. In his letter to Einstein, Schwarzschild declared that when the mass of the body exceeds a certain limit, the mass crosses the event horizon and produces a *singularity*—the point in spacetime where matter is infinitely dense (figure 1.10). Schwarzschild ended his letter on a personal note: "As you can see, the war has been kind to me by allowing me, despite heavy artillery fire, to walk in the land of your ideas."[5] While serving in the trenches in Russia, Schwarzschild, who was Jewish, contracted pemphigus, a rare (and often fatal) skin disease to which Ashkenazi Jews are especially susceptible. He was invalided home in March 1916 and died two months later.

That same year, the German mathematician David Hilbert, who had been Schwarzschild's colleague at the University of Göttingen, calculated the curved path of light near a black hole. Hilbert understood that a black hole is not an object *in* spacetime but instead a feature of spacetime itself (figures 1.11 and 1.12). Thus, a black hole is part of the structure of reality

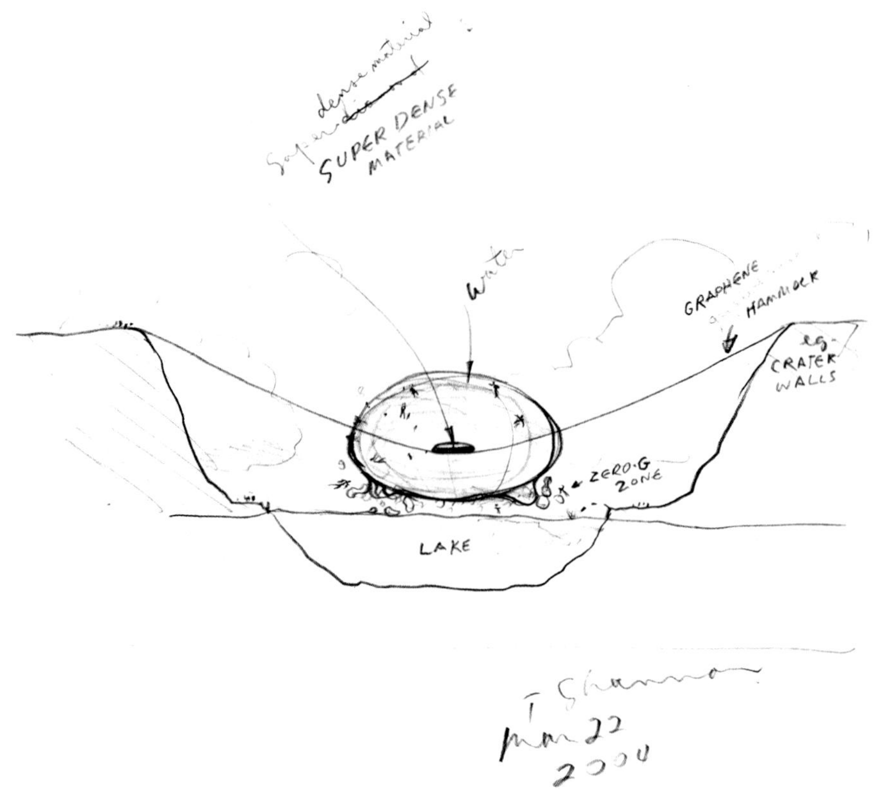

1.12 Tom Shannon (American, born 1947), Thought experiment, 2004. Pencil on paper. Courtesy of the artist.

Shannon is an artist who is interested in extreme gravity, so he imagined a crater with a lake at the bottom. Across the crater is strung a graphene hammock supporting a super-dense object such as a neutron star or a black hole, causing the lake to rise up and surround the super-dense object.

and only seems to be an alien object that distorts reality. In 1919 Einstein's general theory of relativity was confirmed by the shift of starlight during a solar eclipse. Since that time, general relativity has provided the framework for events that happen on a large (astronomical) scale. At both small (atomic) and large (astronomical) scales, nothing can travel faster than light. *Escape velocity* is the speed at which an object leaves the surface of another body to escape its gravity. If light can't escape from a black hole, nothing can.

In 1930 Subrahmanyan Chandrasekhar, while traveling from his native India to Cambridge, England, was thinking about the evolution of stars. He realized that a star's mass determines its final stage.[6] A small star that is very dense is known as a *white dwarf*. Chandrasekhar said that a white dwarf has a maximum mass of 1.4 solar masses (one *solar mass* equals the mass of the Sun); a white dwarf of more than 1.4 solar masses will implode.

A further advance in our understanding of astronomical objects was made by the American physicists J. Robert Oppenheimer and Hartland Snyder, who calculated the fate of a spherical, symmetrical "dust cloud" based on general relativity.[7] At the time, it was thought that the maximum mass for a white dwarf was about the same as for a neutron star, which is the result of stellar collapse. Today it's known that the mass of a neutron star is slightly larger than that of a white dwarf. During the star's collapse, the pressure of the implosion fuses protons into an atomic nucleus the size of a city: protons merge with electrons and become neutrons. What is left is a rotating sphere composed of solid neutrons—a *neutron star*—that has an escape velocity more than half the speed of light. Oppenheimer and Snyder showed that if a spherical, symmetrical dust cloud has around three solar masses, it will collapse into what is today known as a black hole.

The use of the term *black hole* to describe a collapsed celestial object came about in the 1950s when the American physicist David Finkelstein noticed that Schwarzschild had found a region of spacetime from which nothing can escape. Finkelstein described the boundary of a black hole (which he called a "Schwarzschild surface," today known as the *event horizon*) as "a perfect unidirectional membrane: causal influences can cross it, but only in one direction."[8] Hearing such theories, the physicist Robert Dicke remarked that a celestial object from which nothing can escape is like the Black Hole of Calcutta, the infamous eighteenth-century dungeon from which no prisoner escaped (figures 1.13 and 1.14). The term *black hole* was popularized by the physicist John Wheeler, Dicke's Princeton colleague.[9] The term entered the scientific and popular lexicon and strengthened cultural associations with other dark unknowns, such as the infinite nothingness of death (figure 1.15).

Associating black holes with nothingness seemed appropriate since scientists questioned whether they actually existed. Schwarzschild's spherical nonrotating body and Oppenheimer's spherical dust cloud are both perfectly

OPPOSITE

1.13 The Black Hole of Calcutta, in Louis Figuier, *Les merveilles de la science* (Paris: Jouvet, 1870), 353, fig. 236.

The Black Hole was a dungeon within Fort William, a British citadel in Calcutta that protected the East India Company trading corporation. This engraving illustrates an incident during the siege of Fort William in June 1756. The Indian army rounded up British soldiers and merchants and forced the terrified prisoners into the Black Hole, where many died of suffocation and heat stroke.

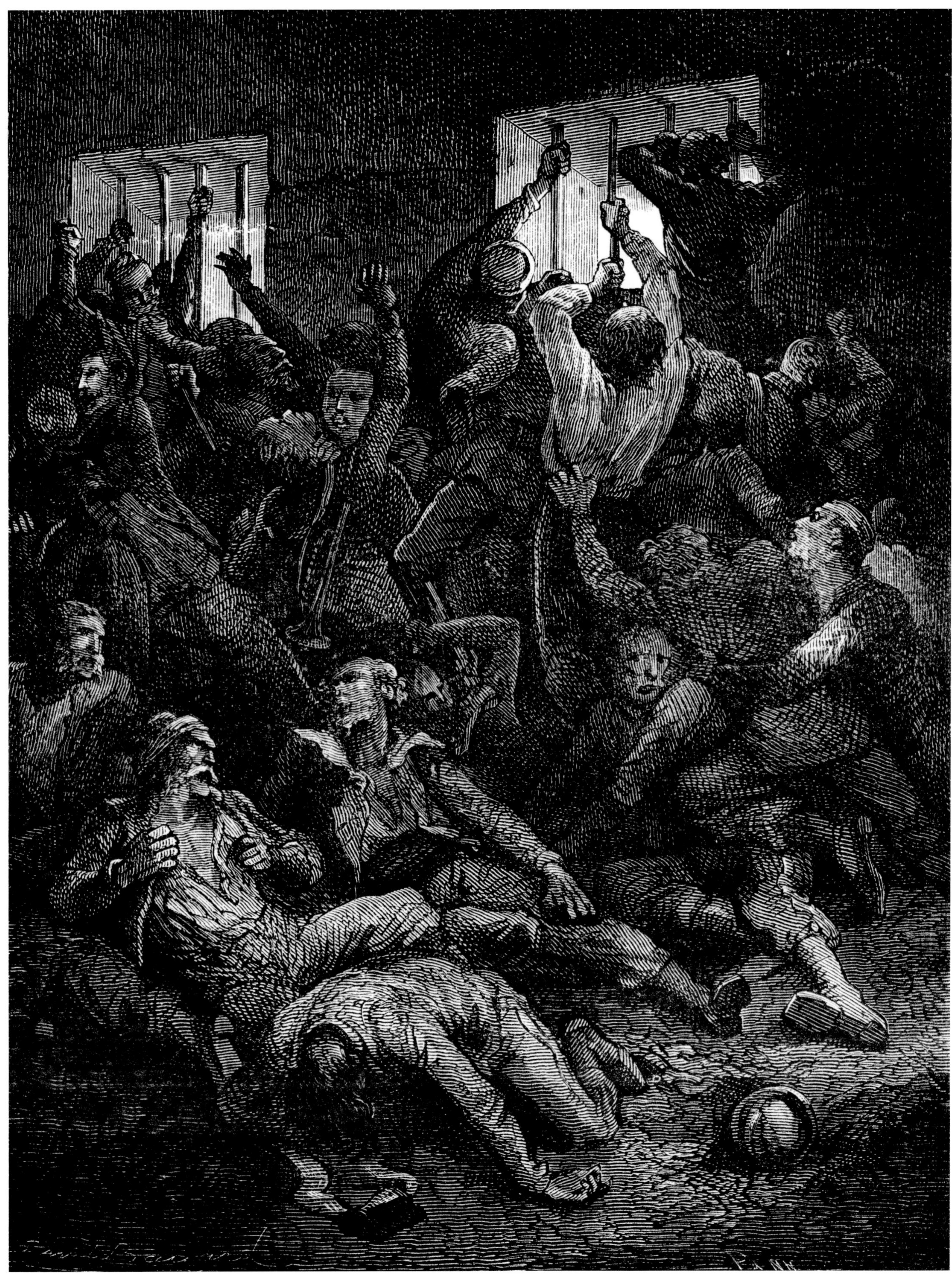

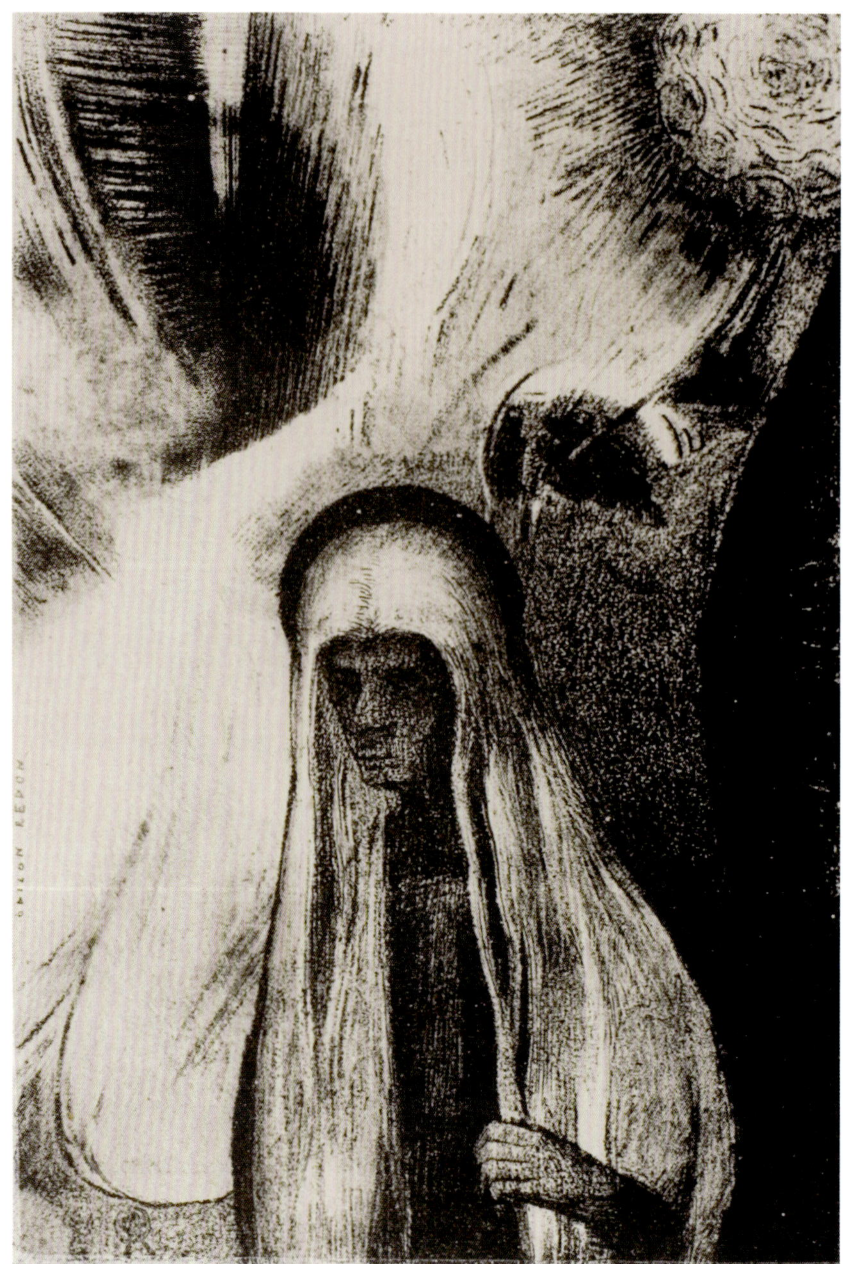

ABOVE

1.14 Tarje Eikanger Gullaksen (Norwegian, born 1973), *Reconstruction*, 2016. Tipp-ex on book, 8¼ × 5½ in. (21 × 13.5 cm). Courtesy of the artist.

In 1966 Noel Barber published *The Black Hole of Calcutta: A Reconstruction*. Gullaksen created this artwork by taking Barber's book and covering some words on the cover with white correction fluid.

RIGHT

1.15 Odilon Redon (French, 1840–1916), *The Old Woman*, 1896. Number 19 in a portfolio of 24 lithographs, image: 6½ × 4¼ in. (16.2 × 10.7 cm). The Art Institute of Chicago.

Redon created a portfolio of lithographs inspired by Gustave Flaubert's 1874 novel *La Tentation de Saint Antoine*, which concerns the torments of an early Christian mystic who lived as a hermit. In the episode illustrated here, Saint Anthony contemplates suicide by falling into an abyss when an old woman (shown here) encourages him to follow through on the act: "What are you afraid of? A wide, black hole! It is a void, perhaps?" Reflecting nineteenth-century secularism, Flaubert suggests that there is no life after death, only a void, which Redon symbolized with the black shapes in the background.

symmetrical. Therefore, these entities were regarded by most scientists as purely theoretical: one finds perfect spheres in geometry but not in nature. Einstein himself denied that a black hole could form in the natural world and treated Schwarzschild's singularity as a mathematical curiosity, concluding a 1939 paper with these words: "The essential result of this investigation is a clear understanding as to why the 'Schwarzschild singularities' do not exist in physical reality."[10]

During the period from the 1930s to the 1950s, most physicists agreed with Einstein that black holes did not exist. But the British mathematician Roger Penrose realized that the key to the formation of a black hole is what Finkelstein called the "Schwarzschild surface" (the event horizon). Penrose developed new mathematical tools to prove that, given a star's sufficiently large mass, general relativity makes the formation of a singularity inevitable. Once a star begins to collapse and an event horizon forms, nothing can stop the collapse from continuing until all matter reaches the singularity (figure 1.16). Penrose's proof is included in a short essay that many scientists consider the most important contribution to the theory of relativity since Einstein.[11]

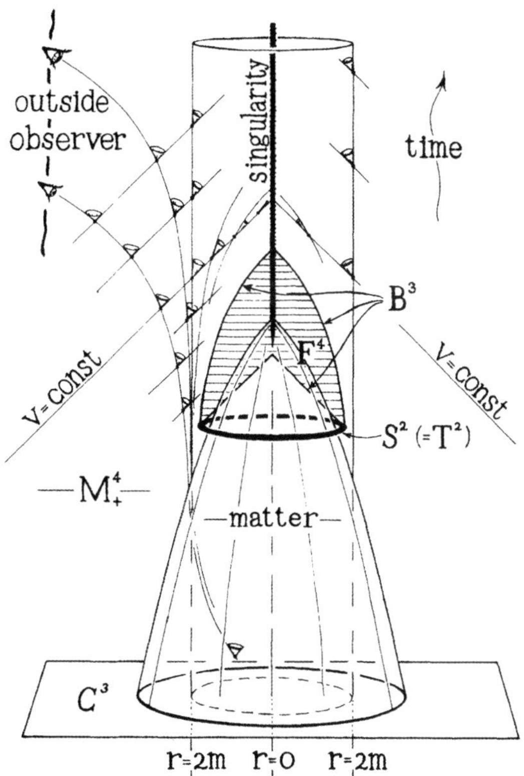

1.16 Roger Penrose, "Gravitational Collapse and Space-Time Singularities," *Physical Review Letters* 14, no. 3 (1965): 58, fig. 1.

Central to Penrose's proof is his diagram of what happens when a black hole is formed, as opposed to a visualization of what a black hole looks like. At the bottom is a collapsing spherical star (represented by a conical volume and labeled "matter"). As the star caves in, it becomes denser, reaching a point of no return: the event horizon (indicated by the thick circle labeled S^2). Over time, all the matter of the star (indicated by lines labeled B^3) continues to fall inward to the singularity. At left, the diagram shows an "outside observer" (represented by eyeballs) viewing starlight (little cones of light) at two points in time. The observer is able to detect the light until it goes inside the vertical lines that indicate the event horizon.

According to Penrose, the singularity is where mass has infinite density, time stops, and the known laws of physics break down. This doesn't make sense—it's literally nonsense—and this nonsense is at the core of what is truly startling about black holes (figures 1.18 and 1.19). Penrose showed that black holes are a direct consequence of the general theory of relativity, but they end in a singularity where general relativity no longer applies. Einstein himself thought his theory of general relativity must be incomplete because it implies a singularity.[12] Perhaps a singularity signals the need for a long-sought theory of quantum gravity—a theory encompassing quantum mechanics and general relativity, both of which are needed to describe what happens inside a black hole.

Penrose then joined with the British physicist Stephen Hawking, and they applied ideas about black holes to the whole universe. They discussed a model of the universe that implies "the existence of an initial (e.g., 'big bang' type) singularity"—in other words, a model in which the universe came into being from a point of infinite density, although they conceded that this model "is virtually unverifiable by observation."[13]

As revolutionary understandings of light and gravity evolved in the twentieth century, the Dutch artist M. C. Escher responded. He used linear perspective, invented in the Italian Renaissance to create rational space, to produce irrational worlds. Amsterdam's museum of modern art, the Stedelijk, organized an exhibition of lithographs by Escher on the occasion of a meeting of the International Congress of Mathematicians in 1954. Penrose saw the exhibit and became a fan. He was struck by a particular print, *Relativity* (figure 1.17), that depicts a scene in which gravity pulls in different directions.[14]

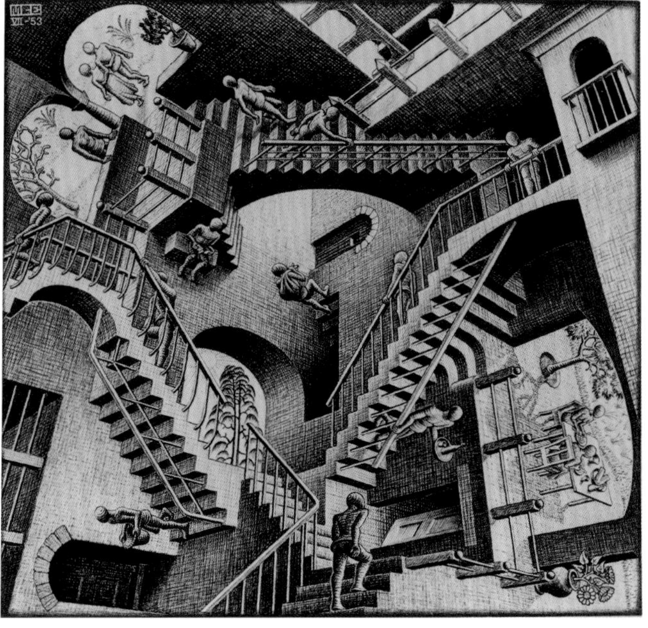

1.17 M. C. Escher (Dutch, 1898–1972), *Relativity*, 1953. Lithograph, 11 × 11½ in. (27.7 × 29.2 cm).

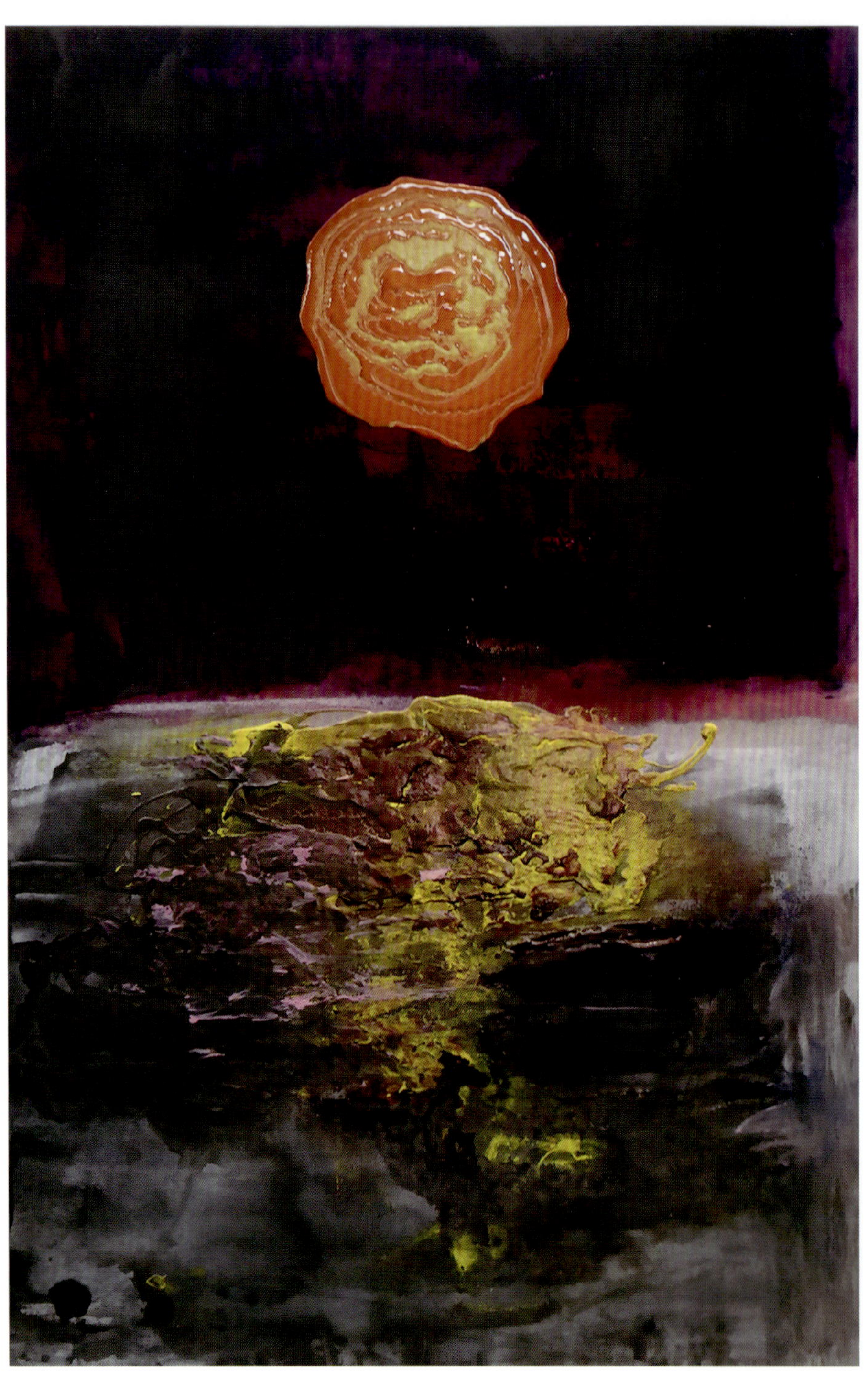

1.18 Darryl Hughto (American, born 1943), *Event Horizon*, 2016, from the series *Heavenly Body*. Acrylic on canvas, 56¾ × 36¾ in. (144.1 × 92 cm). Courtesy of the artist.

Hughto placed Earth in the foreground of this composition, below the night sky and a glowing black hole, perhaps radiating X-rays. The division is not only between terrestrial and celestial but also between near and far, known and unknown, comfortable and terrifying—between the natural world and a faraway place where the laws of physics break down.

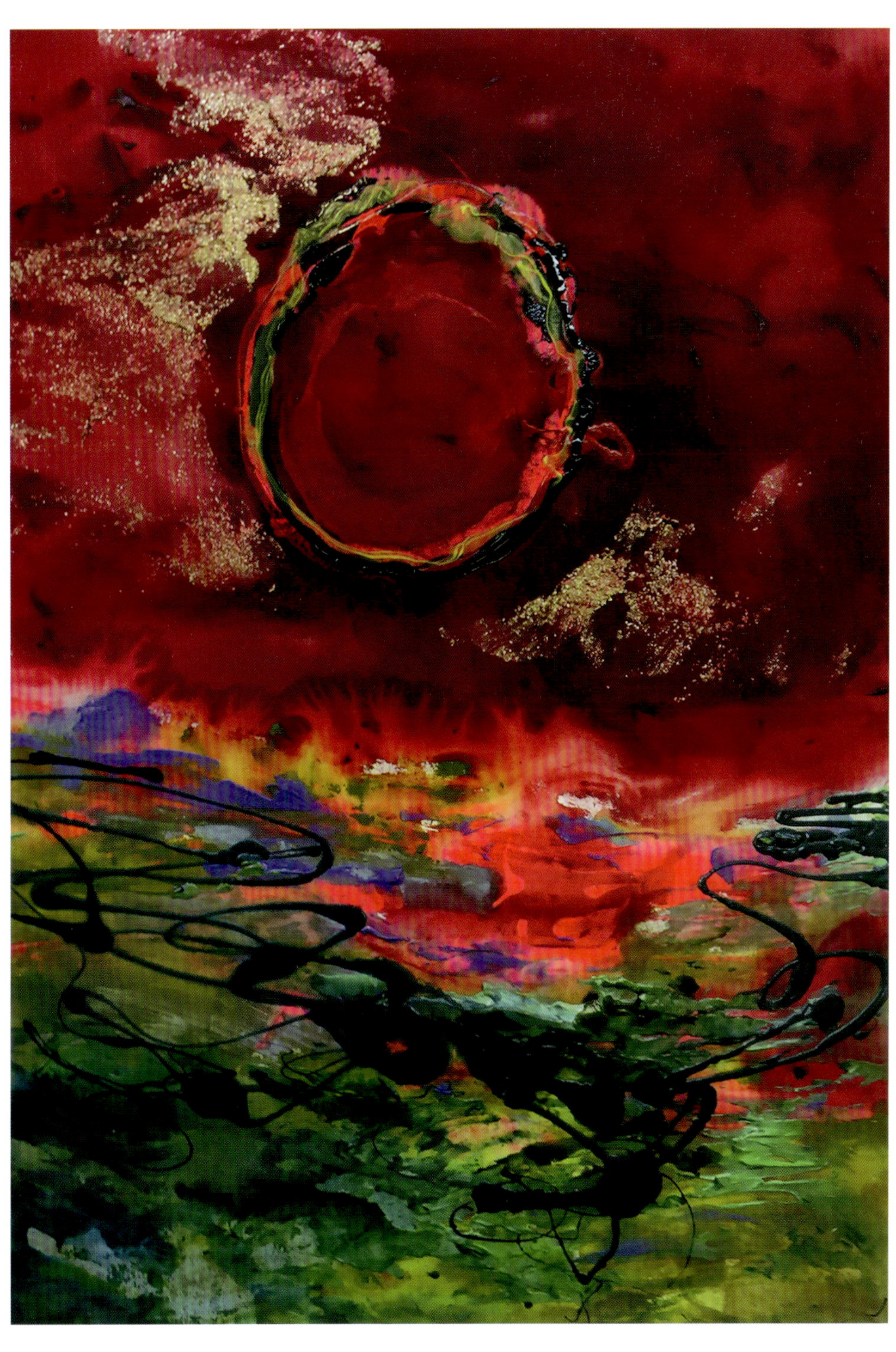

1.19 Darryl Hughto (American, born 1943), *Black Hole*, 2016, from the series *Heavenly Body*. Acrylic on canvas, 66¾ × 45¾ in. (169.5 × 116.2 cm). Courtesy of the artist.

In this painting, a colorful accretion disk surrounds a dark sphere representing a black hole.

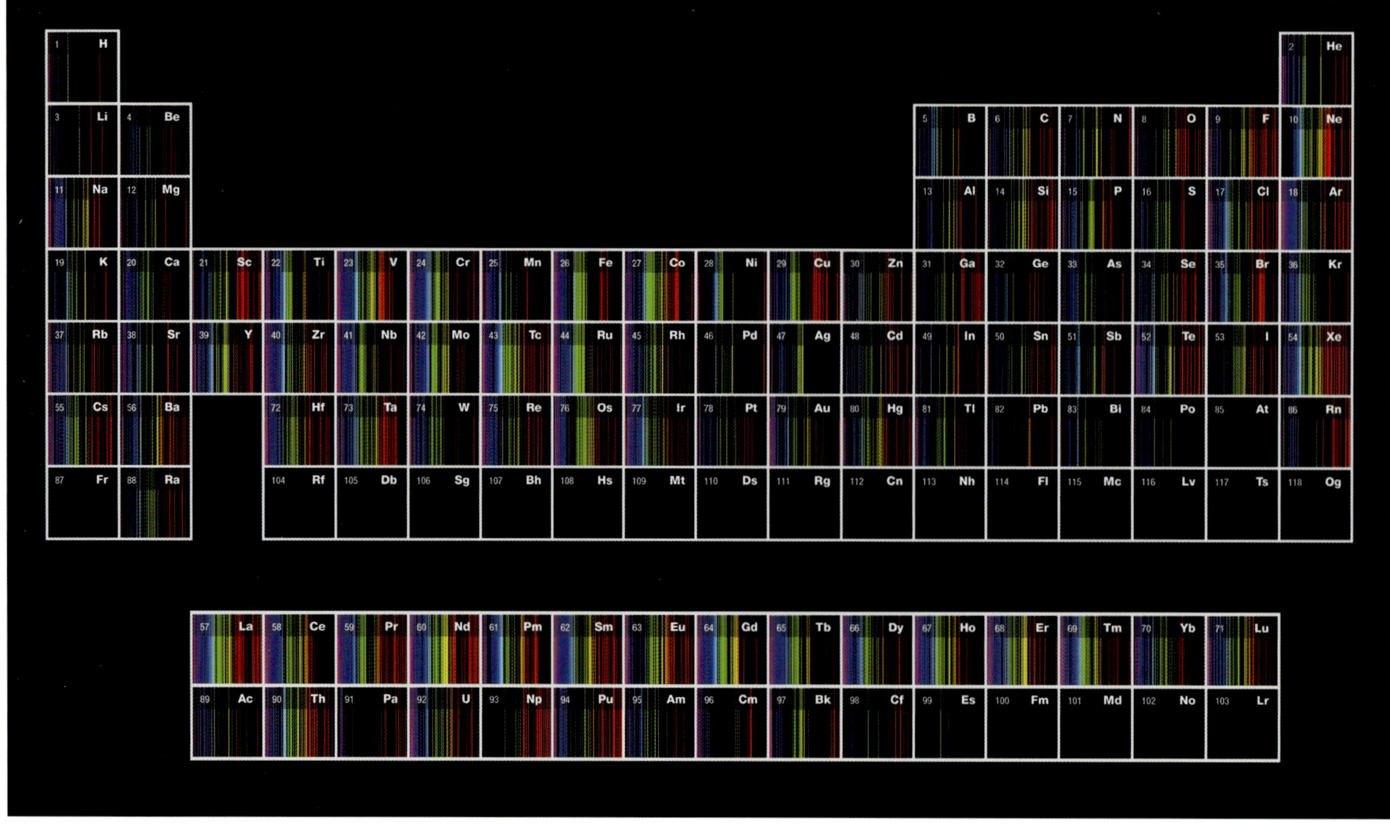

1.20 Periodic table of emission spectra of the elements. Courtesy of Field Tested Systems LLC.

The spectrum of a quasar's light was determined in the following way: In the early nineteenth century, the German physicist Joseph von Fraunhofer passed sunlight through a prism and observed that the colored band of light was not continuous but had dark lines (now called *Fraunhofer lines*). Then, in 1859, the German physicist Gustav Kirchhoff declared that each element has a unique pattern of lines—its *spectrum*. If the element is absorbing energy, one observes dark lines in the colored band; if the element is emitting energy, one observes the reverse, in other words, colored lines in the dark band (as shown here). Astronomers can point a telescope at light coming from a quasar, put that light through a prism, and the pattern of lines they observe is the quasar's spectrum. It determines what element(s) the quasar is made of.

THE DISCOVERY THAT BLACK HOLES EXIST

In 1963 the physicist Roy Kerr of New Zealand found a solution to Einstein's equations for general relativity that describes a rotating black hole, as Schwarzschild had in 1915 for one that is stationary. (Astronomers refer to a stationary *Schwarzschild black hole* and a rotating *Kerr black hole*.) Then, in 1967, the Irish scientist Jocelyn Bell discovered the first *pulsar*, so named because it's a celestial object that pulses, emitting electromagnetic radiation at regular intervals. In the late nineteenth century, James Clerk Maxwell had predicted (correctly) the existence of a whole spectrum of electromagnetism: radio waves, microwaves, infrared waves, visible light, ultraviolet light, X-rays, and gamma rays. Pulsars give off light across the entire electromagnetic spectrum, and they were soon determined to be rapidly rotating neutron stars. Thus, scientists had confirmed the existence of a neutron star that is formed when a star collapses. But what if a star collapses further?

With early radio telescopes, astronomers had detected objects that emit large amounts of radiation in many frequencies but less in the visible range. If the source was visible at all, it appeared as a point of starlight, hence the name *quasi-stellar radio source*, or *quasar*. By the early 1960s scientists had charted the spectrum of a quasar's light (figure 1.20). At first, some astronomers thought the spectrum was from an unknown element, but others suggested that it was the spectrum of ordinary hydrogen redshifted to indicate an enormous receding velocity (figure 1.21). This meant that quasars were traveling at extreme speed at a great distance while emitting enormous radiation. Since the most powerful optical telescope (one that observes in the visible range) could see them only as faint points of light, quasars must also be very small. This defied explanation: a celestial object that is very small, very far away, going very fast, yet the source of enormous energy. What could it be? Scientists confronting this question were like the detectives in *The Invisible Man* (1897) by H. G. Wells who were trying to solve the mystery of the stranger in town. His appearance defied explanation: he wore a hat and gloves indoors, his head was wrapped in bandages, he wore sunglasses at night, and he had a prosthetic nose. Who was he?

While some scientists aimed to crack the mystery of the identity of quasars, others surveyed the sky for X-rays using Geiger counters mounted on high-flying aircraft; unlike optical light and radio waves, X-rays don't penetrate Earth's atmosphere. In the 1960s these missions discovered a powerful X-ray source in the constellation Cygnus. Then, in 1970, the US National Aeronautics and Space Administration (NASA) launched the first X-ray telescope into space: the Uhuru X-ray Satellite. The telescope was named "Uhuru" (Swahili for "freedom") in recognition of the cooperation NASA got from the people of Kenya, where the space telescope was launched. Uhuru performed the first comprehensive survey of the entire sky for X-ray sources, and in 1971 researchers identified a region of space—Cygnus X-1—as the source of X-rays in the constellation Cygnus originally discovered by Geiger counters (figure 1.22). When astronomers pointed an optical telescope to that spot in the sky, they saw a star (HDE 226868) they understood to be incapable of generating such powerful X-rays. Astronomers were familiar with what is known as a *binary system*: two celestial objects that are linked by gravity. Several scientists suggested that Cygnus X-1 and HDE 226868 are a binary system. They proposed that Cygnus X-1 is a black hole pulling matter from its companion star, HDE 226868, forming an *accretion*

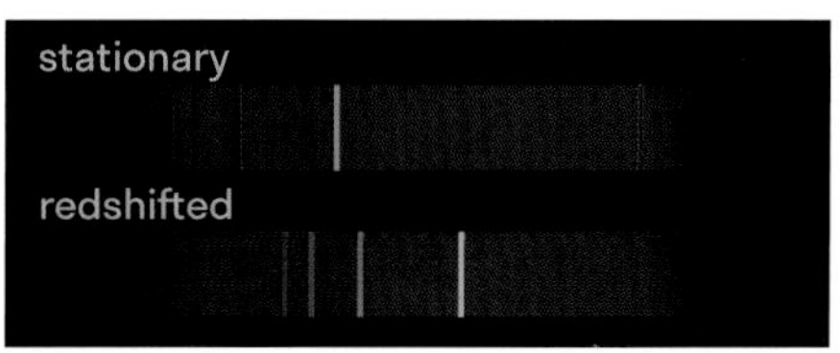

1.21 Emission spectra of stationary and redshifted hydrogen.

When measuring sound, the observed frequency is higher or lower than the emitted frequency depending on whether the source is moving toward or away from the observer; the phenomenon is called the *Doppler effect* after the Austrian physicist Christian Doppler, who explained it in 1842. The same is true for light, which appears closer to the blue or red end of the colored band depending on whether the source is moving toward or away from the observer. Here, the top spectrum of hydrogen has a stationary source; in other words, the source and the observer are at rest relative to each other. The bottom spectrum is redshifted because the source is moving away from Earth. The amount of a light's redshift (or blueshift) allows astronomers to calculate the speed at which the source of hydrogen is moving away from (or toward) the observer on Earth.

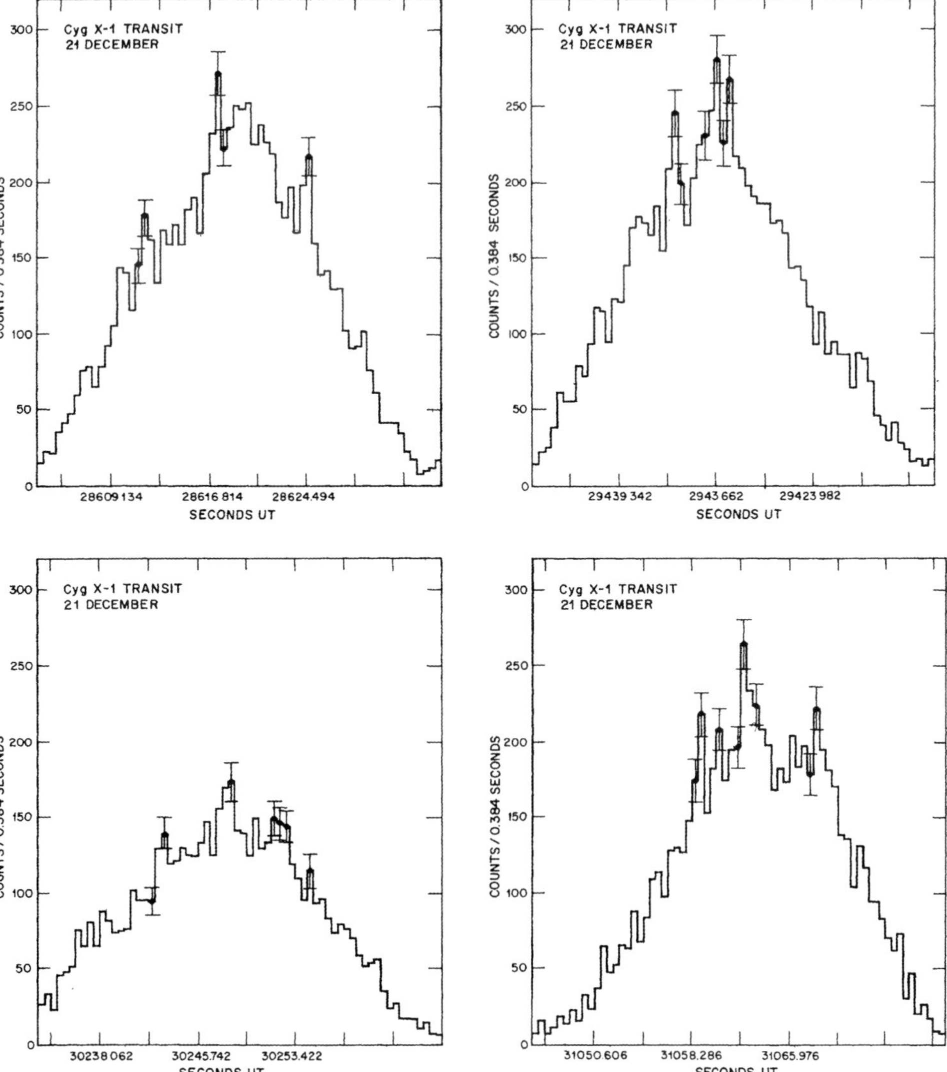

1.22 X-ray pulses from Cygnus X-1 recorded by the Uhuru X-ray Satellite in Minoru Oda, Paul Gorenstein, Herbert Gursky, Edwin Kellogg, Ethan Schreier, Harvey Tananbaum, and Riccardo Giacconi, "X-Ray Pulsations from Cygnus X-1 Observed from *UHURU*," *Astrophysical Journal* 166, no. 1 (June 1971): L2, fig. 1.

The authors of this paper were the first to suggest that Cygnus X-1 might be a black hole.

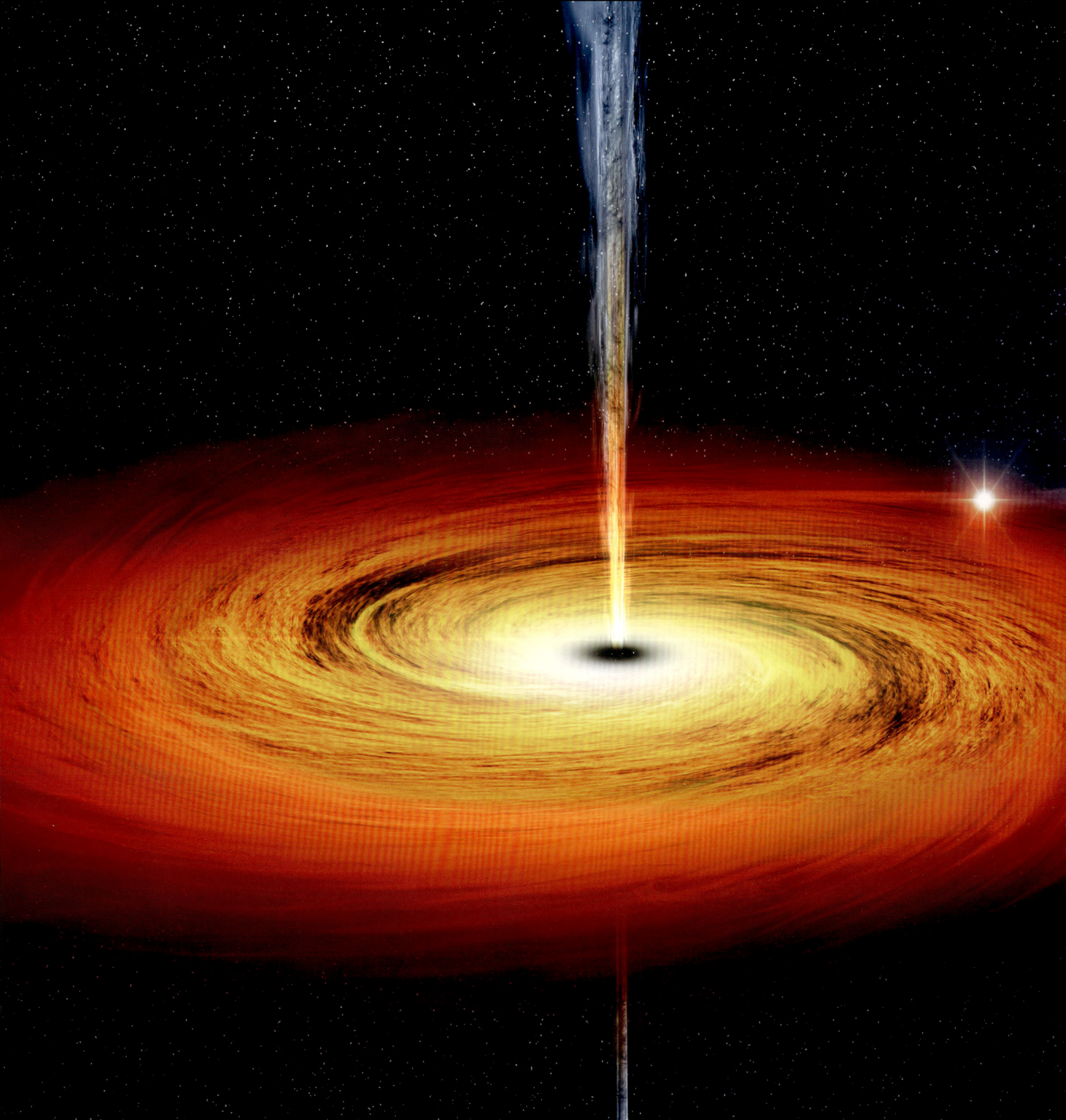

1.23 Artist's impression of Cygnus X-1 and HDE 226868. NASA/CXC/Melissa Weiss.

This illustration shows the black hole's intense gravity pulling matter from its companion star (shown in blue). This matter forms an accretion disk (shown in yellow, orange, and red) that rotates around the black hole.

disk, a cloud of dust and gas that orbits a black hole close to the speed of light, becoming so hot it is incandescent and emits radiation across the entire electromagnetic spectrum, including powerful X-rays (figure 1.23). By the early 1970s, most astronomers agreed that the existence of the first black hole had been confirmed based on X-ray evidence. Since then, NASA launched into space the Chandra X-ray Observatory, named for Subrahmanyan Chandrasekhar. Using X-rays recorded by Chandra, scientists determined that Cygnus X-1 is spinning an astonishing 800 times a

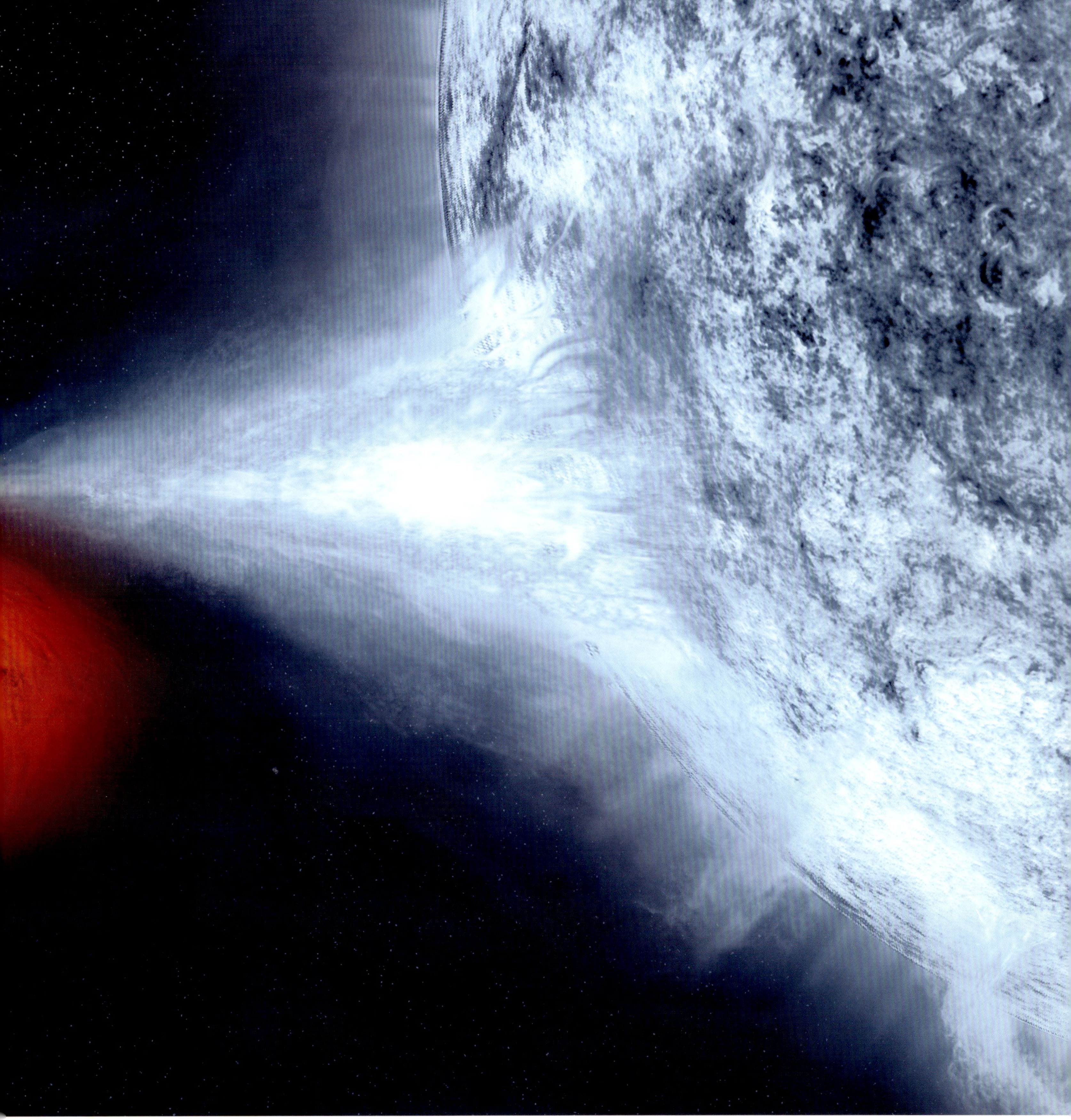

second (figures 1.24, 1.25, and 1.30). Chandrasekhar said, “The black holes of nature are the most perfect macroscopic objects there are in the universe: the only elements in their construction are our concepts of space and time.”[15]

After it had been confirmed that Cygnus X-1 is a black hole, two physicists at Princeton University, Remo Ruffini and John Wheeler, coauthored the essay “Introducing the Black Hole,” published in *Physics Today* in January 1971. For the cover of the magazine, the editor commissioned a painting of a black hole from Helmut Wimmer, who worked as an illustrator at the

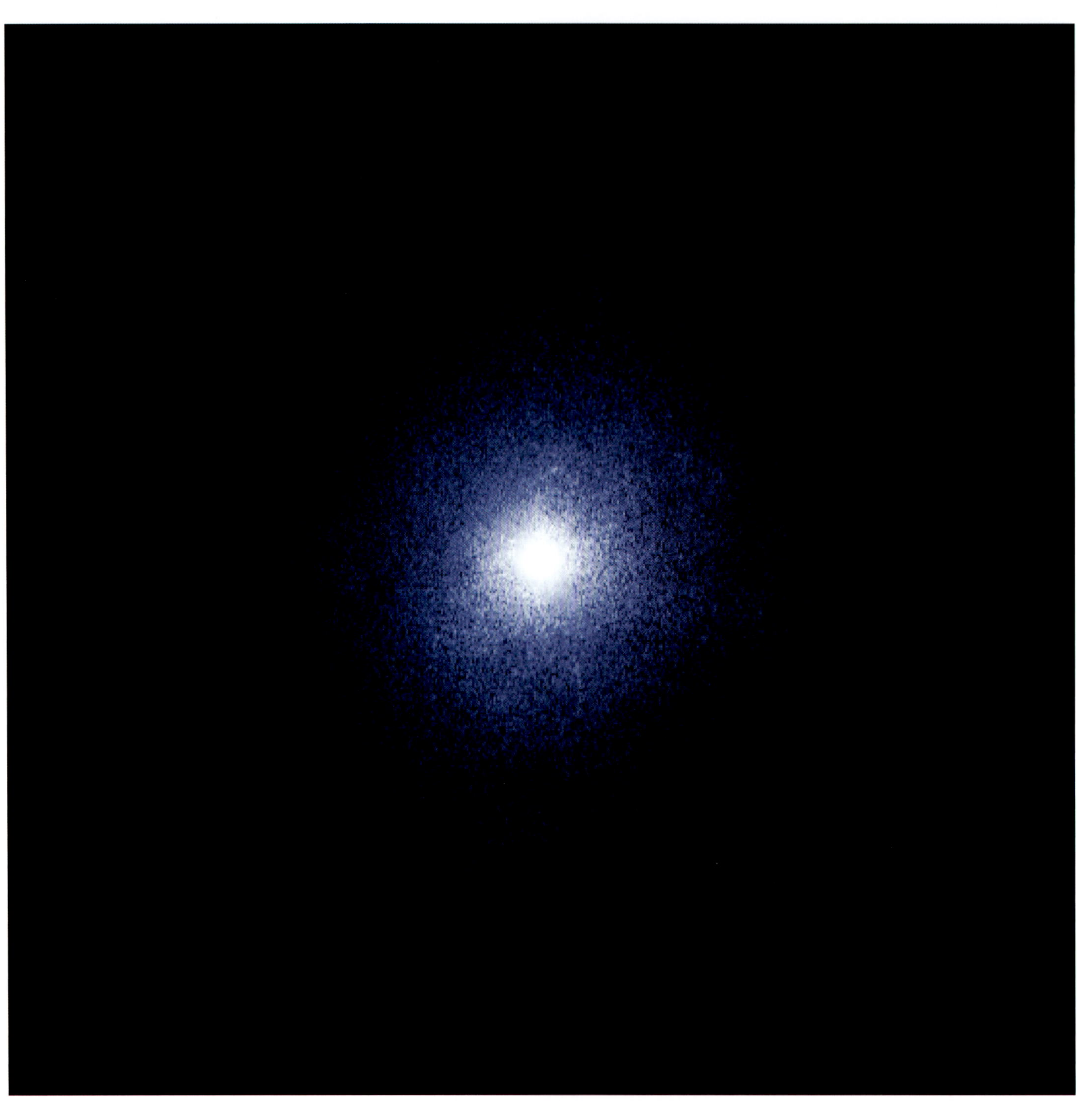

1.24 X-ray pulses from Cygnus X-1 recorded by the Chandra X-ray Observatory in 2001–2003. NASA/CXC.

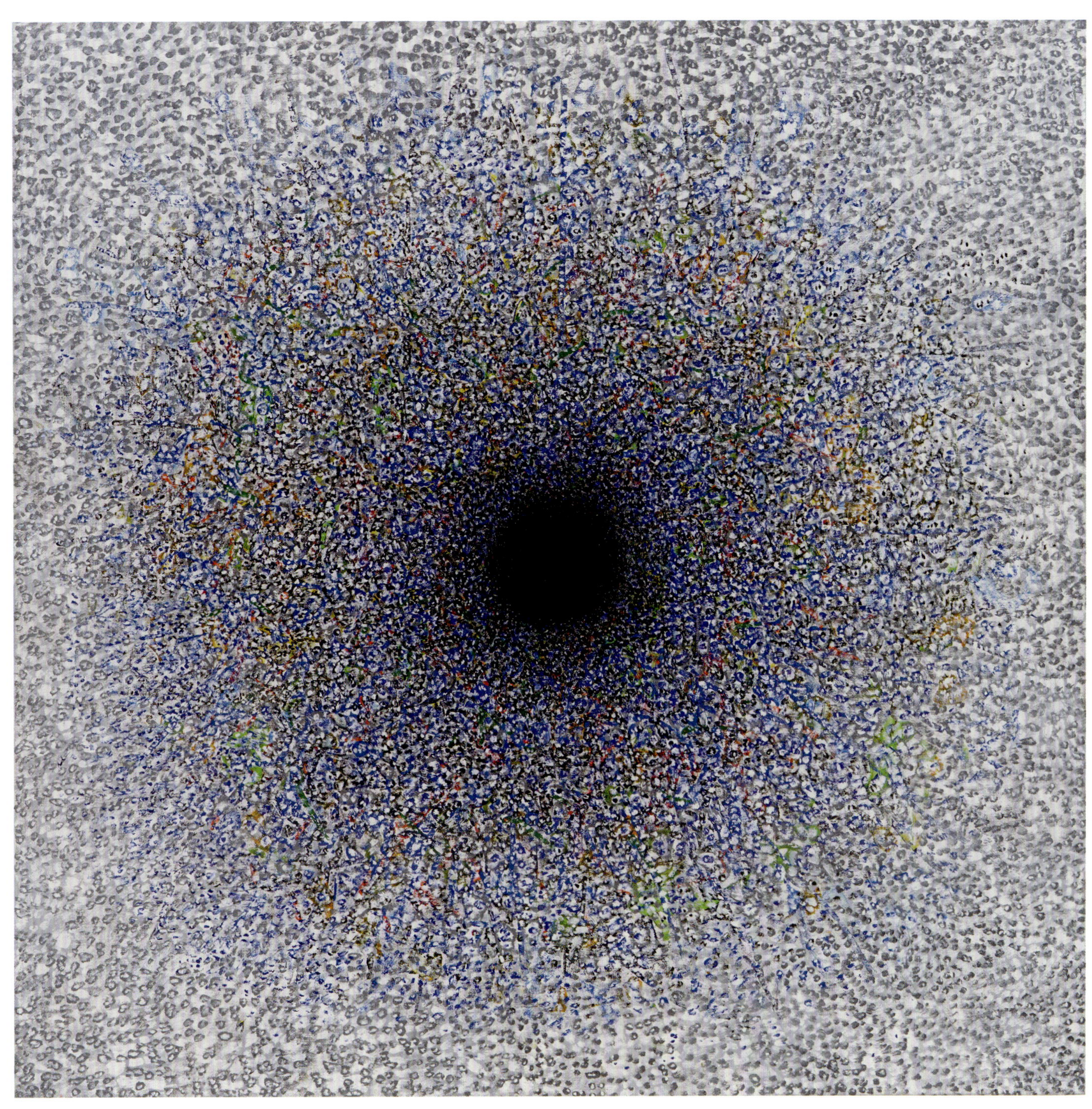

1.25 Richard Pousette-Dart (American, 1916–1992), *Imploding Black*, 1985–1986. Acrylic on linen, 72 × 72 in. (182.9 × 182.9 cm). Courtesy of Pace Gallery, New York.

In this image by Pousette-Dart, who was associated with Abstract Expressionism, our eye is drawn inward by the increasing density of the dabs of color.

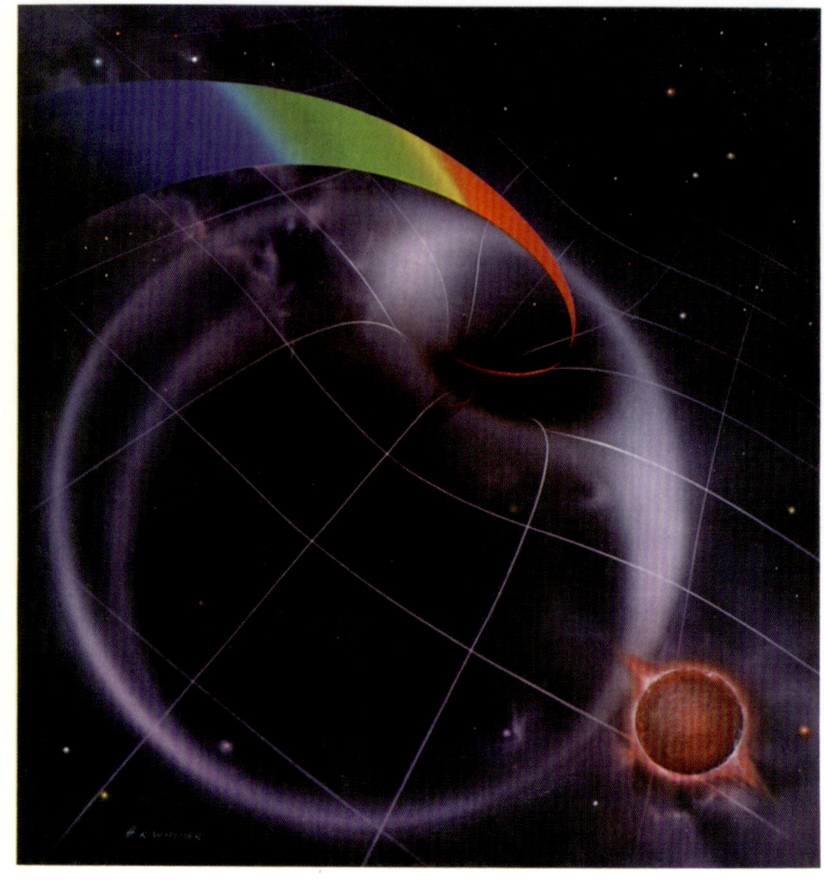

1.26 Helmut Wimmer (American, born Germany, 1925–2006), *Black Hole*, 1971, cover of *Physics Today* 24, no. 1 (January 1971).

Hayden Planetarium of the American Museum of Natural History in New York. Ruffini described Cygnus X-1 for Wimmer: "We had a discussion in New York to explain the concept at the Hayden Planetarium where Helmut Wimmer was working. . . . He listened and then he went home. He could not sleep that night; he woke up at five in the morning and said to himself 'I understand the way it should be' and he did the painting. He called me the day after. I went to New York and saw the painting. It was splendid."[16] Wimmer painted a black sphere with a spectrum of light entering it (figure 1.26). In the painting, a two-dimensional grid shows the deformation of spacetime; at the lower right is the companion star. Science illustrators standardized Wimmer's deformed grid into a diagram of a singularity, which combines a gridded globe with a vanishing point in linear perspective (figure 1.27).

Any object that crosses the event horizon will be gone forever. Ruffini and Wheeler emphasized this with an illustration of an amusing surreal drama: a piece of furniture, a flower, a television, and a potato (or rock) are flying around in outer space and fall into a black hole (figure 1.28). The German artist Björn Dahlem picked up on the physicists' witticism in his sculpture *Black Hole: Cygnus X-1*, in which furniture, lamps, and other found objects are pulled into a metaphorical black hole (figure 1.29).

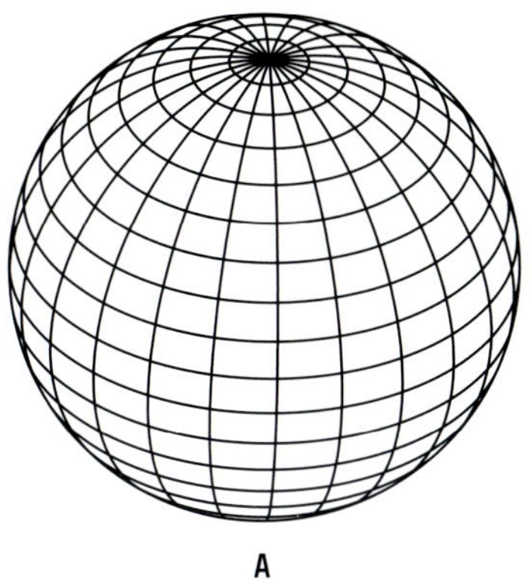

A

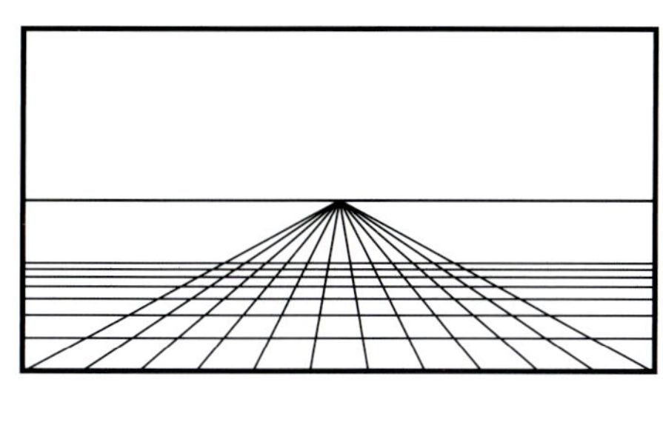

B

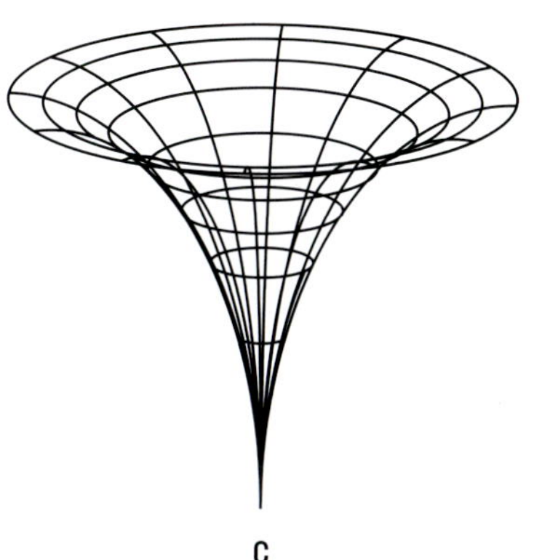

C

1.27 A singularity.

In the third century BCE, the Greek mathematician Eratosthenes first established lines of latitude and longitude on maps of Earth, which he knew to be a sphere. In the second century BCE the Greek astronomer Hipparchus made a gridded globe (A). The viewing position for Hipparchus's globe was the sky, where Zeus was thought to reside. The Renaissance architect Filippo Brunelleschi invented linear perspective after he noticed that receding parallel lines appear to vanish at a point on the horizon (B). Brunelleschi defined a unique viewing position for an image in linear perspective; standing with one eye closed, the viewer is located in front of the vanishing point.

Science illustrators show a black hole as a sphere. Anything that passes the event horizon is drawn into the center of the black hole, where it disappears in a vanishing point of infinite density—a singularity (C). In his 1924–1925 essay "Perspective as 'Symbolic Form,'" the German art historian Erwin Panofsky had the insight that Renaissance artists used linear perspective to express their concept of the world and only secondarily to imitate their visual experience. Similarly, scientists use the diagram of a singularity to communicate their understanding of a black hole and only secondarily to imagine how a singularity might actually appear. Like Hipparchus's globe, the diagram of the singularity represents a god's-eye viewpoint, which, in our modern secular age, is called "the view from nowhere," a phrase coined in 1986 by the philosopher Thomas Nagel in *The View from Nowhere*.

BELOW

1.28 "Figurative representation of a black hole in action," in Remo Ruffini and John Wheeler, "Introducing the Black Hole," *Physics Today* 24, no. 1 (January 1971): 31, fig. 1.

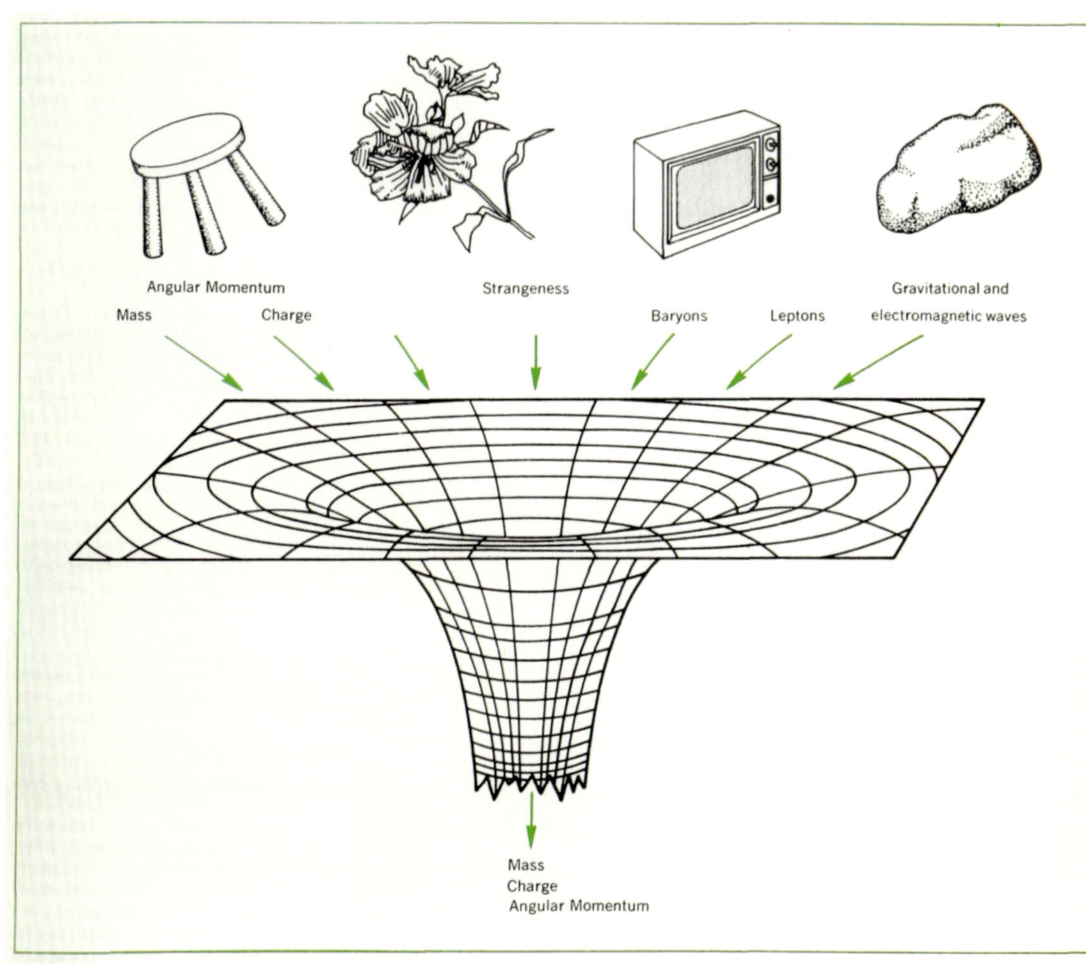

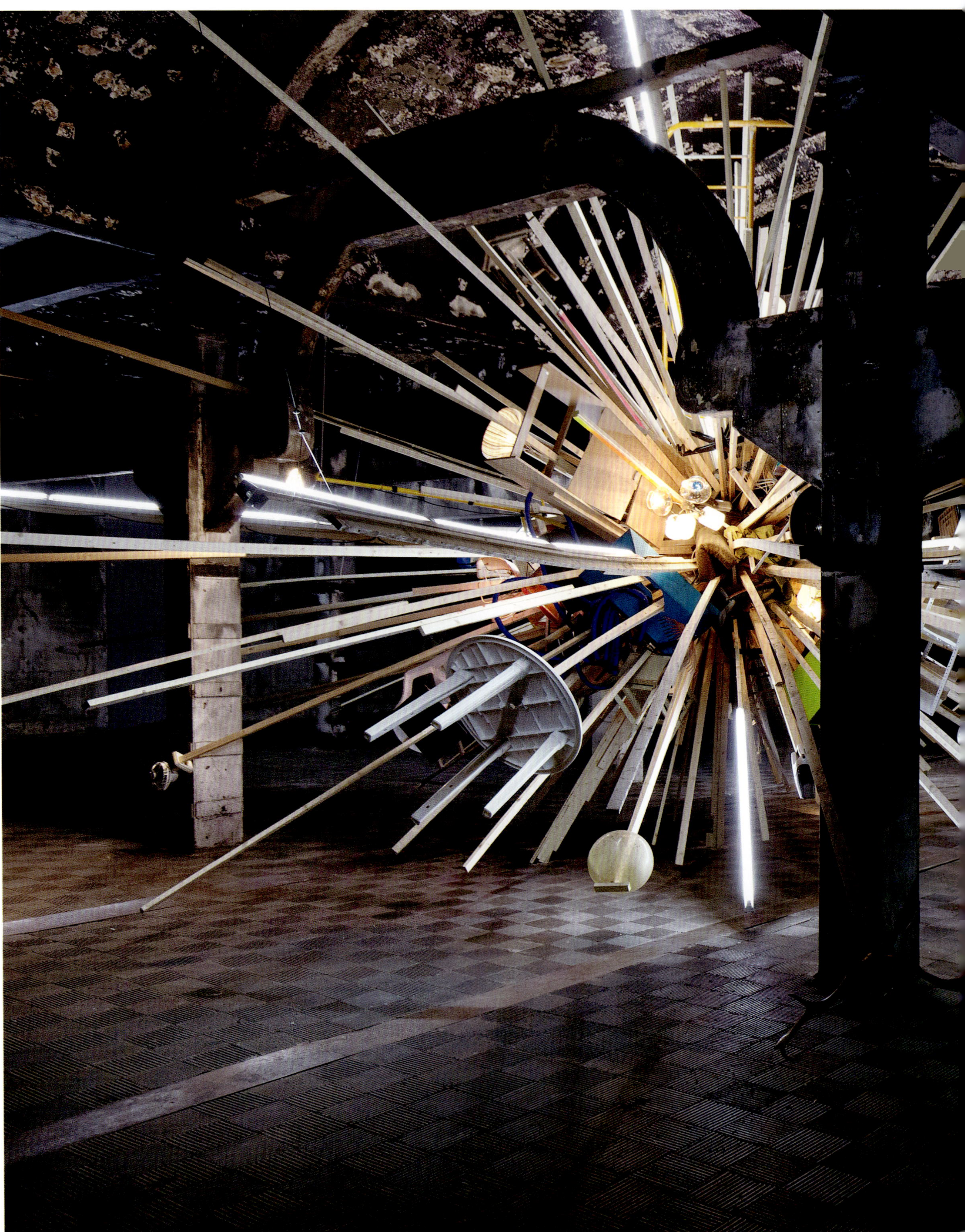

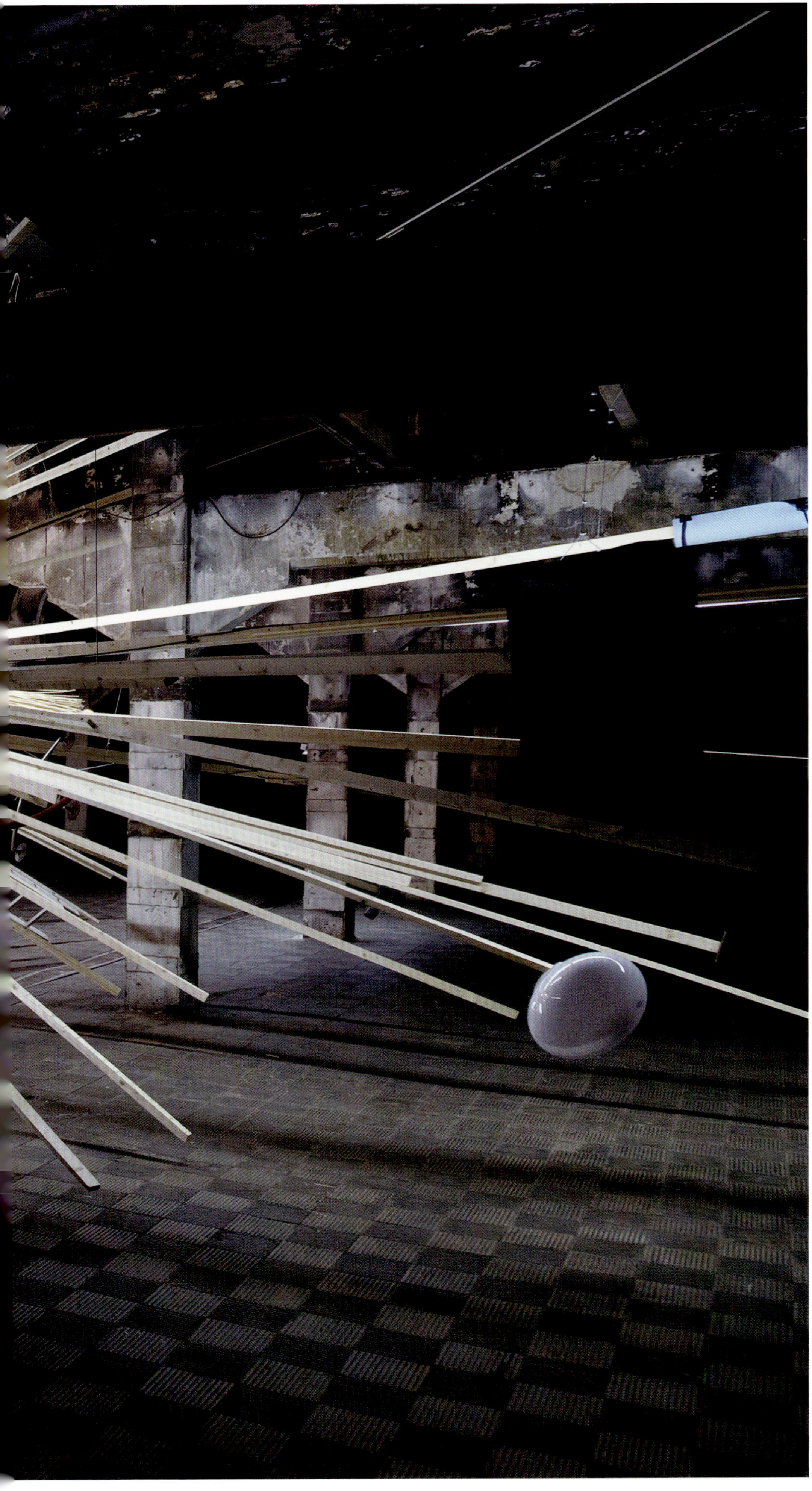

LEFT
1.29 Björn Dahlem (German, born 1974), *Black Hole: Cygnus X-1*, 2014. Wood, fluorescent lamps, various found objects, 19 ft. 8 in. × 49 ft. 2½ in. × 49 ft. 2½ in. (6 × 15 × 15 m). Installation view at Matadero Center for Contemporary Art, Madrid. Courtesy of Ehrhardt Flórez Gallery, Madrid, Sies + Höke, Dusseldorf, and Guido W. Baudach Gallery, Berlin.

FOLLOWING PAGES
1.30 Artist's impression. Chandra X-ray Observatory orbiting Earth. NASA/CXC/J. Vaughan.

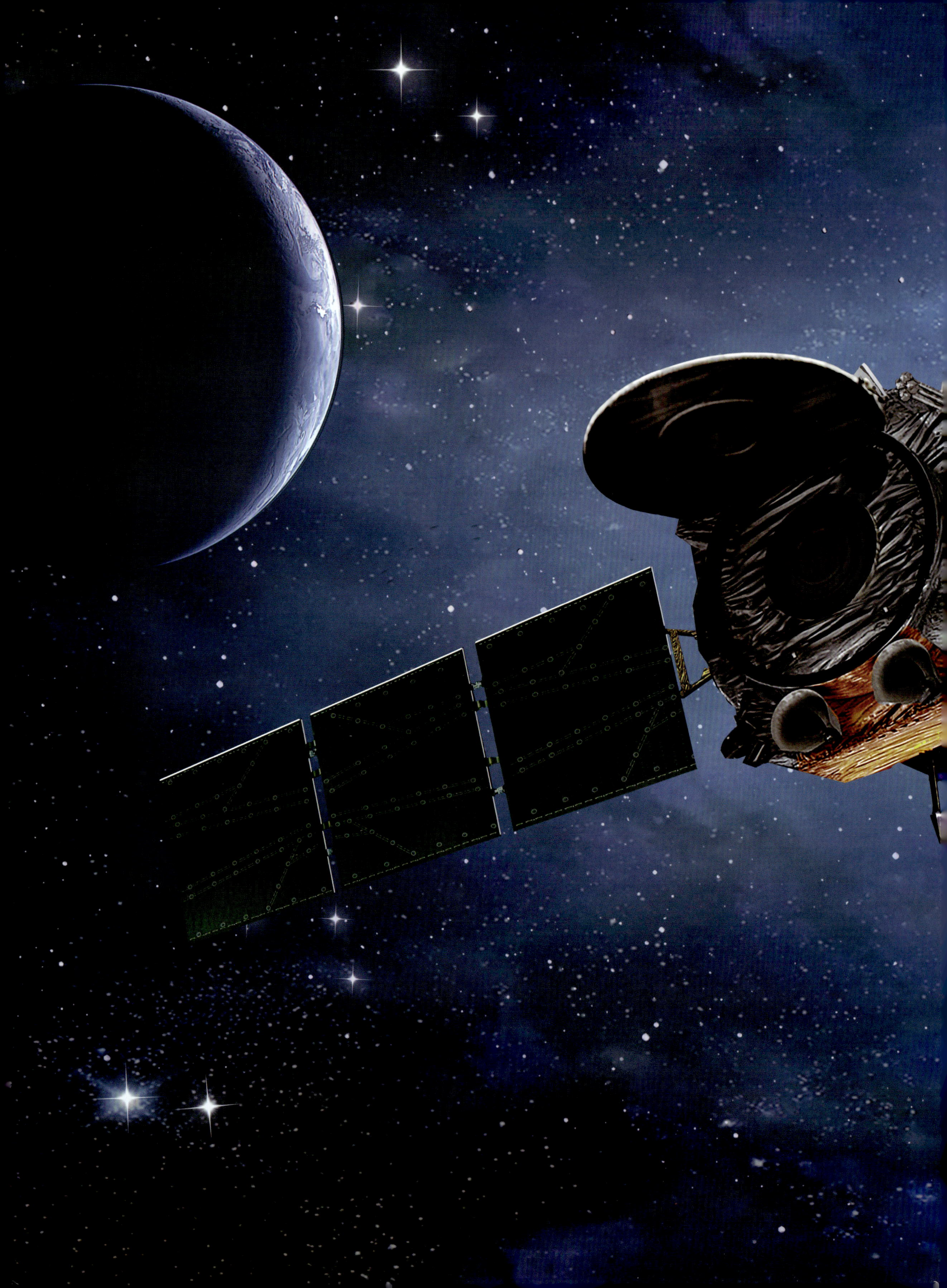

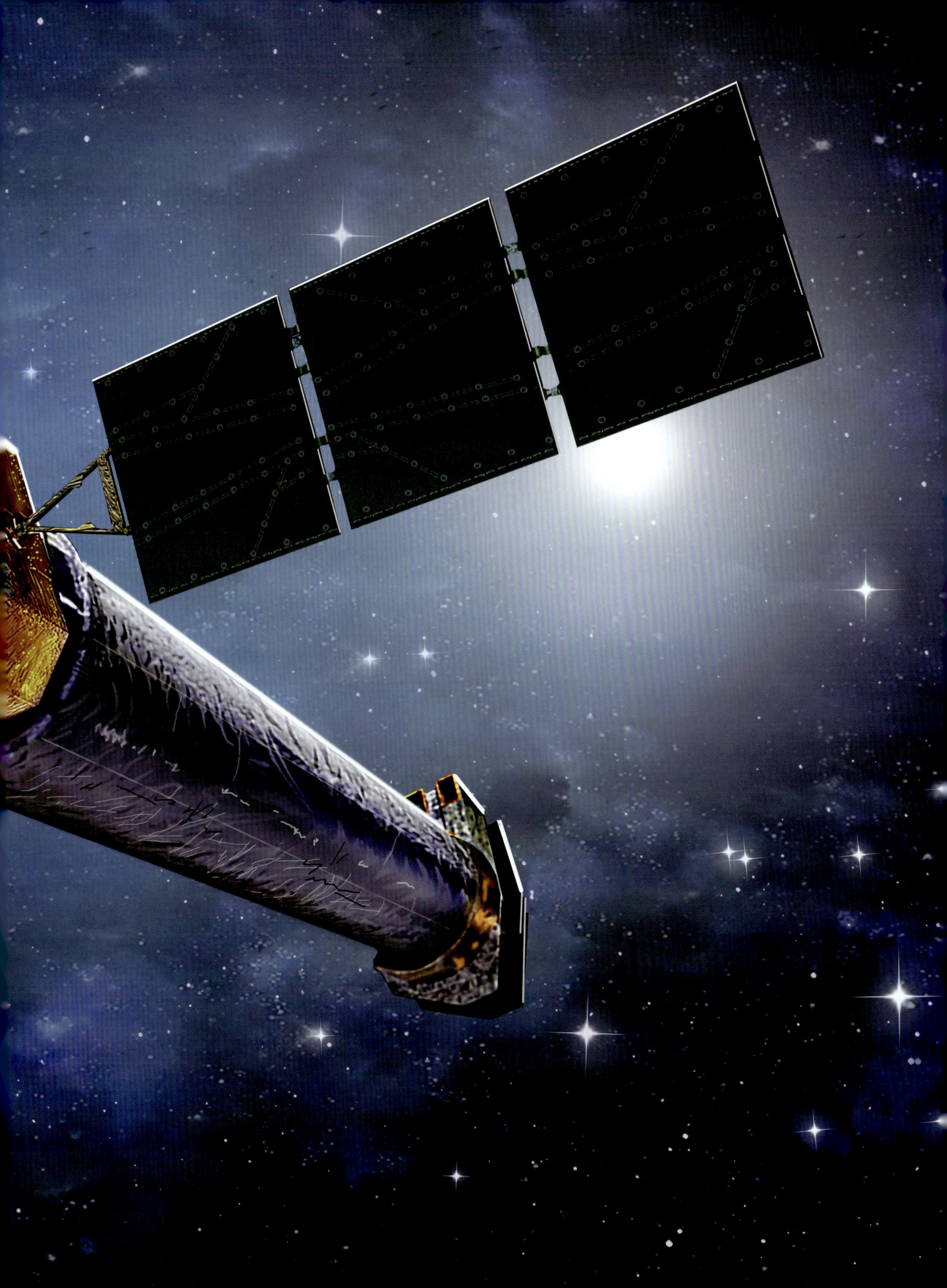

古巳上合氣象有卅八條臣冒考有驗故錄之也未曾占考不敢
輙掄入此卷臣不揆庸賓見敢縉愚情輟而錄之具如前件謹
陳階庭弥加戰越死罪死罪謹言

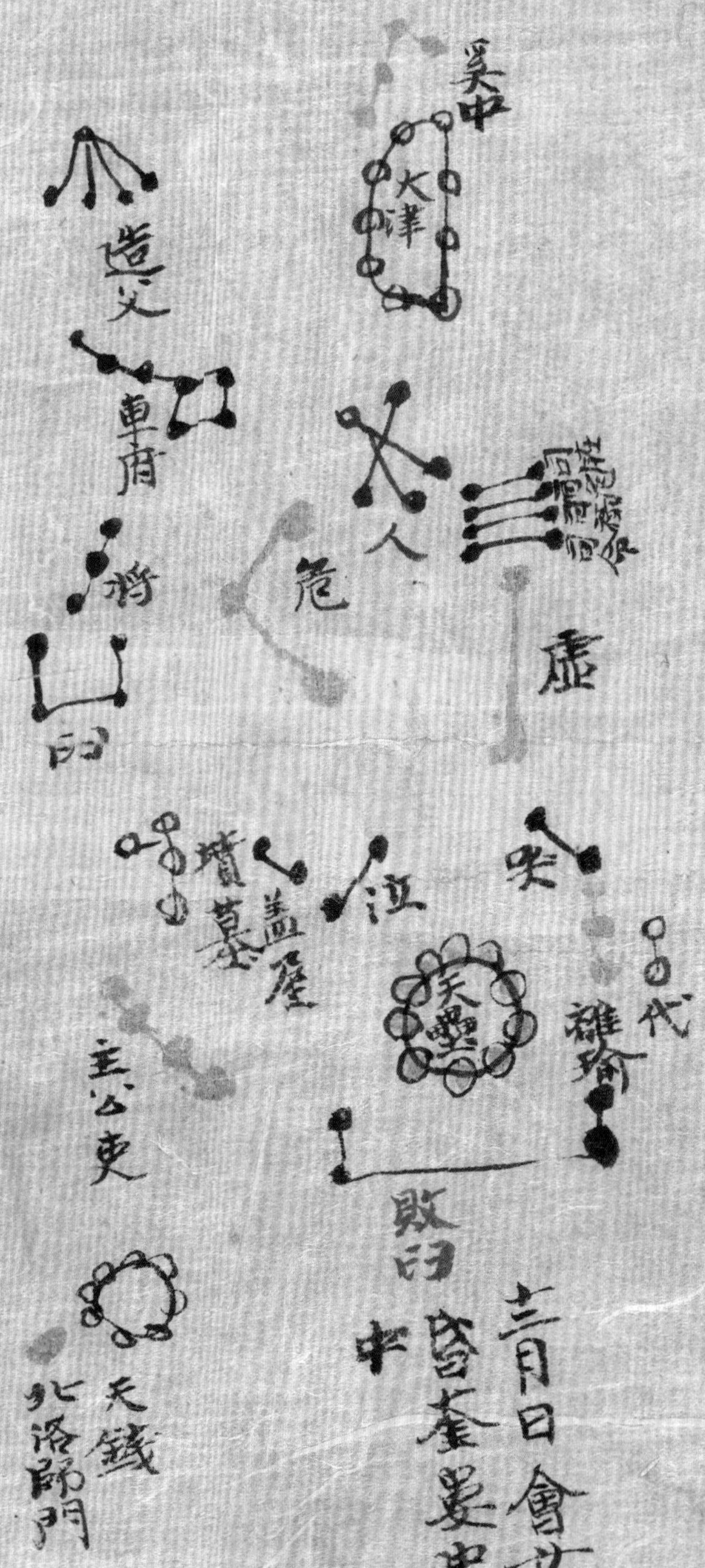

十二月日會女虛
昏奎婁中旦氐
中

2

HISTORICAL AND PHILOSOPHICAL BACKGROUND TO IMAGING BLACK HOLES

The foundations of how modern artists and scientists understand black holes are rooted in historical and philosophical traditions that began in antiquity with creation myths and the search for meaning in the world. As science developed in the West, artists responded to discoveries of unseen realms by creating abstract art. After 1945, Eastern philosophical traditions and Western modern art merged, forming a single global outlook. Scientists discovered the existence of mysterious black holes, and international artists symbolized these strange invisible objects.

ANCIENT OUTLOOKS

Black holes paradoxically embody the concept of nothingness while they are simultaneously a kind of totality—for lack of a better term, an everythingness. Anything that crosses the event horizon disappears into a void; within the black hole is a singularity where everything that has crossed the point of no return is crushed into infinite density—the most extreme concentration of matter in the universe. Asian art about black holes focuses on this dichotomy because the paradoxical pair of absence/totality is embedded in ancient Eastern cosmology. In Chinese thought, the natural world came into being by self-assembly following the *Tao* (Chinese for "way") of nature, and according to the legendary fourth-century BCE founder of Taoism, Laozi (figure 2.2): "All things under heaven sprang from It as existing (and named); that existence sprang from It as nonexistent (and not named)."[17] In other words, everything came from nothingness/emptiness/void. This quotation from Laozi is an ancient echo of Roger Penrose and Stephen Hawking's suggestion that the Big Bang was

2.1 The constellation Emptiness in the Dunhuang Star Chart (detail), 649–684. Handscroll; ink on paper, 9½ in. × 12 ft. 11 in. (24.4 cm × 3.94 m). The British Library, London.

Drawn in China during the Tang dynasty (618–907), the Dunhuang Star Chart is the oldest star atlas in the world. It records stars visible from the Northern Hemisphere. The constellation Emptiness (虛 *xu*)—which has two stars, one directly above the other—is colored gold and located in the center of this section of the chart. Chinese astronomers divided the sky into twenty-eight regions or "mansions." The mansion in which the constellation Emptiness is located is also called "Emptiness." The text on the right can be translated as: "Dark means black, this is the color of north; emptiness means exhaustion. In the eleventh month the *yang* vapors descend, and the *yin* vapors ascend, the myriad things are dominated by darkness and death, there is no living thing around, Heaven and Earth are bare and empty. This is why it is called 'Dark Emptiness.' This division corresponds to the Kingdom of *Qi*." The word *qi* is Chinese for "energy" or "life force."

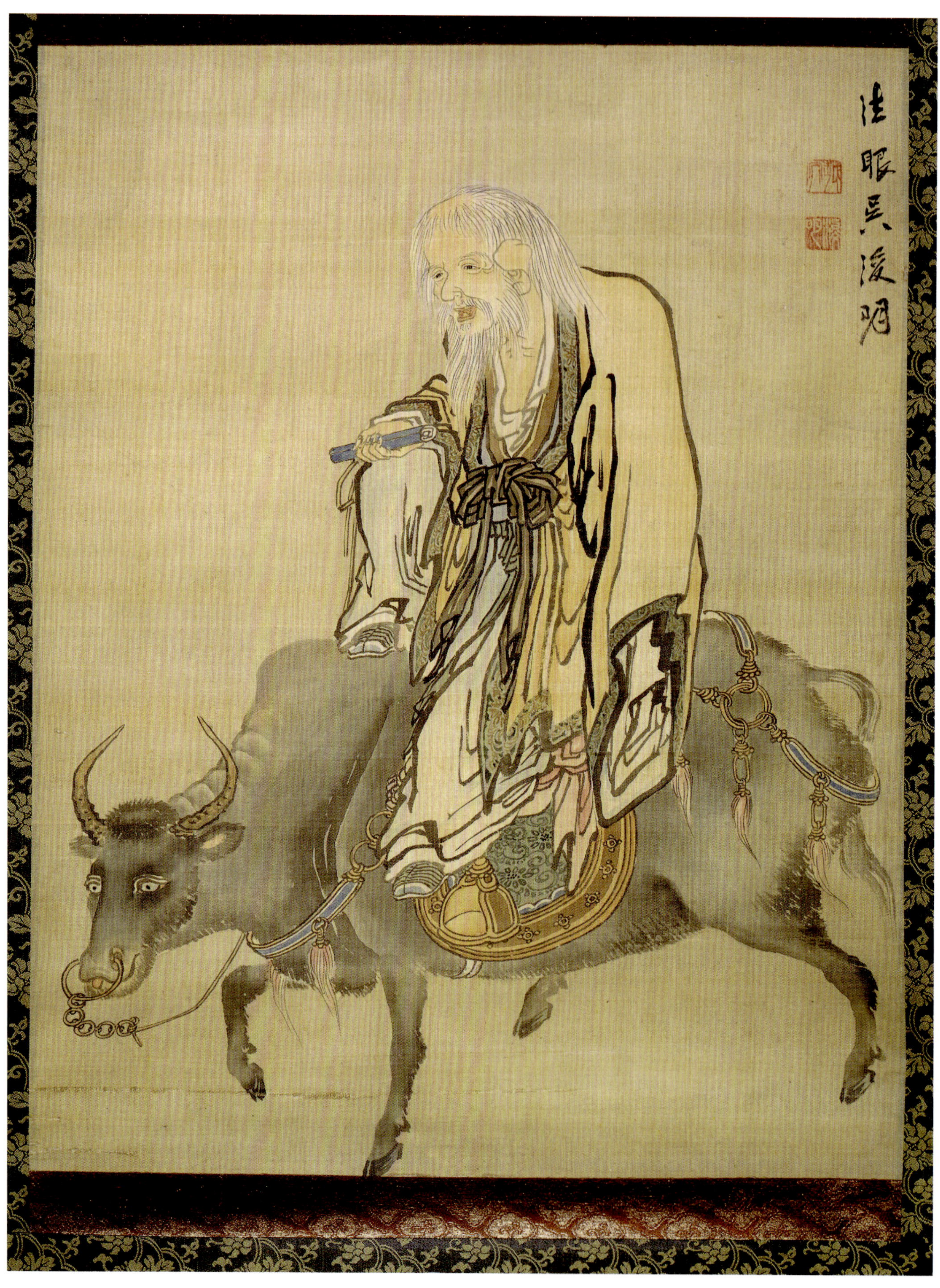

2.2 Igarashi Shunmei (Japanese, 1700–1781), *Laozi on Water Buffalo*, 18th century. Paint and ink on silk, 17¼ × 13 in. (43.8 × 33.5 cm). British Museum, London.

an "initial . . . singularity"—a point of infinite density that is the primordial source of everything.

Western creation myths feature a divine person (such as the Hebrew God of Abraham or Plato's Craftsman) who imposed order on chaos and decreed laws of nature, but Asian creation myths lack such a divine person. Laozi said, "The Tao that can be trodden is not the enduring and unchanging Tao. The name that can be named is not the enduring and unchanging name."[18] In other words, not only is the way of nature not a person but, furthermore, the Tao is incapable of being expressed in words. When Chinese astronomers looked at the night sky, they saw patterns in the stars and named one constellation "Emptiness" (figure 2.1). Hinduism began a thousand years earlier in the Indus Valley, and, as Mahatma Gandhi said, "Hinduism is a persistent pursuit after truth."[19] Hindus want to reach nirvana: the absence of attachments, aversions, and delusions. The highest form of reality is a realm of pure mind—the Brahma—which humans know by thought alone. A person can achieve enlightenment by uniting eternally with the Brahma through meditation and ascetic living. Born in Nepal, Buddha (late sixth and early fifth centuries BCE; figure 2.3) declared that reaching enlightenment is open to all. Many Buddhists emigrated from Nepal to China, where Buddhism merged with Taoism—to reach nirvana, one follows the way of nature—and this Buddhist-Taoist mix spread throughout Southeast Asia as Chan (Japanese Zen) Buddhism.

On the ancient shores of the Mediterranean Sea and the Yellow River, scholars asked different questions. In the West, Greek philosophers asked: "What is being?" In the East, Chinese philosophers asked: "What is nothingness?" Plato said that being is built from sentient particles—or *monads* (from the Greek for "unity"). During the eighteenth century, the French philosopher René Descartes countered that there are two kinds of being—mind and matter. This dualism culminated in the German Idealism of Immanuel Kant, who wrote in *Critique of Pure Reason* (1781) that one knows only mind. Eastern philosophy rejects all dualisms. Nothingness is neither *being* nor *nonbeing*. To consider nothingness as nonbeing would be to make this abstract concept into a concrete thing. To experience nothingness, one needs to follow the advice of Zhuangzi (late fourth century BCE) to forget everything and focus on absolute nothingness so that one's finite existence can be in harmony with the infinite Tao: "Be empty, that is all. The perfect man uses his mind like a mirror—going after nothing, welcoming nothing, responding but not storing."[20]

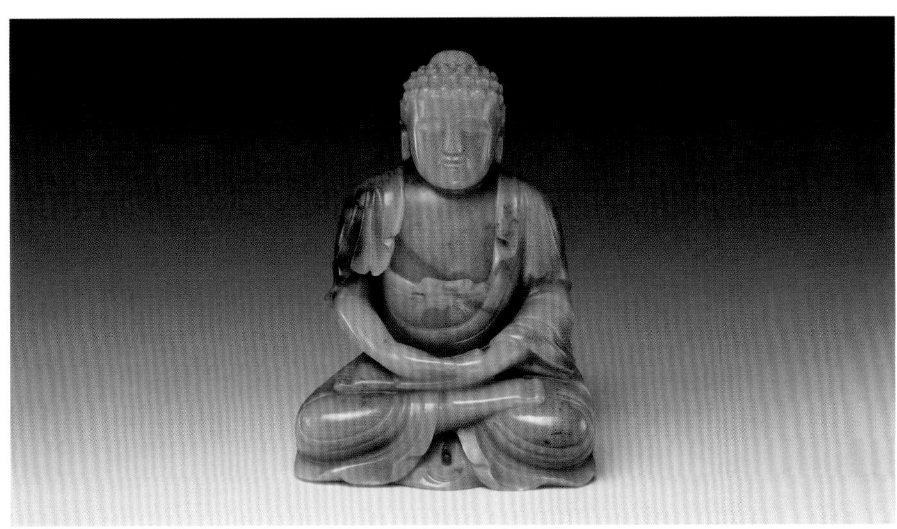

2.3 *Seated Buddha*, 18th century. Jade, 5¼ × 3¾ × 2¾ in. (13.5 × 9.5 × 7 cm). Metropolitan Museum of Art, New York, gift of Heber R. Bishop, 1902.

THE RISE OF SCIENCE

Black holes are investigated by the precision tools and creative minds of science, a discipline that began in the West during the Enlightenment and was adopted by Asian cultures. Many in the West, in turn, adopted the Asian view that nature is a sacred place that came into being by self-assembly, without a divine person.

In the West some people looked for the order imposed by Plato's Craftsman or the God of Abraham, leading them to gain knowledge of the laws of nature. The Greek astronomer Ptolemy computed the position of the planets in the second century CE, and Isaac Newton discovered the law of universal gravitation in 1687. According to the twentieth-century British scholar Joseph Needham, a key reason science didn't develop beyond a rudimentary level in Asia is that Eastern creation myths do not feature a divine person who imposed order on chaos; therefore, there never arose an assumption that there is an imposed order to discover. In the Asian understanding of things, nature spontaneously made itself.[21]

As Western thinkers explored nature, they inevitably also confronted its majesty and power, which was not exclusively creative. The *sublime* was defined by the British philosopher Edmund Burke as "whatever is fitted in any sort to excite the ideas of pain and danger, that is to say, whatever is in any sort terrible . . . is a source of the sublime."[22] According to Burke, we appreciate beauty because it has aesthetic appeal, but we are amazed and terrified by the sublime because it has the power to destroy us. Burke was referring to the power of nature as seen in Joseph Wright's painting of Vesuvius (figure 2.4). Dark and dangerous—like a black hole—the sublime has been part of modernism since its inception.

The rise of science in the West entailed the diminution of the Abrahamic religions—Judaism, Christianity, and Islam. Widespread atheism developed in the mid-nineteenth century after the English biologist Charles Darwin presented overwhelming evidence that mankind was not created by the God of Abraham in his image but "descended from a hairy, tailed quadruped."[23] Psychology emerged as a medical specialty in the late nineteenth century, and in *The Future of an Illusion* (1927) Sigmund Freud gave an atheistic account of belief in God: Putting one's trust in a divine father is an infantile overvaluation of the parent that has survived into adulthood. More and more features of the human body and soul yielded to the explanatory power of science, causing a seismic shift from belief in the God of Abraham to denial of the very Creator whose existence had driven Western science since the Renaissance. Because Taoism, Hinduism, and Buddhism posit no celestial lawgiver, they merged seamlessly with the scientific worldview. Western science confirmed the traditional Asian outlook—the natural world *did* develop spontaneously by self-assembly. Eastern outlooks, especially Zen Buddhism, provided an alternative to atheism after the steep decline of organized religions in the West.

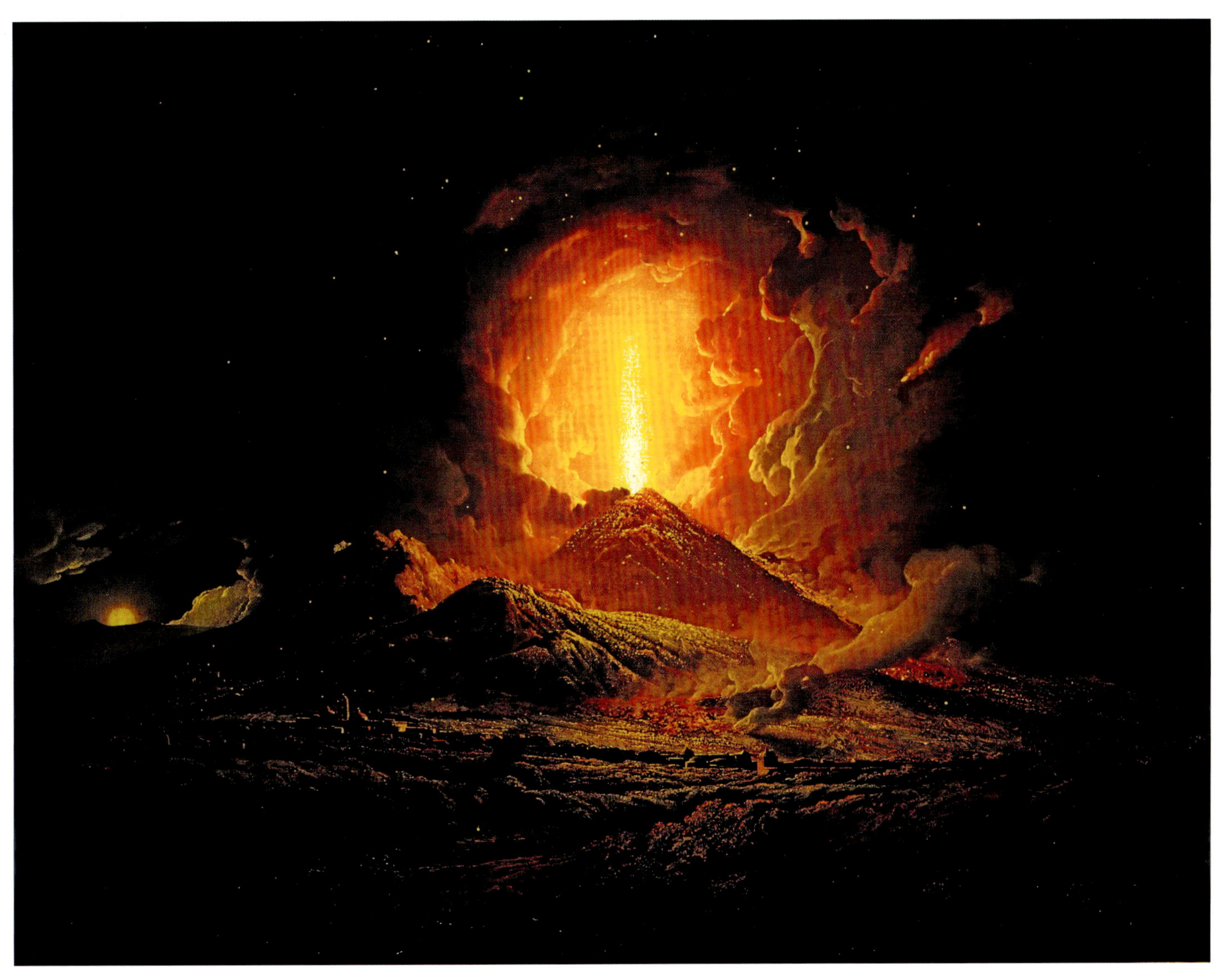

2.4 Joseph Wright of Derby (English, 1734–1797), *Vesuvius from Portici*, ca. 1774–1776. Oil on canvas, 39¾ × 50 in. (101 × 127 cm.). The Huntington Library, Art Museum, and Botanical Gardens, San Marino, California, funds from Frances Crandall Dyke Bequest.

When Wright traveled to Italy in 1773, he heard dramatic stories about Mount Vesuvius, whose last major eruption had been six years earlier. From Portici, near Naples, the artist saw constant smoke and an occasional burst of cinders still spewing out of the volcano. In this painting, Wright imagined a force of nature so strong that it sent white-hot lava shooting straight up into the sky; the buildings in the lower part of the painting give a sense of scale. On the left, the rising moon is obscured by smoke and clouds. The terrifying volcano is immense, and, like Edmund Burke's sublime, it has the power to destroy.

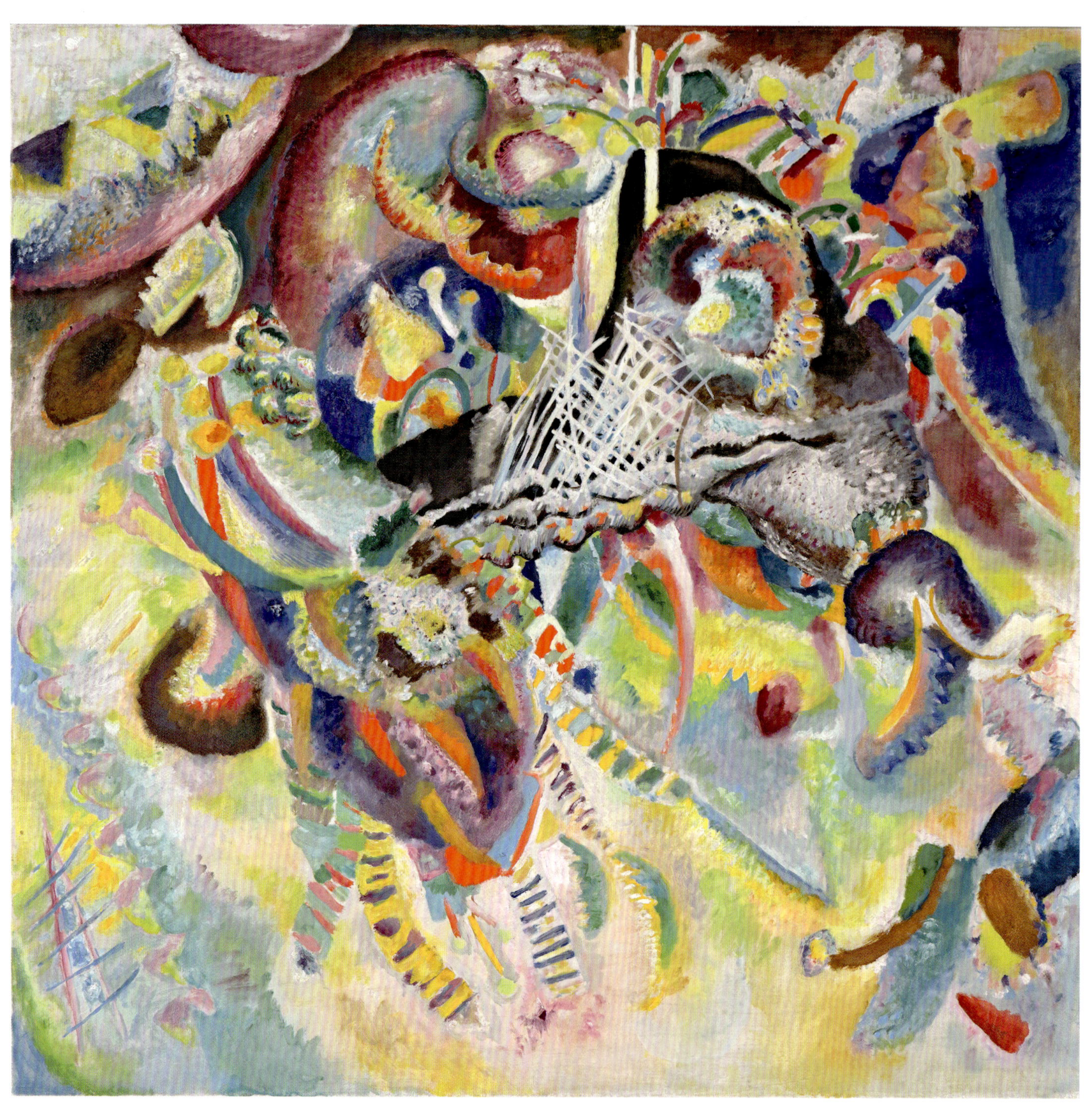

2.5 Wassily Kandinsky (Russian, 1866–1944), *Fugue*, 1914. Oil on canvas, 51⅞ × 51⅞ in. (129.5 × 129.5 cm). Fondation Beyeler, Riehen/Basel, Switzerland.

THE CREATION OF ABSTRACT ART

Artists express the worldview of their times: Greek artists carved statues of Zeus, Chinese artists painted nature to express the Tao, and modern artists created abstract art to convey a view of human experience shaped by science. During the Enlightenment, telescopes and microscopes expanded human vision to worlds that were too distant or small to see with the unaided eye. Nineteenth-century scientists discovered invisible radiation, and they viewed nature as a web of dynamic forces without a predetermined purpose or meaning. Artists created abstract art to capture these unseen realms and forces. From the start, abstract art entailed the concept of nothingness—it was not a picture of anything—but also its opposite because abstract art could express everything: infinite possibilities, reality itself. Today, most art about black holes is abstract because they are invisible and can't be pictured.

Abstract art was invented in Germanic lands, where scientists asked theoretical questions inherited from Kant's German Idealism. In Munich and Moscow, where German was the common language and German Idealism was the shared intellectual heritage, the artists who called themselves Der Blaue Reiter (The Blue Rider) and Russian avant-garde artists reflected Germanic theoretical science by symbolizing invisible realms. In 1898, the Munich artist August Endell declared: "We are at the beginning of a totally new art, an art with forms that mean nothing and remind one of nothing and represent nothing." He was quick to add, however, that this nothingness was filled with potentiality: "Yet these forms are able to move our souls so deeply, so strongly, as before only music could do."[24] Wassily Kandinsky, a member of Der Blaue Reiter, was another artist who compared abstract art to music and saw that infinite possibilities could be symbolized by something invisible and immaterial. In 1911 he attended a performance of Arnold Schoenberg's atonal String Quartet No. 2, which the artist felt was a spiritual expression.[25] Kandinsky stated that such music is a profound expression because it is immaterial—made of nothing at all: "An artist who sees that the imitation of natural appearances, however artistic, is not for him—the kind of creative artist who wants to, and has to, express his own inner world—sees with envy how naturally and easily such goals can be attained in music, the least material of the arts today."[26] After Kandinsky moved into complete abstraction, he gave his paintings musical titles (figure 2.5).

As science developed in the West, so did resistance to its core values: objectivity and rationality. This began in the 1880s when the German mathematician Georg Cantor, in Romantic rebellion against soulless science, developed his philosophy of the infinite, which culminated in a divine "Absolute Infinity" (*Absolut Unendlichen*).[27] Cantor's ideas about infinity were popularized in Moscow by the Russian Orthodox monk and mathematician Pavel Florensky, who inspired Suprematist poets and painters to symbolize infinity. The poet Alexei Kruchenykh invented the language of *zaum* (Russian for "beyond reason"), which inspired Kazimir Malevich to develop a painterly counterpart that would describe the artist's vision of a realm beyond reason. This followed in the tradition of negative theology, which originated in the third century CE with the Desert Fathers—Christian mystics who lived in North Africa. Around 500, the Eastern Orthodox monk and mystic Pseudo-Dionysius wrote that God is so transcendent that he can be approached only indirectly by way of negation. In other words, one can only say what God is *not*. Malevich felt kinship with painters of Eastern

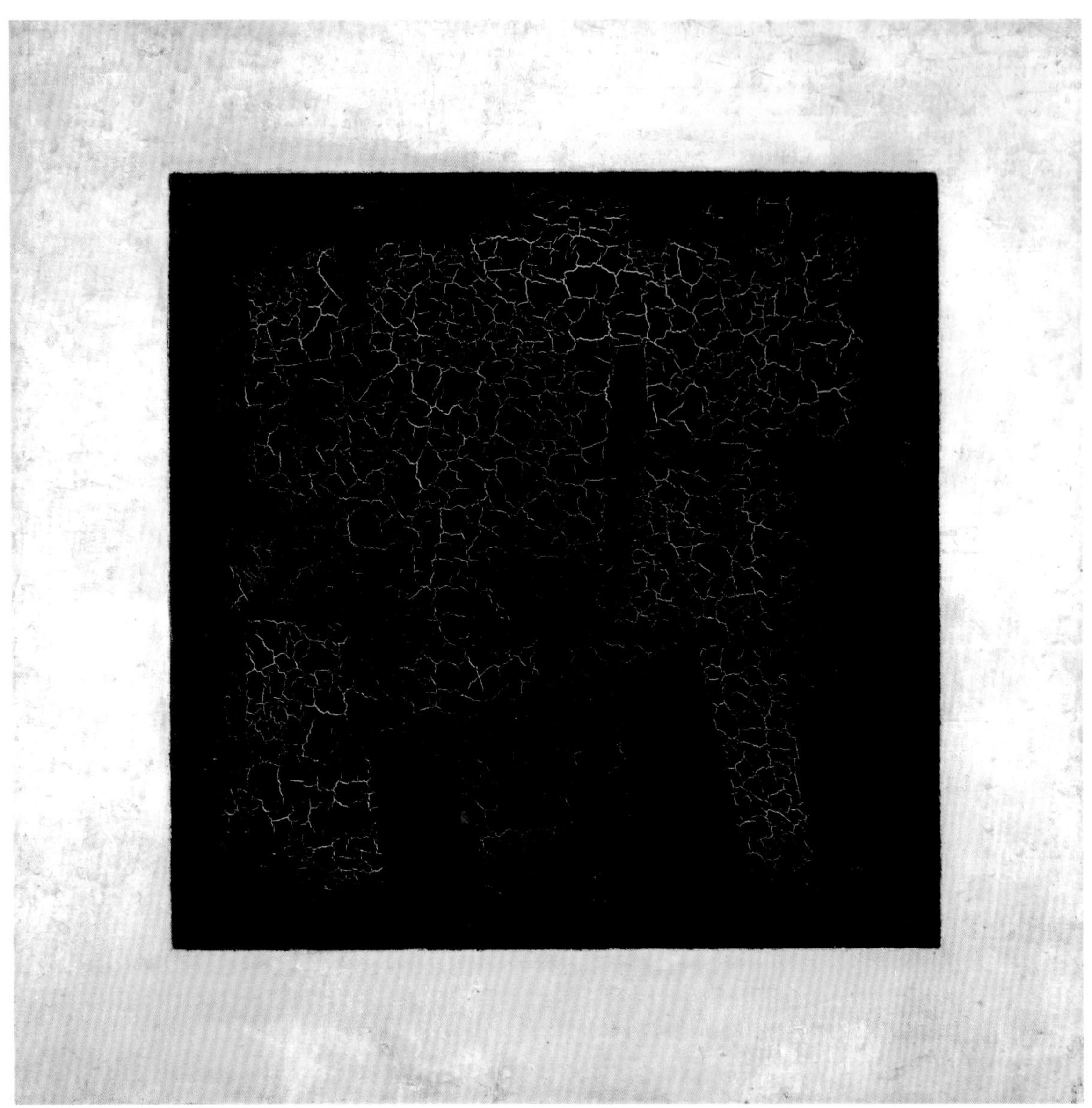

2.6 Kazimir Malevich (Russian, 1879–1935), *Black Square*, 1915. Oil on canvas, 31¼ × 31¼ in. (79.5 × 79.5 cm). State Tretyakov Gallery, Moscow.

Malevich painted *Black Square* over an earlier work; the surface has cracked over the years, revealing colors underneath.

Orthodox icons: "I was strongly moved by icons. I felt some kinship and something splendid in them. I saw in them the entire Russian people with all their emotional creativeness."[28] Malevich expressed the irrationality of *zaum* in his 1915 nonobjective painting *Black Square* (figure 2.6), and he described being attracted to nonobjective art: "A blissful sense of liberating non-objectivity drew me forth into the desert, where nothing is real except feeling."[29] Once scientists discovered that the laws of physics no longer applied within a black hole, abstract artists drawn to irrationality were also drawn to black holes.

A reference to the nothing/everything dichotomy was made by Malevich's contemporary, the Russian Constructivist artist Aleksandr Rodchenko, who created artwork that was both meaning-free and filled with meaning. In 1921 Rodchenko painted a meaning-free colored rectangle called *Red* (figure 2.7) and then *gave* the color and the form meaning by

2.7 Aleksandr Rodchenko (Russian, 1891–1956), *Red*, 1921. Oil on canvas, 24½ × 20¾ in. (62.5 × 52.5 cm). © A. Rodchenko and V. Stepanova Archive, Moscow.

applying them to practical tasks for the new Communist society after the Russian Revolution of 1917.

In 1915, while artists in Germany and Russia were developing abstract art, Karl Schwarzschild was postulating black holes to Einstein. The creative efforts of these artists and scientists were highly theoretical and owed their intellectual lineage to German thought and culture. Later in the twentieth century, the art world and science community outgrew these Germanic roots to become truly international in scope.

THE POST-1945 MERGER OF EASTERN THOUGHT AND WESTERN MODERNISM

Both Eastern thought and Western modernism entail the relationship of nothing and everything, and after World War II ended in 1945, these outlooks merged. Western modernism was attractive to Eastern artists as their countries adopted the scientific worldview from the West, and Eastern Zen Buddhism appealed to Western agnostics who felt there was neither meaning nor purpose in the world after the war.

World War II ended after the American military dropped atomic bombs on Hiroshima and Nagasaki—a human-created sublime—prompting Japan's surrender to the Allied powers. As the Allies occupied Germanic lands, the world realized the full horror of the Holocaust.

After decades of isolation from Europe, Japanese artists were eager for contact with Western avant-garde art movements. By 1970 Japan had made tremendous advances in technology and boasted one of the world's largest economies. Many Japanese artists, however, felt their art had become too Westernized and that Japan's economic gains had come at a loss of traditional spirituality. There were many attempts to forge modern art styles with a distinct Japanese identity, such as Kazuo Ohno's *ankoku butoh* (Japanese for "dance of darkness"; figure 2.8) and Morita Shiryū's *bokujinkai* (Japanese for "people of the ink"; figure 2.9).

The politics of the era was dominated by the Cold War between the United States and the Communist states in China and the Soviet Union, which flared into proxy wars fought on Asian soil in the Korean War (1950–1953) and the Vietnam War (1964–1975). After the Soviet Union developed nuclear weapons in 1949, it and the United States refined rocket technology with the intention of annihilating each other with nuclear bombs. As Einstein remarked in that year: "I know not with what weapons World War III will be fought, but World War IV will be fought with sticks and stones."[30] Many other countries have since developed nuclear weapons. Soviet and American scientists appropriated rocket technology to launch satellites into space, beginning with Sputnik (1957).

The bleak atmosphere of the Cold War was reflected in art about blackness and nothingness. Mordecai Ardon left his native Poland for Germany to study art, attending the Bauhaus from 1921 to 1925 and teaching at the Swiss designer Johannes Itten's school in Berlin. In 1933 Ardon fled Germany and settled in Jerusalem, where he joined the faculty of the New Bezalel School of Arts and Crafts, becoming its director in 1940. After World War II, he took up subjects related to the Holocaust. In his painting *Landscape with Black Sun* (figure 2.10), the sky is the color of blood and the sun is black. Theodor Adorno, a German Jew who spent World War II in exile, returned to his homeland after the war and wrote: "If works of art are to survive in the context of extremity and darkness, which is social reality, and if they are to avoid being sold as mere comfort, they have to assimilate themselves to that reality. Radical art today is the same as dark art; its background color is black."[31]

The American Abstract Expressionist Ad Reinhardt studied Asian art from 1944 to 1952 at the Institute of Fine Arts in New York. During these years he traveled throughout Asia and became a Taoist, writing: "The *tao* is through and through dark and mysterious."[32] Like a black hole, Reinhardt's paintings may appear to show nothing, but something is definitely there (figure 2.11). The artist structured this painting on the format of the Taoist

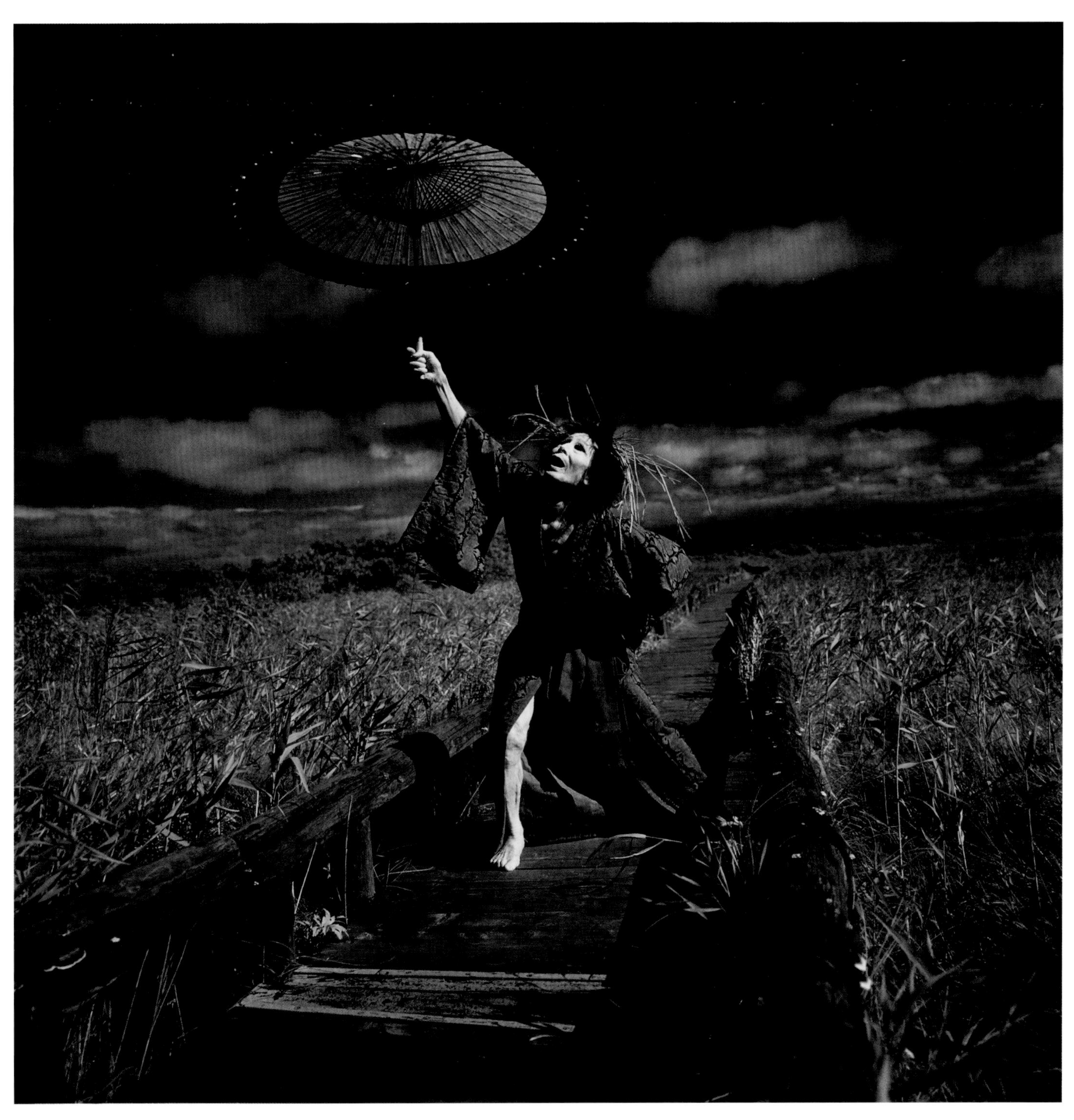

2.8 Eikoh Hosoe (Japanese, 1933–2024), *Kazuo Ohno, Dancing in the Kushiro Marshland IV*, 1994. Gelatin silver print. Courtesy of Taka Ishii Gallery Photography/Film, Tokyo.

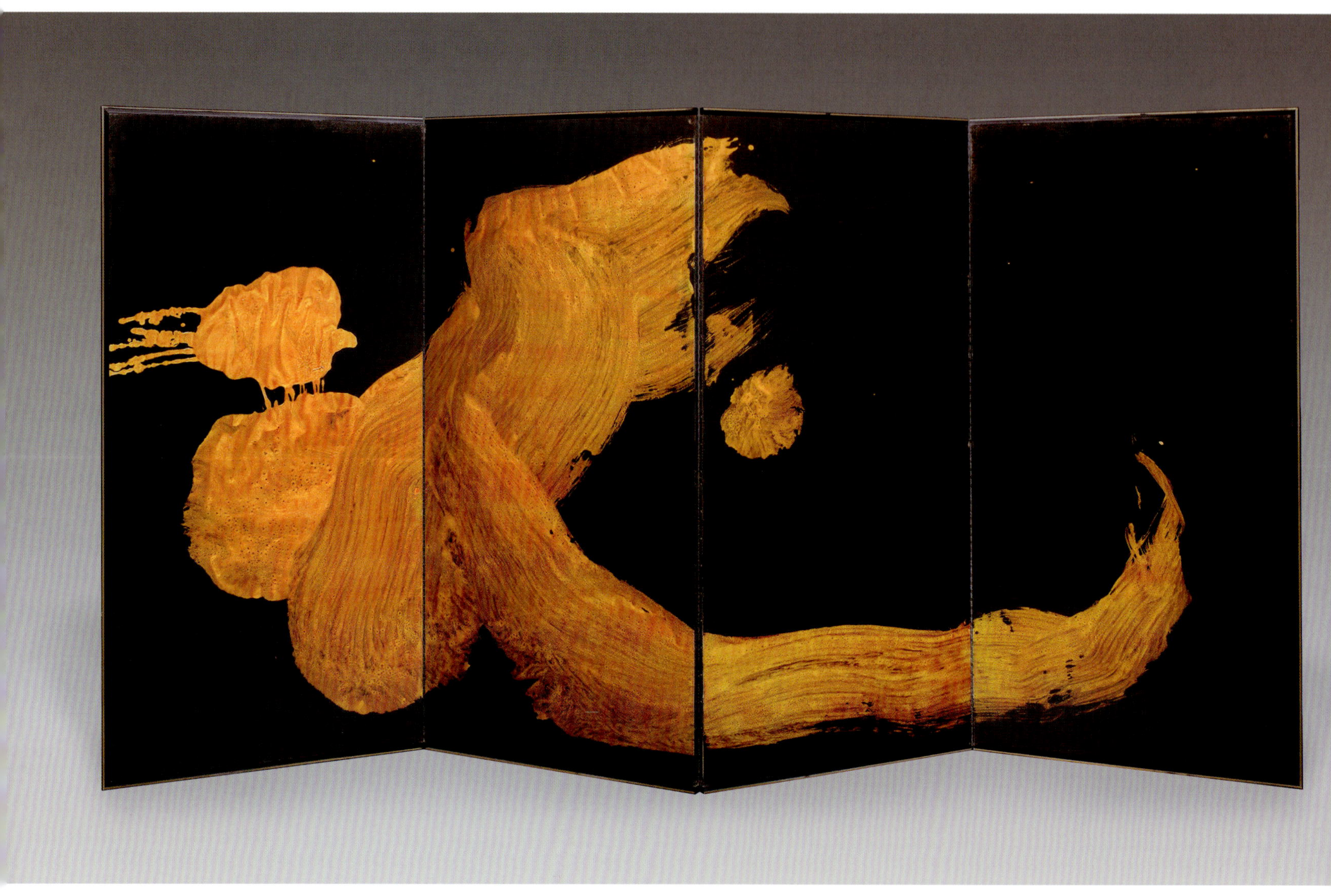

2.9 Morita Shiryū (Japanese, 1912–1998), *Ryu* (Dragon), 1965. Four-panel screen, aluminum-flake pigment in polyvinyl acetate medium and yellow alkyd varnish on paper, 63⅝ × 122 in. (161.5 × 310 cm). The Art Institute of Chicago, gift of Charles C. Haffner III.

Morita Shiryū was a founder of *bokujinkai*, which combined Eastern calligraphy with Western gestural abstract painting. Unlike traditional calligraphy in which the glyphs have a set meaning, Shiryū painted made-up characters such as this.

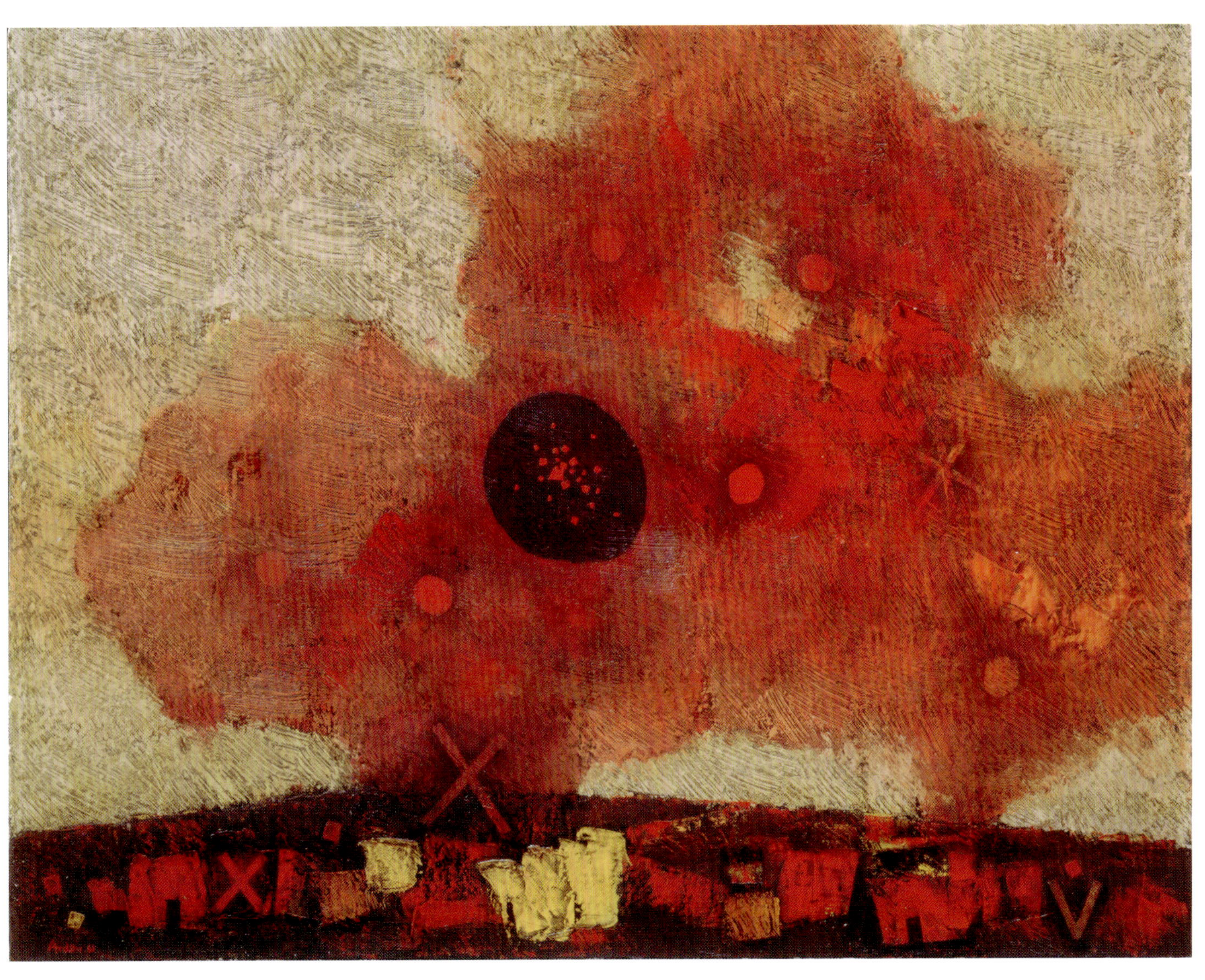

2.10 Mordecai Ardon (Israeli, born Poland, 1896–1992), *Landscape with Black Sun*, 1961. Oil on canvas, 31 × 39¼ in. (78.7 × 100 cm). © Ardon Estate. The Jewish Museum, New York, Beren Foundation Fund.

2.11 Ad Reinhardt (American, 1913–1967), *Abstract Painting*, 1963. Oil on canvas, 60 × 60 in. (152.4 × 152.4 cm). Museum of Modern Art, New York.

magic square, a three-by-three grid that symbolizes the creation of the cosmos, which he then painted in different shades of black.

The Japanese scholar D. T. Suzuki was the leading spokesman for Zen Buddhism in the West for half a century, and his contact with the Abstract Expressionist circle was part of the merger of Eastern thought and Western modernism. While teaching philosophy at Columbia University in New York in the late 1940s and 1950s, Suzuki began a conversation with the Trappist monk Thomas Merton. Merton was, in turn, a good friend of Ad Reinhardt, who did a painting for his friend's monastic cell (*Small Painting for T. M. [Thomas Merton]*, 1957).[33] Merton thought that much of the mystical spirituality of the Desert Fathers had been lost in the modern Christian church, and he vowed to combine the tradition of negative theology with Zen Buddhism.[34] The American composer John Cage was a student of Suzuki's at Columbia for three years, attending his lectures on Zen Buddhism. Cage adopted the ancient Chinese magical system known as the *I-Ching*, which is described in the divination manual *Zhouyi* (Book of Changes). Cage used the *I-Ching* to determine the pattern and rhythm of the notes when composing music such as *Music of Changes* (1951).

Like Suzuki and the Abstract Expressionists, artists in India and China also merged Eastern and Western influences in their art. Early twentieth-century Indian artists were taught to speak English and paint in Western styles at the British-run Government College of Art and Craft in Kolkata. After Mahatma Gandhi led the nonviolent Indian Independence Movement and India won its independence from Britain in 1947, Indian artists merged Western modern art with native traditions. In more recent decades, Chinese artists have begun visiting the West, merging modernism with China's ancient traditions and language.

As we have seen, in Western creation myths a divine person imposed order on chaos, while in Hinduism, Buddhism, and Taoism nature came into being by self-assembly following the mysterious ways of nature. Western philosophers discussed being, while Eastern philosophers focused on nothingness—which, enigmatically, is the source of everything. Westerners sought to understand the order the Creator had imposed, and Newton posited the law of universal gravitation, thus beginning modern science. As Westerners searched further, they found evidence, such as Darwin's theory of evolution, that undermined their own religious traditions. Furthermore, Western scientists concluded that the East had been right all along—nature *did* come into being without the aid of a divine being. Modern art developed in the West in response to scientific breakthroughs, as both artists and scientists tried to make sense of the world. As Asian countries adopted a scientific worldview from the West, they also imported modern art. Japanese artists forged modern art styles with a distinct Japanese identity, Indian artists merged modern art with native traditions, and artists in China fused modernism with Chinese philosophy and language. It was in this historical and philosophical context that black holes were discovered, and it was on this stage that artists produced visual works born from the need to grapple with this profound shift in our understanding of the universe.

FOLLOWING PAGES

2.12 Several radio telescopes in the Atacama Large Millimeter Array (ALMA) in Chile, observing the Milky Way. ESO/José Francisco Salgado.

The dark bands across the plane of the galaxy are clouds of interstellar dust that are opaque to visible light but transparent to electromagnetic waves that are in the millimeter (radio) range.

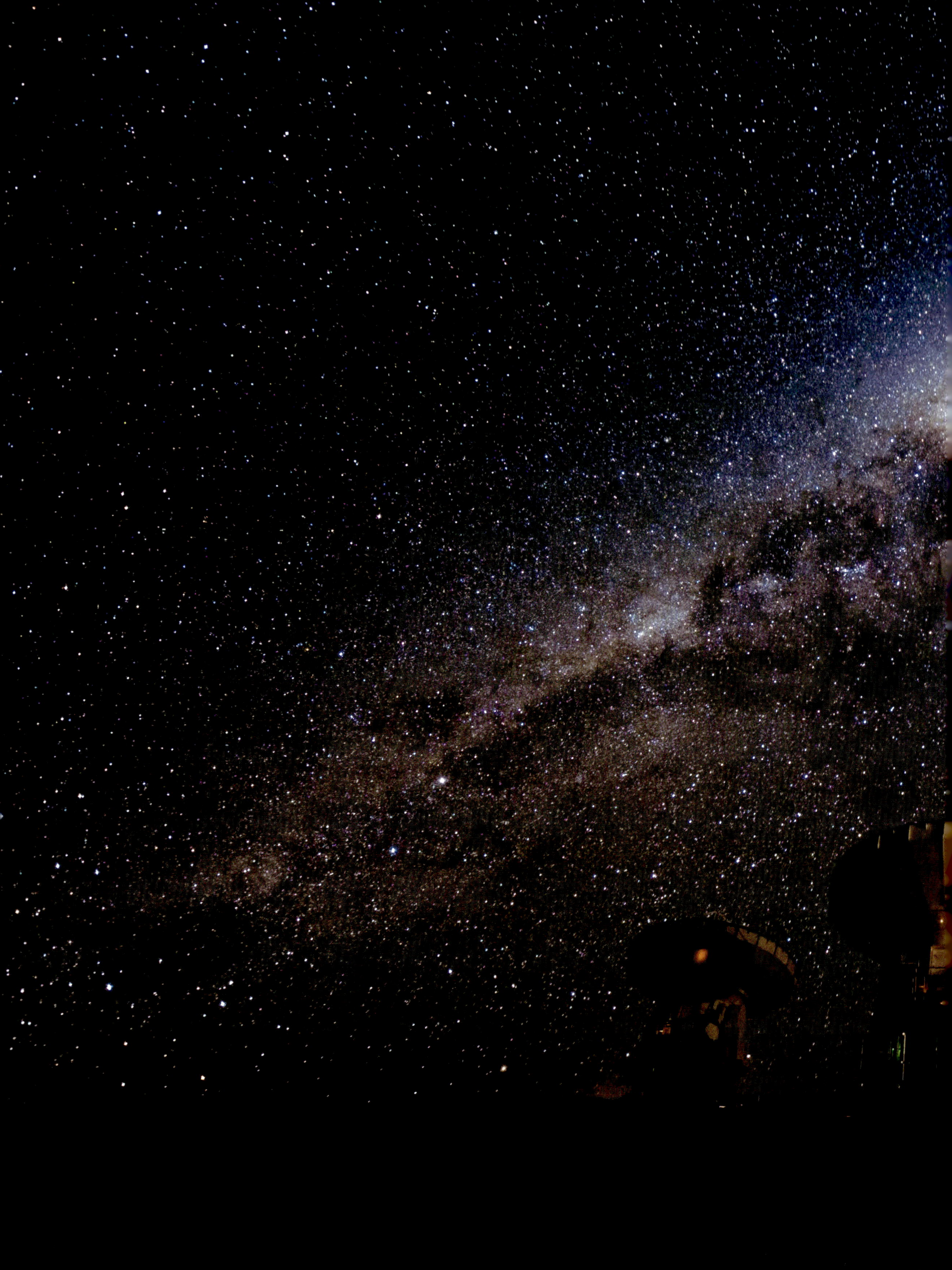

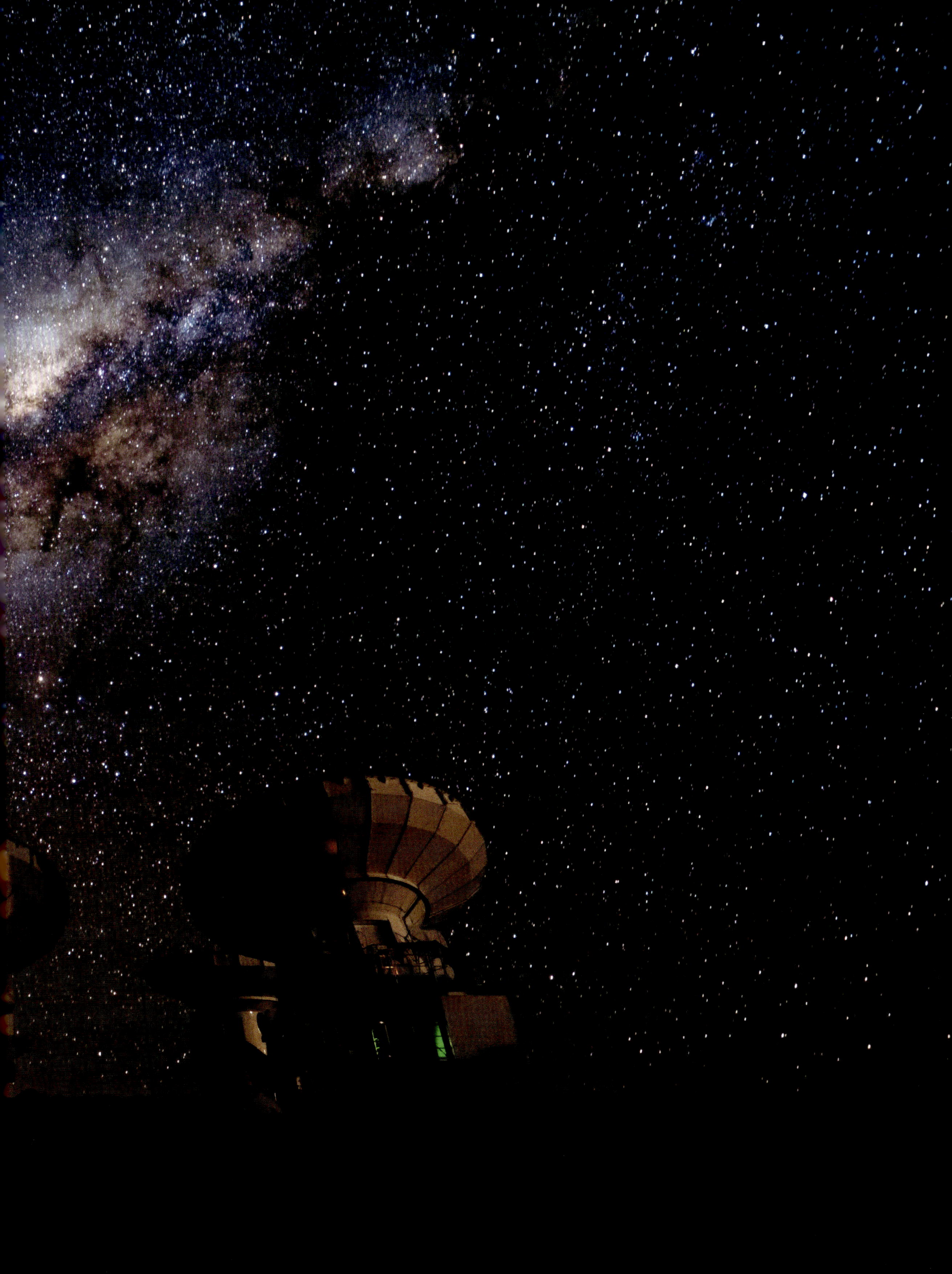

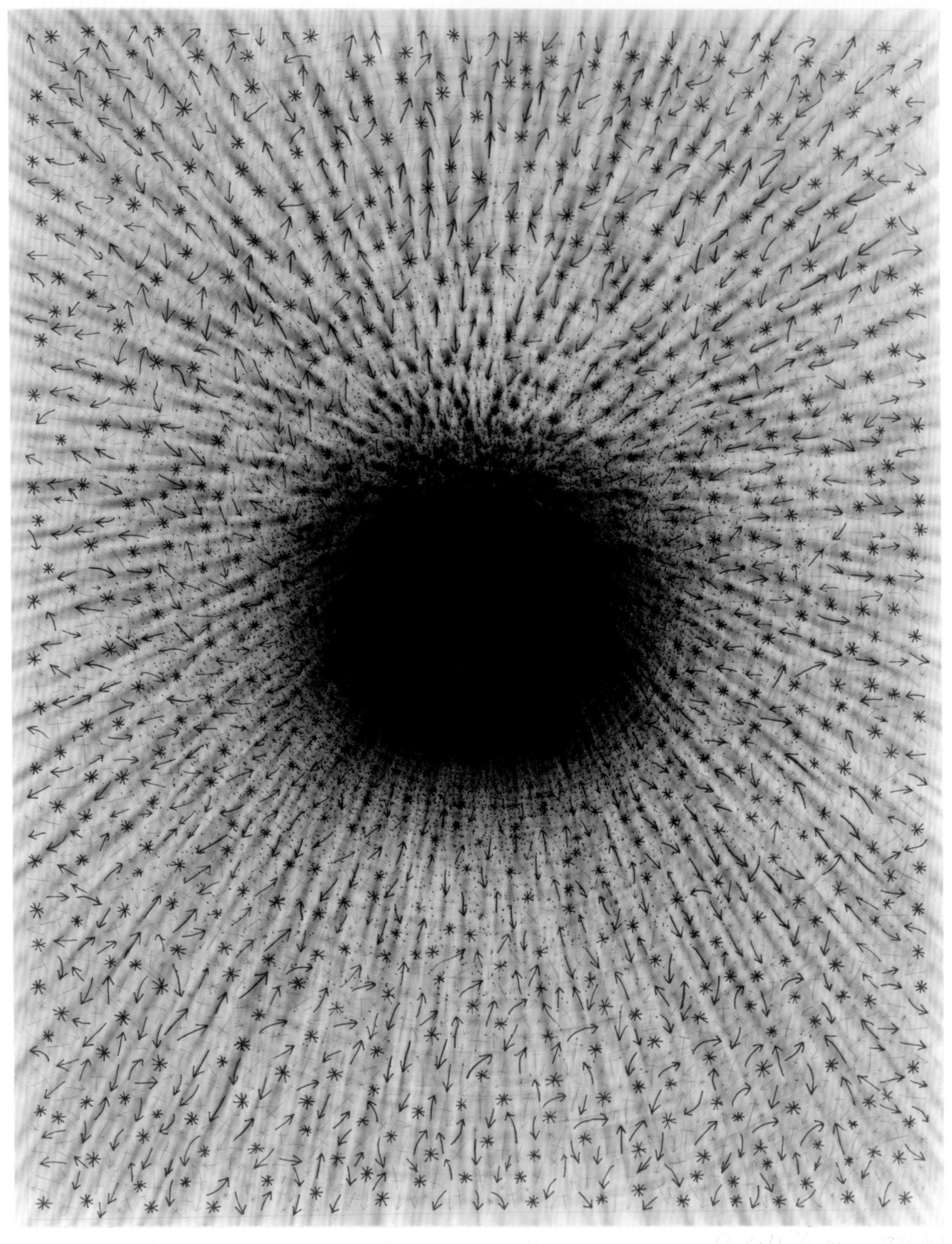

3

ARTISTIC AND SCIENTIFIC IMAGES OF INVISIBLE OBJECTS

In the early 1970s the existence of black holes was reported in scientific papers and newspapers around the world, introducing the phenomenon to the culture's imagination. Scientists symbolized data in charts, graphs, and mathematical formulae and attempted to make images of black holes. But seeing an object requires light, so rather than depicting a black hole itself, scientists imagined what matter surrounding it would look like. Artists, in turn, subjected scientific data to the transformation of the imaginative process and created something completely new: artworks.

EARLY ARTISTIC IMAGES OF BLACK HOLES

In the decades before scientists showed that black holes exist, several artists in the West—the American Barnett Newman, the Argentine-Italian Lucio Fontana, the American Lee Bontecou, and the Englishman John Latham—made abstract art about dark voids. Newman described a post-1945 sublime: "We now know the terror to expect. Hiroshima showed it to us."[35] In 1946 he painted *Pagan Void* (figure 3.2), a dark abyss in the middle of a shape that resembles a mushroom cloud. Abstract Expressionists were familiar with Asian traditions in which nothingness is sacred, but the year after the atomic bomb, Newman painted a *pagan* void. Beginning in 1949 Fontana stretched a canvas and punctured it with holes to create passageways between the terrestrial world and outer space—a metaphorical infinite void beyond the canvas (figure 3.3)—saying: "Einstein's discovery about the cosmos is the infinite, endless dimension."[36]

When Bontecou began her career in the 1950s, she learned to weld steel frameworks over which she stretched canvas until it was taut, creating relief

3.1 Rudolf Sikora (Slovak, born 1946), *Black Hole II*, 1976–1978, from the series *Concentration of Energy*. Photograph, 14½ × 11¼ in. (37.0 × 28.5 cm). Slovak National Gallery, Bratislava, Slovakia.

The artist drew symbols on a manipulated photograph and then rephotographed the image, making the final artwork a black-and-white photograph.

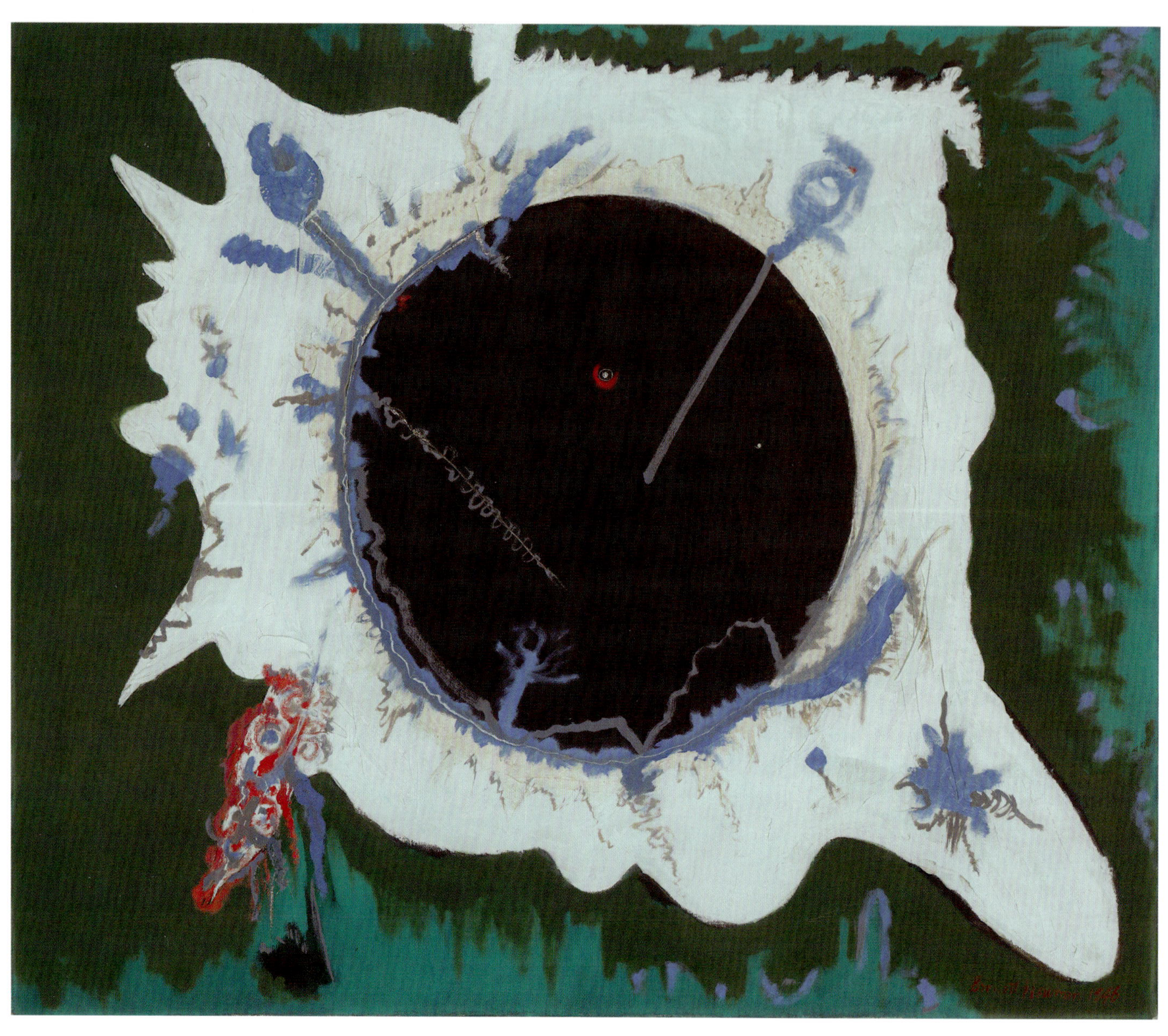

3.2 Barnett Newman (American, 1905–1970), *Pagan Void*, 1946. Oil on canvas, 33 × 38 in. (83.8 × 96.5 cm). National Gallery of Art, Washington, DC, gift of Annalee Newman, in honor of the 50th anniversary of the National Gallery of Art.

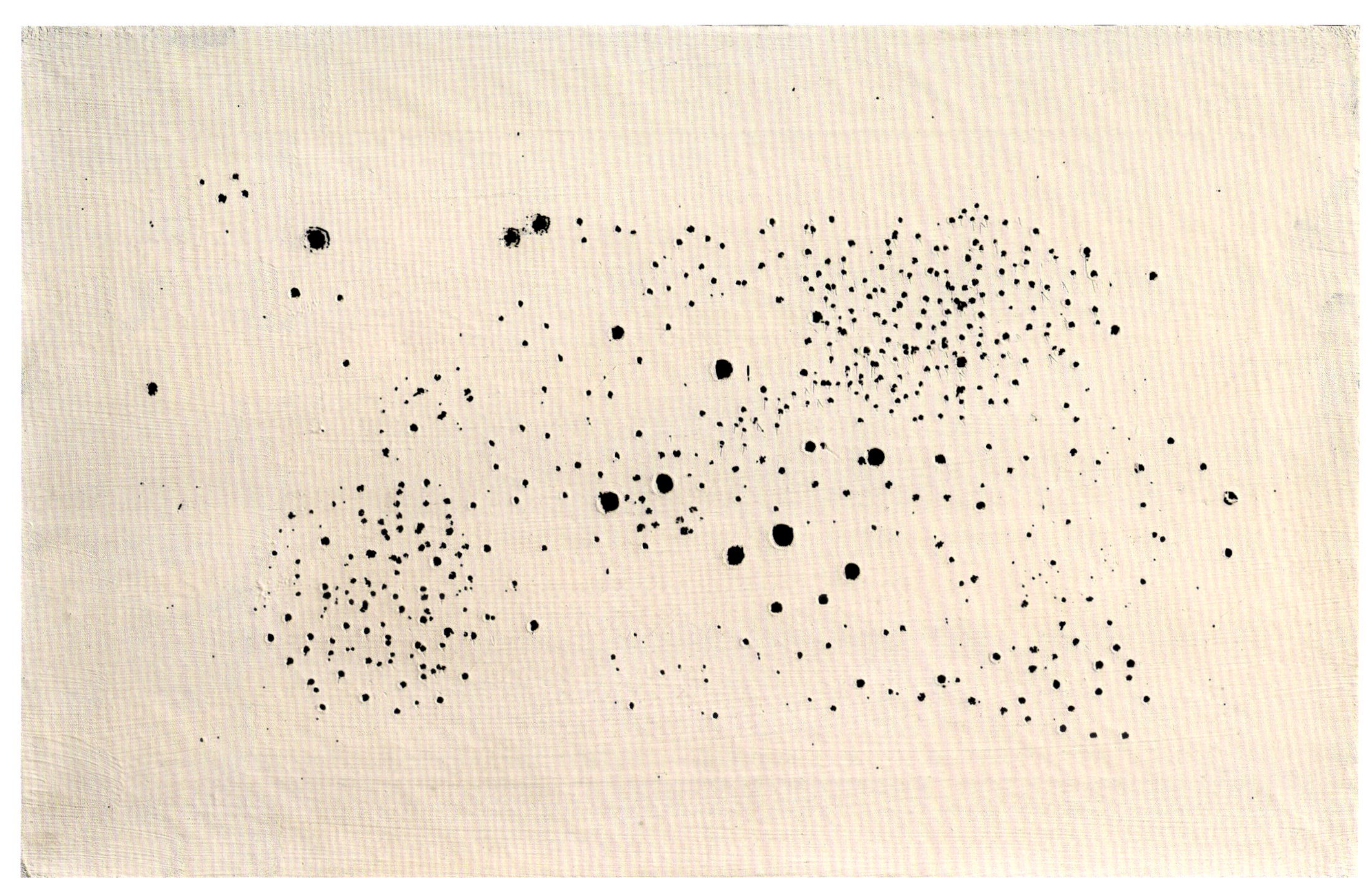

3.3 Lucio Fontana (Argentine and Italian, 1899–1968), *Concetto spaziale* (Spatial concept), 1952. Oil on canvas, 14 × 21½ in. (35.5 × 55 cm). Private collection. Courtesy of Christie's Milan. © Fondazione Lucio Fontana, Milan.

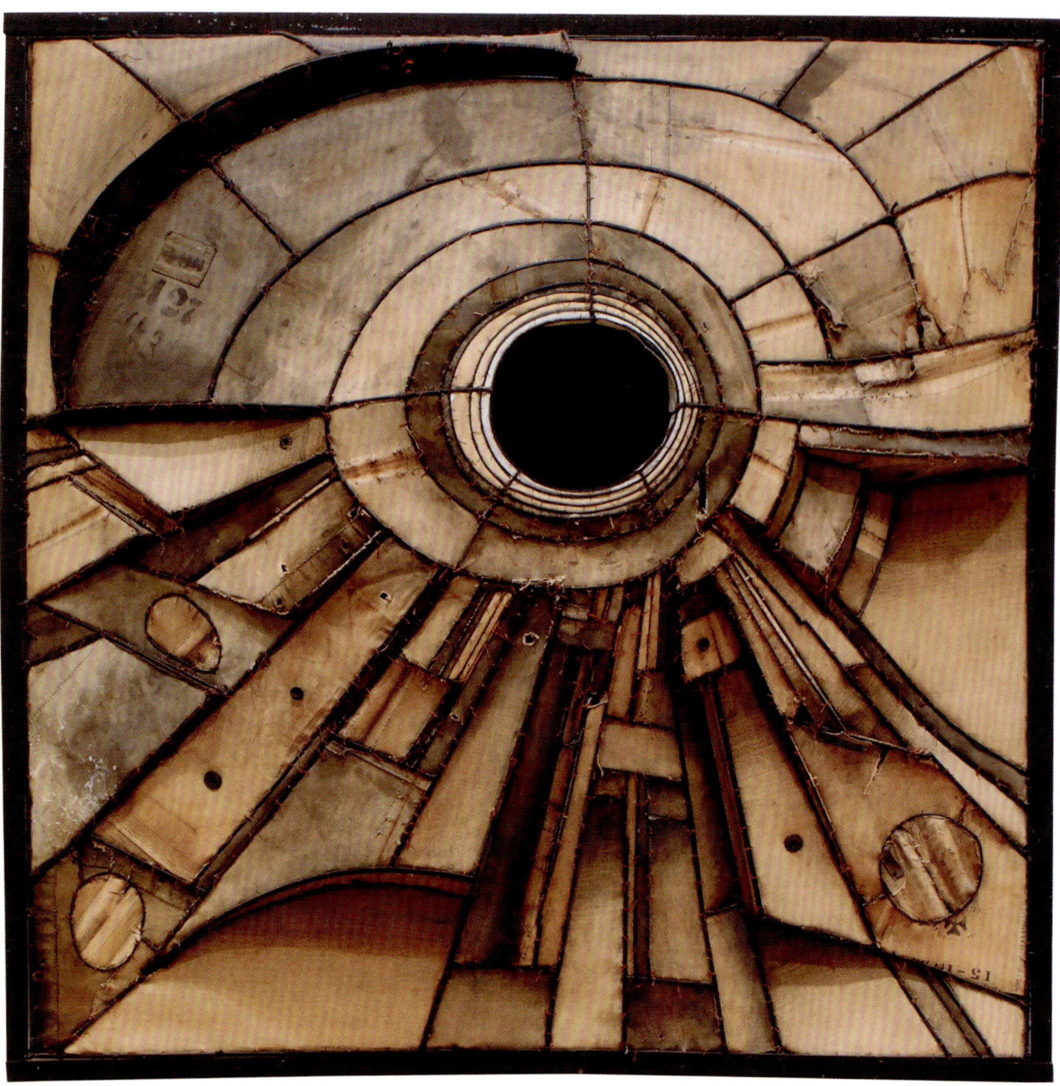

3.4 Lee Bontecou (American, 1931–2022), Untitled, 1960. Steel, canvas, and copper wire, 72 × 72 × 18 in. (182.9 × 182.9 × 45.7 cm). The Art Institute of Chicago, gift of Society for Contemporary American Art.

sculptures centered around a hole that she lined with black velvet so that it wouldn't reflect light (figure 3.4). In the late 1960s she was at work in her studio when she heard on the radio that astronomers were searching the sky for enigmatic objects they called "black holes." Having been making black holes for years, she recalled her reaction: "I thought: Oh yes! Thank you!"[37] In Bontecou's later work, the galactic references are clear. In a large suspended sculpture, forms move dynamically around a still center that resembles a black hole (figure 3.5).

Latham's early work in the 1950s and 1960s expressed the anxiety of the Cold War era; he staged, for example, book-burning events to signal the end of culture. Latham made the very large ethereal painting *Full Stop* (figure 3.6) with a spray gun, shooting black acrylic paint onto unprimed canvas. The title refers to a period at the end of a sentence, but the phrase can also be said in anger, signaling that the conversation is over. Long interested in cosmology and physics, Latham wrote about black holes after Cygnus X-1 was confirmed.[38]

As scientists were confirming the existence of black holes, Frederick Eversley was imagining sculptures of them. He graduated in 1963 from the Carnegie Institute of Technology (now Carnegie Mellon University) in Pittsburgh with a degree in engineering and worked in the aerospace industry building acoustic laboratories for NASA. In 1967 he transitioned to being an artist, creating abstract sculptures in cast polyester. With his background in science, Eversley understood the significance of the

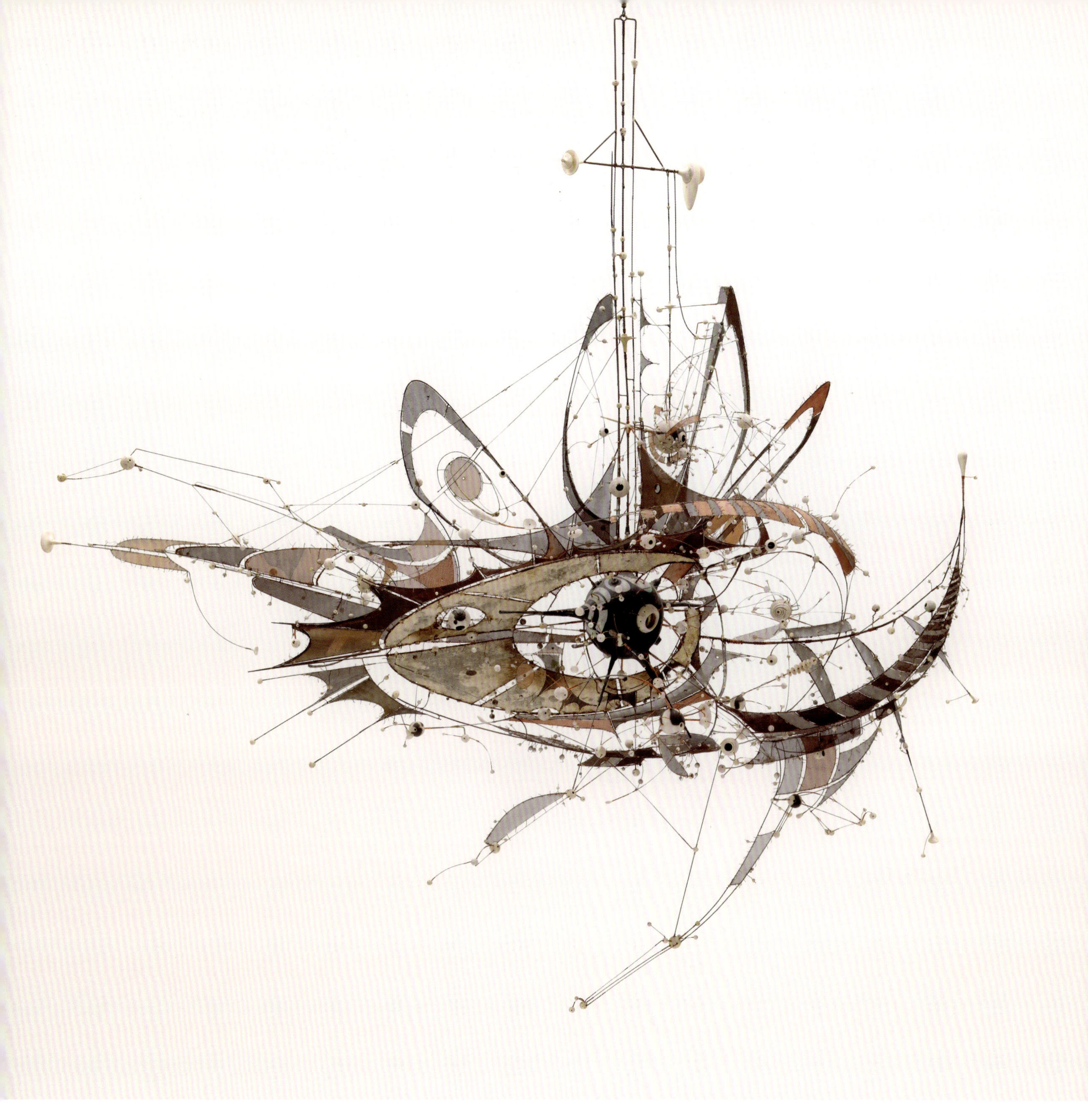

3.5 Lee Bontecou (American, 1931–2022), Untitled, 1980–1998. Welded steel, porcelain, wire mesh, canvas, wire, 7 × 8 × 6 ft. (2.13 × 2.43 × 1.82 m). Museum of Modern Art, New York, gift of Philip Johnson (by exchange) and the Nina and Gordon Bunshaft Bequest Fund.

3.6 John Latham (English, 1921–2006), *Full Stop*, 1961. Acrylic on canvas, 9 ft. 10¾ in. × 8 ft. 5½ in. (3.01 × 2.58 m). Tate, London. © The John Latham Foundation. Courtesy The John Latham Foundation and Lisson Gallery, London.

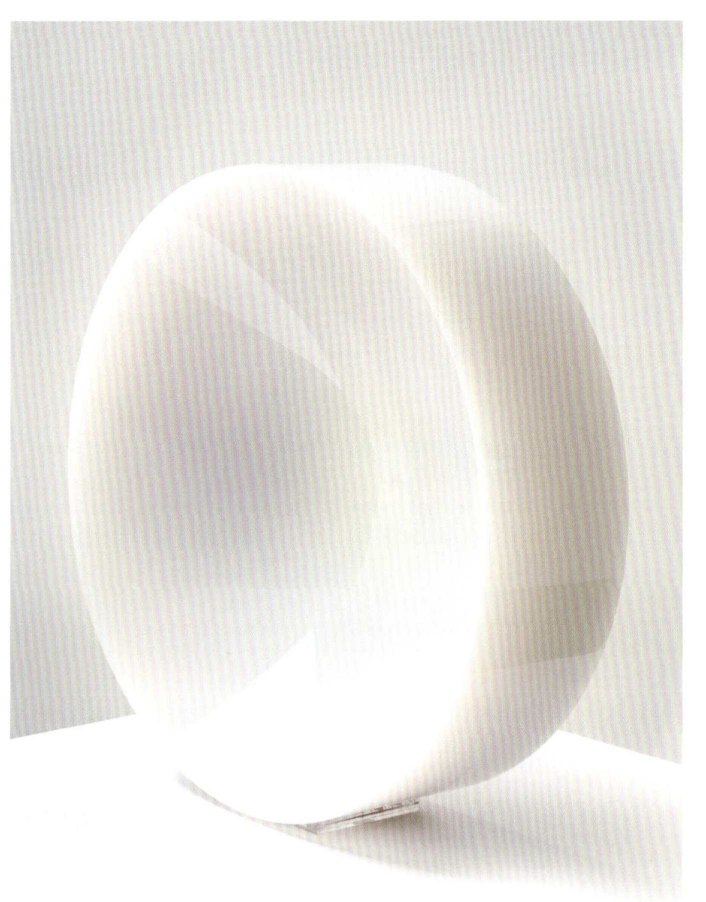

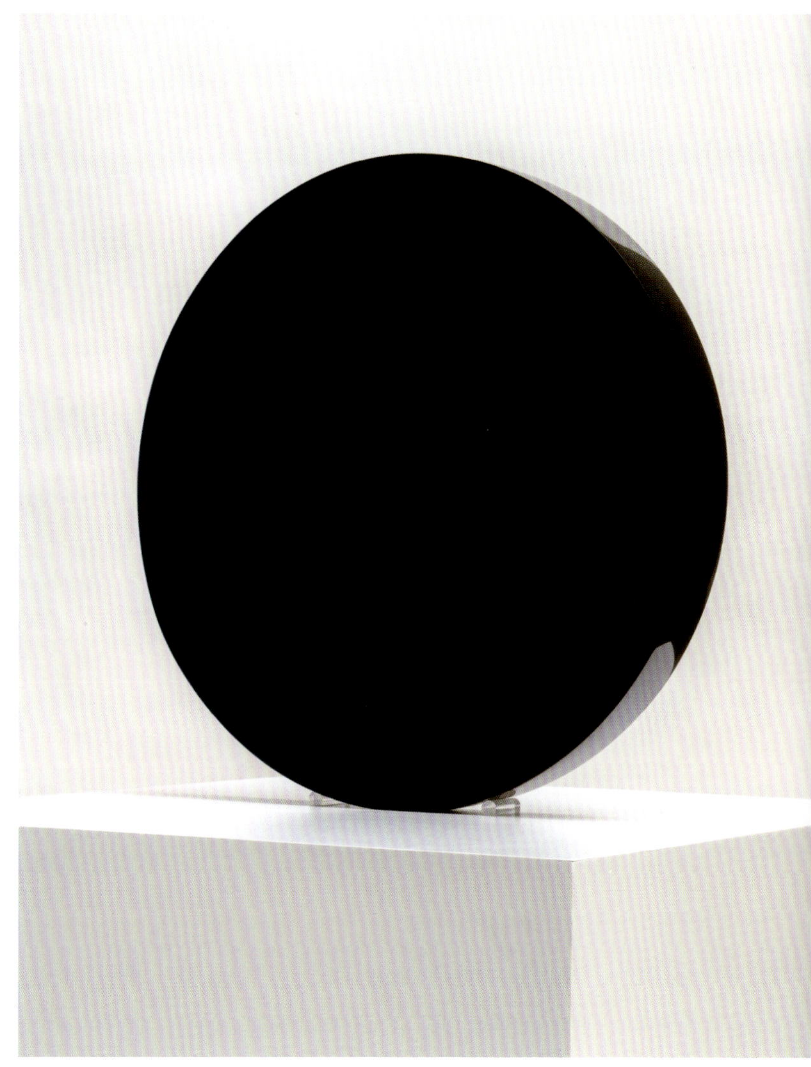

LEFT
3.7 Frederick Eversley (American, 1941–2025), Untitled (*White Dwarf*), 1974. Cast polyester, 20 × 20 × 6 in. (50.8 × 50.8 × 15.2 cm). Courtesy of the artist and Art + Practice, Los Angeles.

RIGHT
3.8 Frederick Eversley (American, 1941–2025), Untitled (*Black Hole*), 1974. Cast polyester, 20 × 20 × 2 in. (50.8 × 50.8 × 5.1 cm). Courtesy of the artist and Art + Practice, Los Angeles.

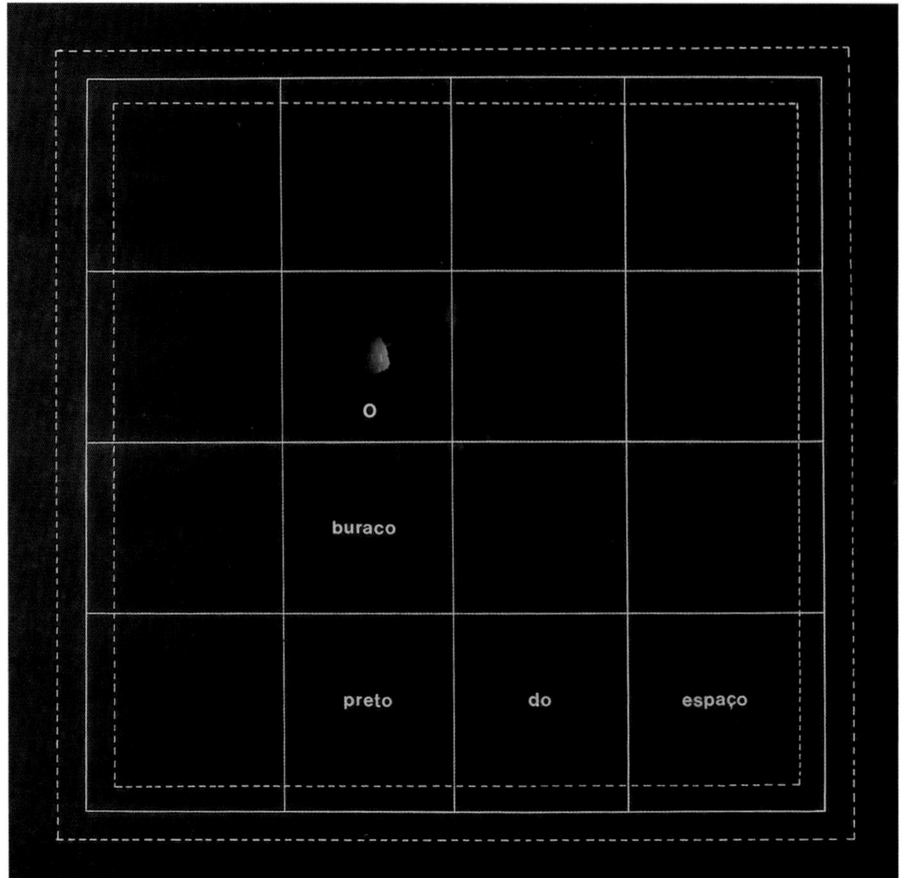

3.9 Anna Maria Maiolino (Brazilian, born 1942), *O buraco preto do espaço* (The black hole of space), 1973, from the series *Mental Maps*, 1971–1976. Acrylic, transfer lettering, graphite, and clipping on paper, 21¼ × 21¼ in. (54 × 54 cm). Museu de Arte Moderna do Rio de Janeiro. Courtesy of Hauser & Wirth, Zurich.

discovery of Cygnus X-1 in 1971. Two years later he created *White Dwarf* and *Black Hole* (figures 3.7 and 3.8). In an artist's statement, he wrote: "Black holes, white dwarfs, and neutron stars are important components of modern cosmological studies," and he went on to say that his sculptures are "literal representations of these phenomenon." Eversley concluded by stating that he was "representing the very transcendental nature of the concept of energy as postulated by Einstein's special theory of relativity."[39]

The year Cygnus X-1 was confirmed, the Brazilian artist Anna Maria Maiolino began a series of artworks about her life called *Mental Maps*. At the time, she was in her thirties and surrounded by political danger triggered by Brazil's military dictatorship. In that series, she made a work titled *O buraco preto do espaço* (The black hole of space; figure 3.9). Maiolino associated black holes not only with the astronomical phenomena with but also with the voids in the work of Lucio Fontana.[40] Whereas most artists in the early 1970s didn't pay much attention to black holes because there were no visualizations of them to fire their imaginations, Maiolino became fascinated with holes filled with darkness. In 1974 she created an artwork called *Black Hole* (figure 3.10) consisting of torn edges, dark shadows, and concentric circles that seem to close in on one another. Perhaps she was drawn to this theme because—like the prison in Calcutta—nothing can escape from black holes, and they became a metaphor for the inescapable danger surrounding her in Brazil in the 1970s.

Black holes were a metaphor for resistance to political repression in the work of Rudolf Sikora—in his case, from the Communist government of Czechoslovakia in the early 1970s. In the wake of World War I and the dissolution of the Austro-Hungarian Empire, Czechoslovakia was formed in 1918 by joining Czech lands with Slovakia. After World War II, the Communist Party seized power in Czechoslovakia in a coup backed by the

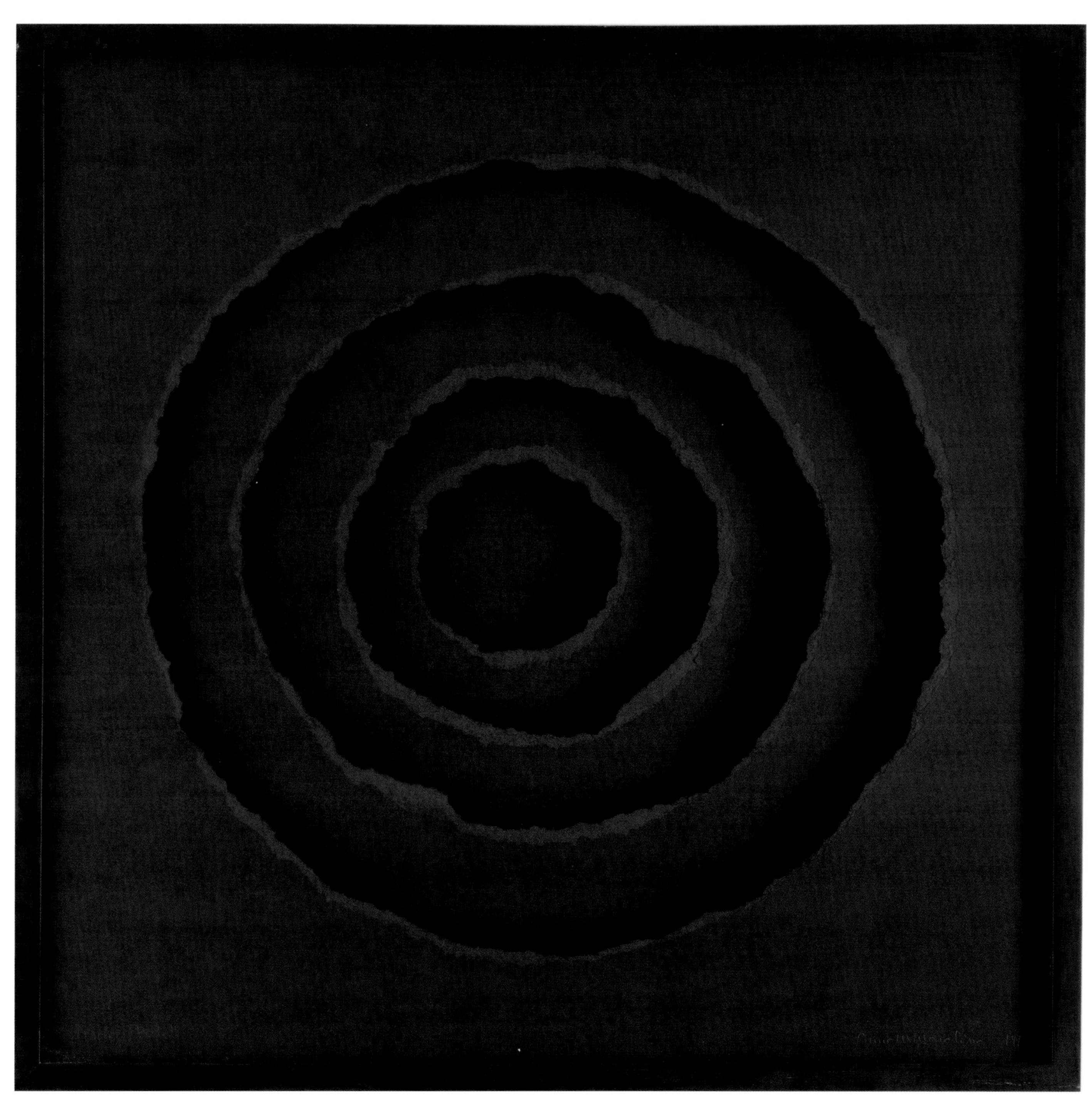

3.10 Anna Maria Maiolino (Brazilian, born 1942), *Black Hole*, 1974, from the series *Holes/Drawing Objects*. Torn paper, 27 × 27 in. (68.6 × 68.6 cm). Museum of Modern Art, New York. Courtesy of Hauser & Wirth, Zurich.

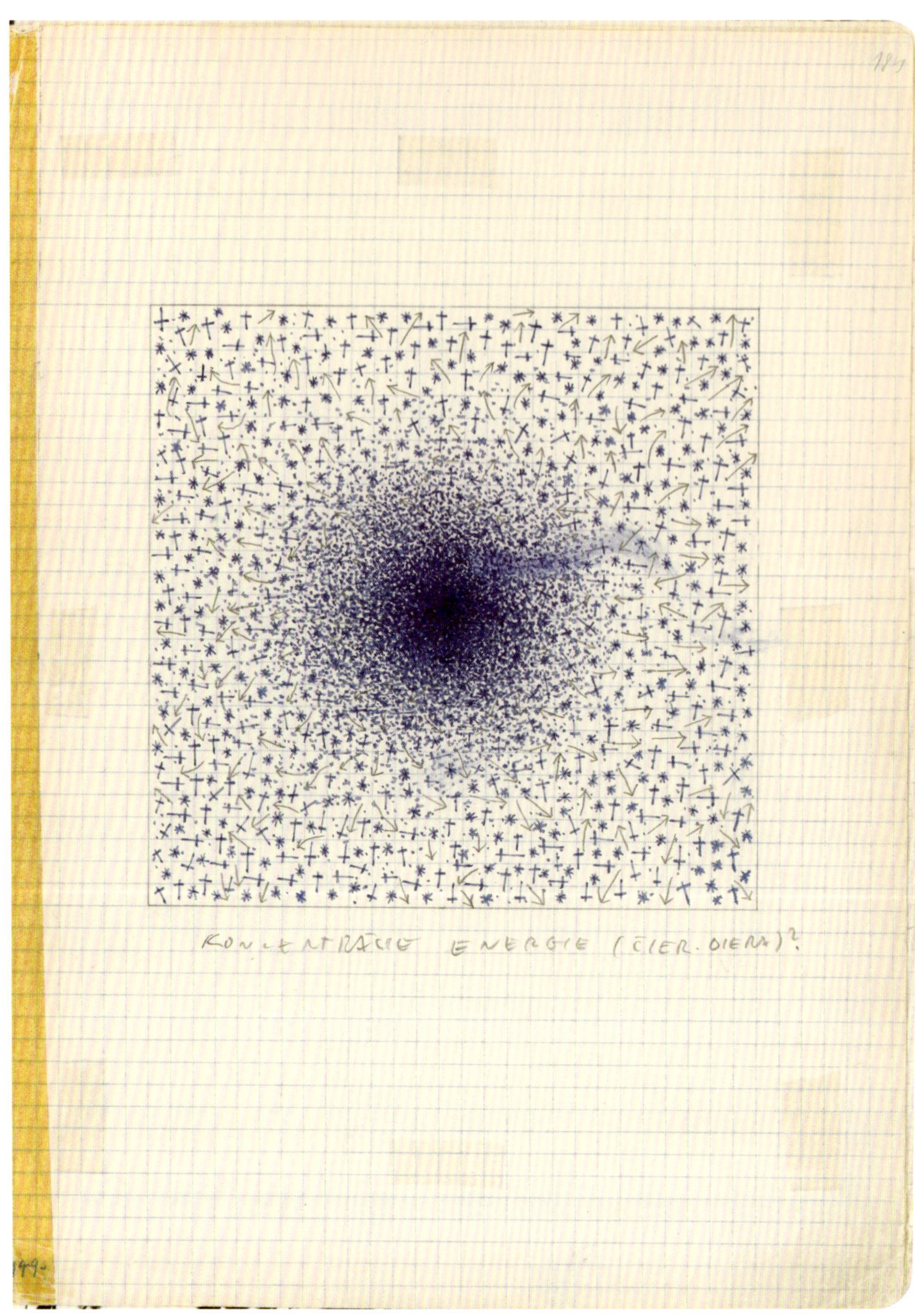

3.11 Rudolf Sikora (Slovak, born 1946), *Concentration of Energy (Black Hole)*, 1970–1975. Pencil and ink on graph paper, 11¼ × 8 in. (28.8 × 20.2 cm). Slovak National Gallery, Bratislava, Slovakia.

Soviet Union. In 1968 a brief liberal period—the Prague Spring—ended violently when Soviet troops invaded Czechoslovakia. Two years later, Sikora organized an exhibition in his studio of artwork protesting the Communist government. The next day the police interrogated him, and he was officially blacklisted. Thus, in 1970 Sikora became part of the Czech underground art scene and developed a metaphorical vocabulary. As scientists were confirming that Cygnus X-1 is a black hole, Sikora began a series called *Concentration of Energy* featuring black holes (figures 3.1 and 3.11). In these works black holes are surrounded by symbols representing origin (*), motion (→), and death (†). As a powerless citizen in an authoritarian state, Sikora saw black holes as a symbol of resistance. He did artwork related to black holes throughout the 1970s and 1980s until, in 1989, the citizens of Czechoslovakia peacefully rid themselves of the Communist government in the Velvet Revolution. No longer blacklisted, Sikora was appointed a professor at the Academy of Fine Arts and Design in Bratislava and the Academy of Fine Arts in Prague. In 1992 Czechoslovakia peacefully split into the Czech Republic and Slovakia (the Velvet Divorce).

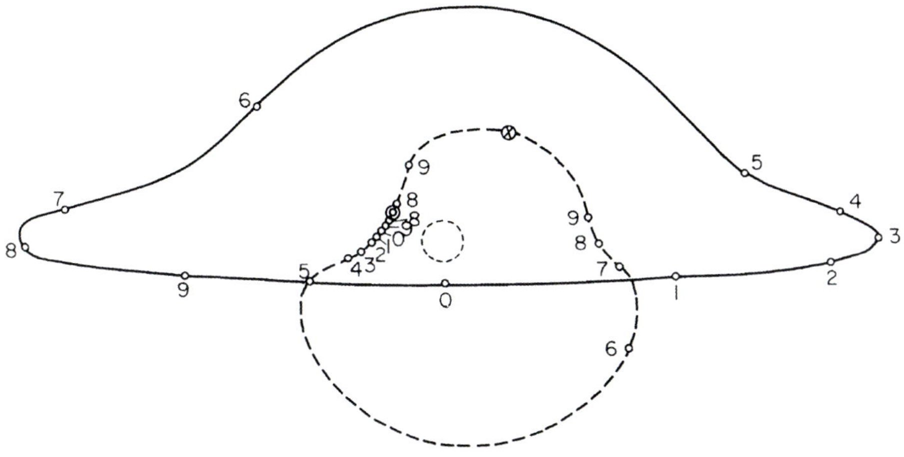

3.12 C. T. Cunningham and James M. Bardeen, "The Optical Appearance of a Star Orbiting an Extreme Kerr Black Hole," *Astrophysical Journal* 183 (1973): 253, fig. 8a.

This illustration shows a star orbiting around a spherical rotating black hole. The small dashed circle indicates the location of the black hole. The solid line indicates the initial path of the star, beginning at 0 and moving to the right; gravitational lensing deforms the star's circular orbit between 4 and 7, making the star's path appear to bulge upward. The larger dashed line indicates the path of the star after it has made one orbit around the black hole; at one point (between the two 9s on the left) the star appears to briefly orbit in retrograde (backward).

EARLY SCIENTIFIC IMAGES OF BLACK HOLES

While Eversley, Maiolino, and Sikora were in their studios making artworks about black holes, the American physicists C. T. Cunningham and James Bardeen were in their laboratory creating an illustration of the deformations in spacetime around a black hole. They imagined a distant observer seeing a star orbiting a black hole at a uniform distance. They knew that the rapidly rotating black hole's gravity affects light passing through its gravitational field in a manner similar to a powerful lens, hence the observer would see light that is distorted by what astronomers call *gravitational lensing*. Cunningham and Bardeen calculated these optical deformations and in 1973 produced the first scientific visualization of spacetime around a black hole (figure 3.12). Melissa Walter's sculpture in figure 3.13 is a metaphor for gravitational lensing. Light passes through cut paper that sways and curves, distorting the light like a gravitational lens. Walter, unlike many artists, understands the crucial distinction between a science illustration and an artwork. Under her maiden name, Melissa Weiss, she works for NASA, executing science illustrations of how a black hole might actually appear, such as the image of Cygnus X-1 and its companion star (figure 1.23). Under her married name, Melissa Walter, she creates artworks such as this. Speaking about the development of her oeuvre, she said: "Abstraction has been the common thread throughout that evolution as it relates to humanity's place in the cosmos."[41]

What would gravitational lensing do to the cloud of dust and gas that orbits a black hole called the accretion disk? The French astrophysicist Jean-Pierre Luminet wanted to make a realistic picture of an accretion disk. Associating realism with photography, he imagined the black hole "as seen by a distant observer" taking a "photograph" from a stationary, authoritative viewpoint.[42] In Luminet's diagram, the accretion disk forms a flat, circular disk of dust and gas that he labeled the "equatorial plane (disk)" (figure 3.14). Friction and magnetic forces heat the accretion disk to hundreds of billions of degrees until it becomes an incandescent plasma emitting radiation. The observer looks down on the disk from a slightly elevated position (at a 10-degree angle, labeled "observer's direction"). While the accretion disk and stars emit light in all directions, for simplicity's sake

3.13 Melissa Walter (American, born 1976), *Gravitational Lensing*, 2018. Cut paper, magnets, steel, and paint, height approx. 12 ft. (3.6 m). Courtesy of the artist.

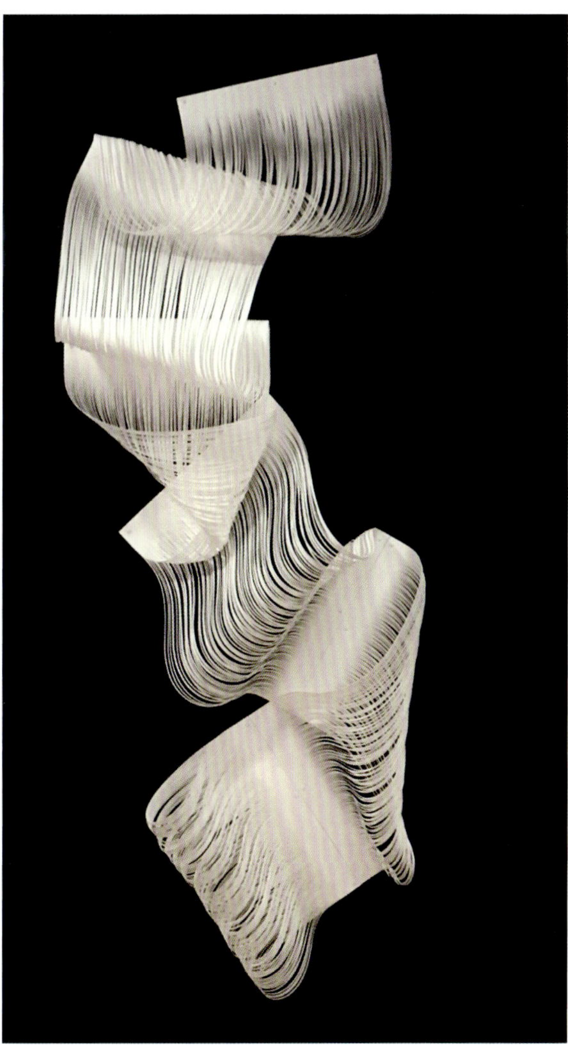

3.14 Diagram of light in Jean-Pierre Luminet, "Image of a Spherical Black Hole with Thin Accretion Disk," *Astronomy and Astrophysics* 75, nos. 1–2 (1979): 231, fig. 4.

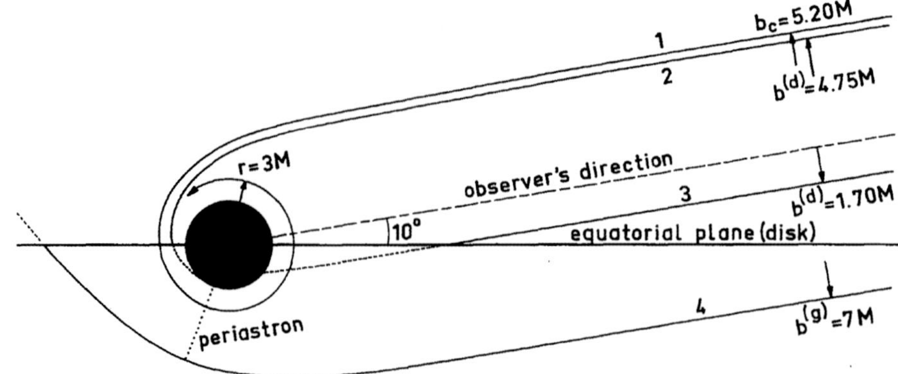

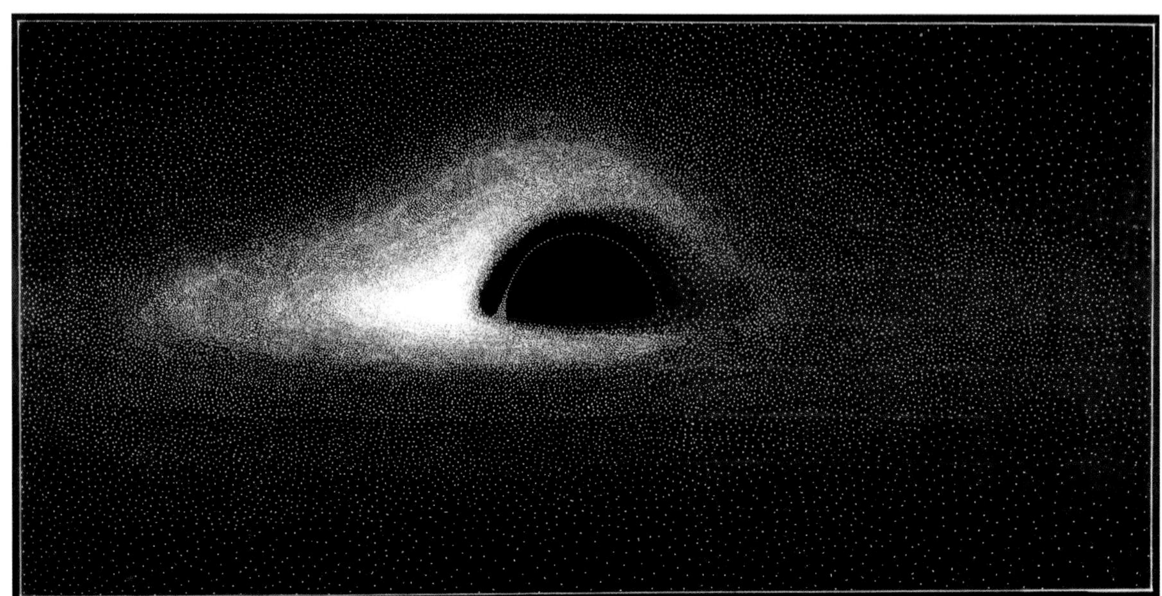

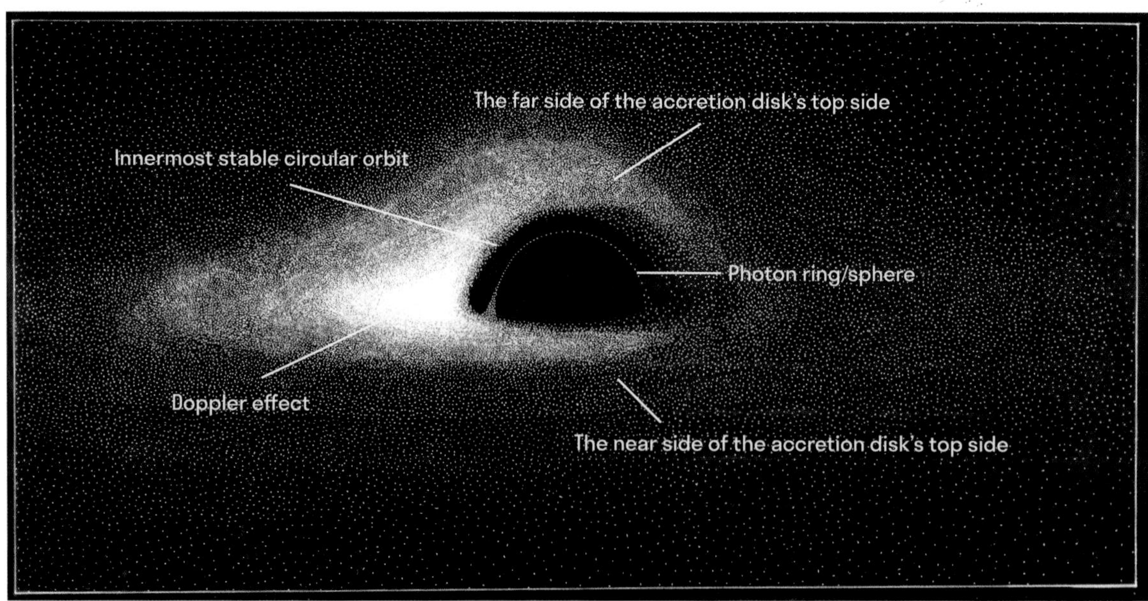

3.15 Drawing of a black hole by Jean-Pierre Luminet. Ink on paper, reversed photographically. "Image of a Spherical Black Hole," 235, fig. 11. For a detail of this image, see figure 0.1 on pages 8–9.

3.16 Drawing of a black hole by Luminet (with labels).

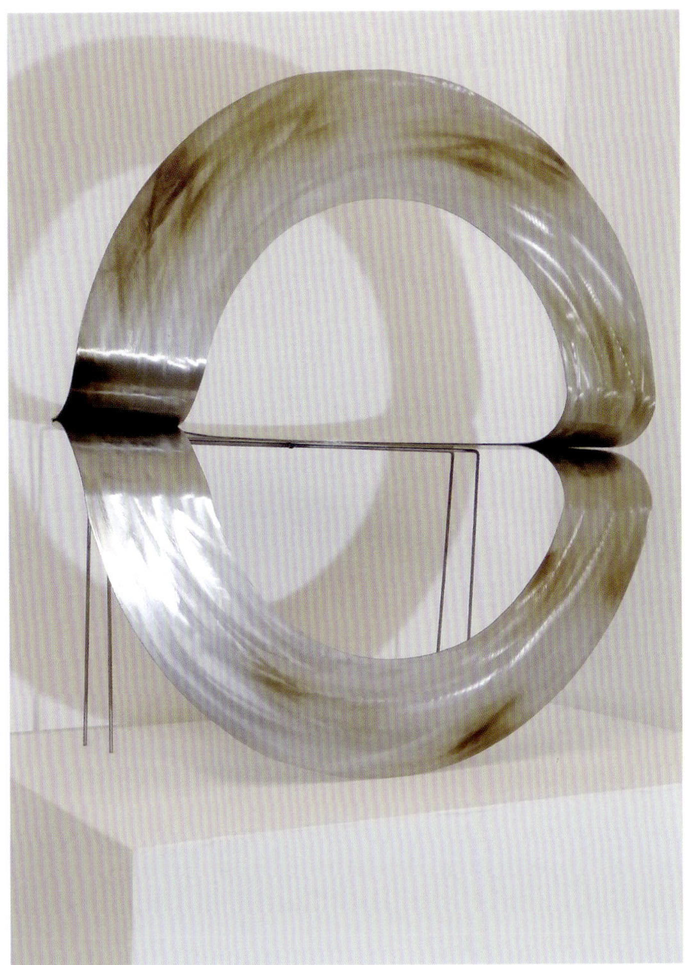

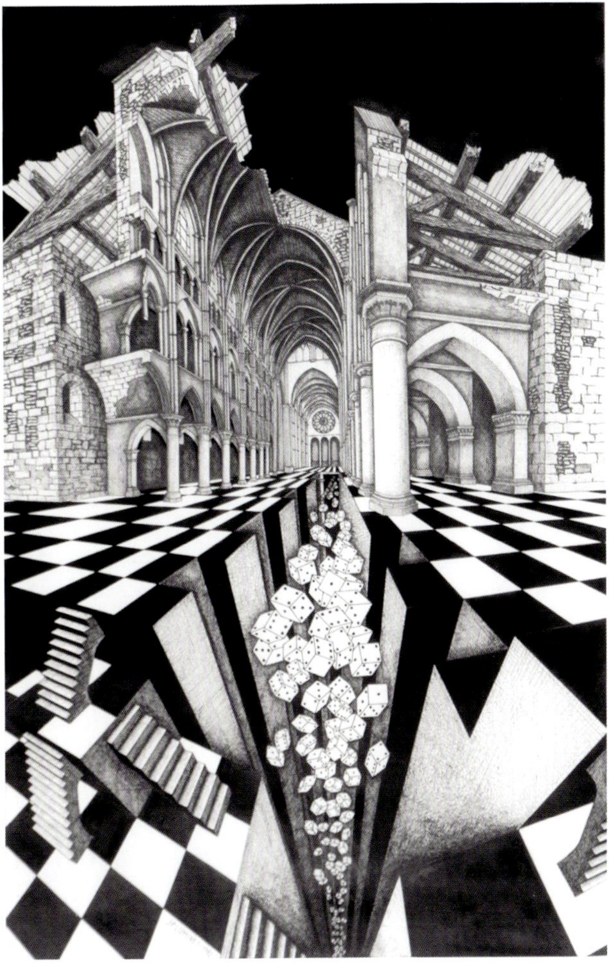

Luminet imagined parallel light rays (numbered 1, 2, 3, and 4 in his diagram) coming from the observer's direction. The curvature of spacetime in the vicinity of the black hole causes light ray 1 to bend around the black hole in a circular path. Light ray 2 goes behind the black hole and strikes the top of the accretion disk, so the observer sees the far side of the top of the disk, as shown in a drawing Luminet made of a black hole (figures 3.15 and 3.16). Light ray 3 strikes the accretion disk in front of the black hole, which Luminet also shows in his drawing. Light ray 4 strikes the underside of the accretion disk behind the black hole, so the observer sees the far side of the accretion disk's underside. (To view this part of the accretion disk, not illustrated in Luminet's drawing, see figure 3.19, top row, third from the left, and figure 3.20).

Luminet made his drawing with tiny dots of black ink on white paper and then photographically reversed the image (see figure 0.1 on pages 8–9) so that it reads white against a black background to create a "simulated photograph" of a luminous object in the darkness of space. His drawing shows one additional optical deformation lacking in Cunningham and Bardeen's line drawing. The accretion disk displays a dramatic Doppler effect (see figure 1.21) since it's rotating close to the speed of light. Recall that light appears closer to the blue or red end of the spectrum depending on whether the source is moving toward or away from the observer. In Luminet's drawing, the disk's left side appears to be moving toward the observer, so the observed frequency (hence the energy) of the electromagnetic waves is very high. Since Luminet's image is black and white, he shows all radiation in the electromagnetic spectrum in what photographers call a *bolometric photograph*.

LEFT
3.17 Silvina Maestro (Uruguayan and British, born 1978), *Photon Sphere 1*, 2023. Steel, 19¾ × 19¾ × 15¾ in. (50 × 50 × 40 cm). Courtesy of the artist.

Maestro imagined an abstract sculpture based on the concept of a photon sphere.

RIGHT
3.18 Jean-Pierre Luminet (French, born 1951), *Black Hole*, 1982. Ink on paper, 26¼ × 16½ in. (67 × 42 cm). Courtesy of the artist.

3.19 A black hole accretion disk seen from various angles, by Jean-Alain Marck and Jean-Pierre Luminet, 1989.

Beginning at the upper left (with a view that reverses left and right in Luminet's drawing in figure 3.15), Marck calculated what an accretion disk would look like from sixteen viewpoints.

In Luminet's image, the *innermost stable circular orbit* is the smallest circular orbit in which matter can stably orbit the black hole; it's the inner edge of the accretion disk. If matter goes inside that orbit, it quickly falls past the black hole's event horizon. Since light has no mass, it can orbit within the innermost stable circular orbit. If light crosses the event horizon it will not escape, but some photons circle on a narrow path between the innermost stable circular orbit and the event horizon. Scientists call this structure a *photon ring* (some call it a *photon sphere* because it's three-dimensional; figures 3.16 and 3.17).

Luminet published his work in 1979 and concluded with these prophetic words: "Thus our picture could represent many relatively weak sources, such as for instance the supermassive black hole whose existence in the nucleus of M87 has been suggested recently."[43] Forty years later, the black hole in the center of galaxy M87 was imaged by the Event Horizon Telescope (figure 3.138).

Luminet is also an artist who works in the style of M. C. Escher. In *Black Hole* (figure 3.18), the space above the checkerboard floor is described by

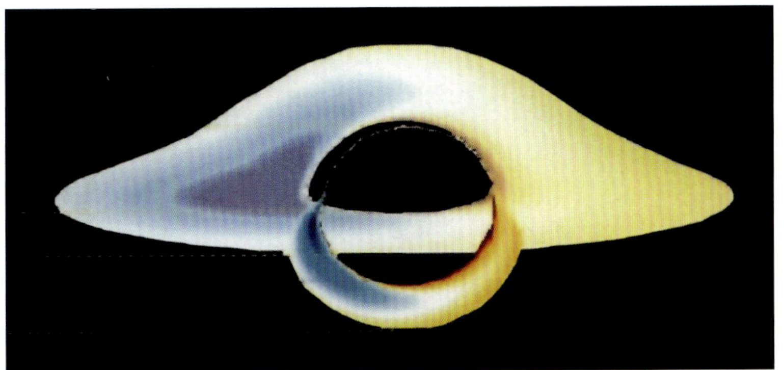

3.20 Edge-on view of a black hole in C. Fanton et al., "Detecting Accretion Disks in Active Galactic Nuclei," *Astronomical Society of Japan* 49, no. 2 (April 1997): detail of fig. 7.

Newton's classical mechanics and depicts the nave of the cathedral of Notre-Dame drawn in linear perspective. The floor opens to reveal a metaphorical black hole whose description requires quantum mechanics, so below he drew tumbling dice and tipping planes to suggest probability theory.[44]

Jean-Alain Marck—Luminet's colleague at the Paris-Meudon Observatory—was an expert in general relativity, computer programming, and calculating geodesics around a black hole.[45] A *geodesic* is the shortest distance between two points on a curved plane. (A line is the shortest distance between two points on a flat plane.) In 1989 Marck calculated the geodesics describing the accretion disk in Luminet's drawing from various angles and, for dramatic effect, added color (figure 3.19). An image of a black hole from 1997 shows the far side of the accretion disk's top side and underside (figure 3.20). Marck and Luminet's image had shown this view earlier (figure 3.19, top row, third from left), but it remained unpublished.

In the early 1990s Marck and Luminet collaborated on a sequence about black holes for a television documentary that was broadcast across Europe. Luminet had drawn his image by hand in the late 1970s because computer graphics programs were not available, but by the 1990s the technology had advanced and Marck was able to write the animation program himself. Marck's calculation is unusual because it shows what a moving observer—riding a magic carpet and wearing a bow in her hair—would see flying past a Schwarzschild black hole on an elliptical trajectory (figure 3.21).

3.21 Flying around a black hole, stills from *Infiniment courbé* (Infinitely curved), 1994, directed by Laure Delesalle.

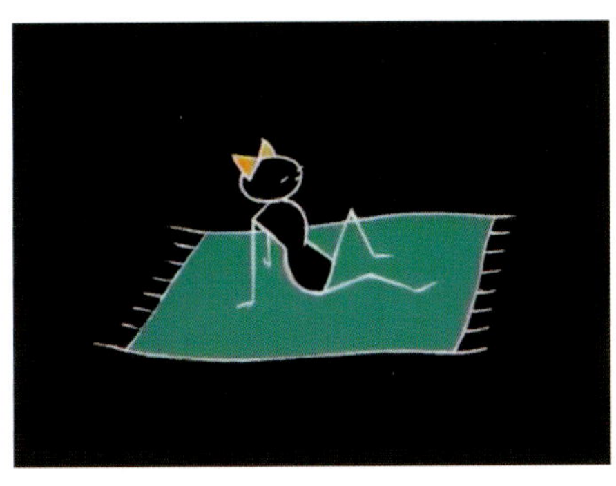

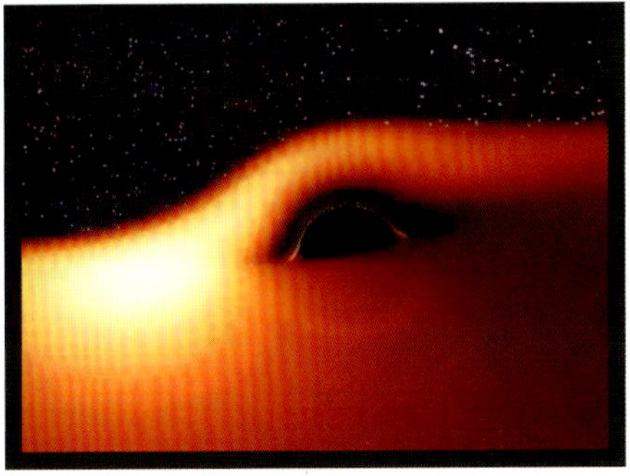

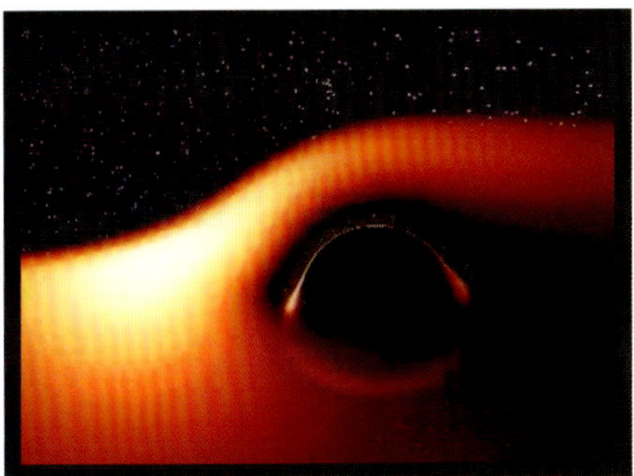

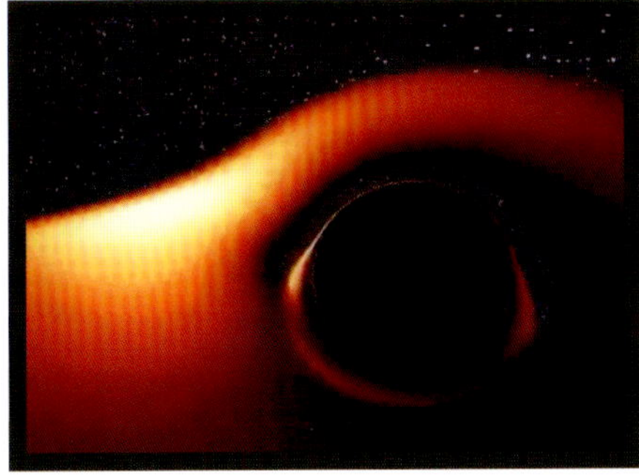

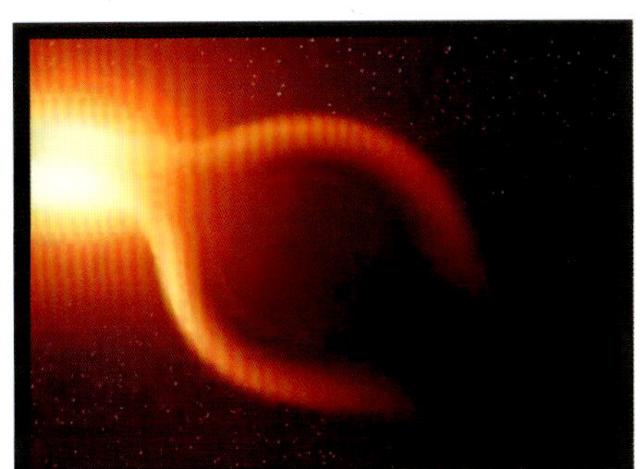

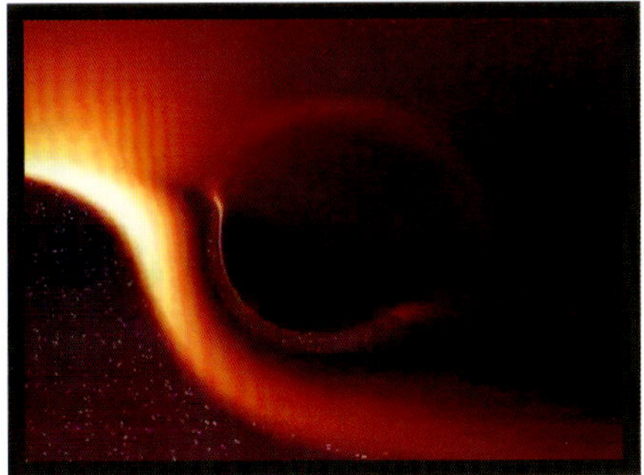

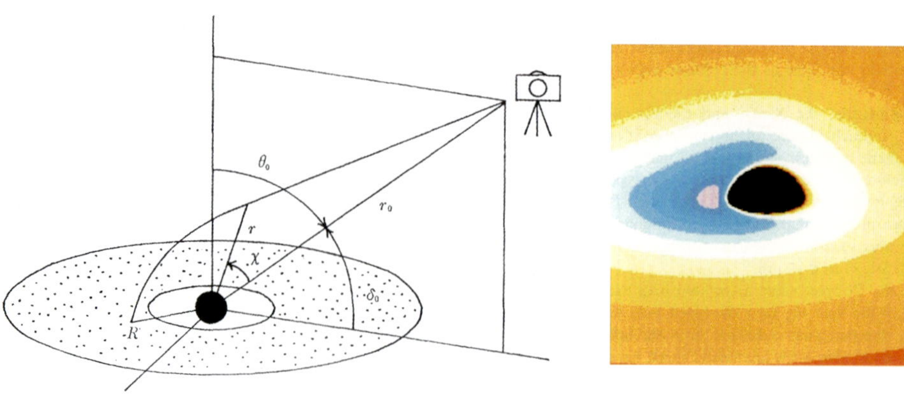

LEFT
3.22 Diagram of an imaginary camera set up on a tripod to capture the black hole and its accretion disk in Jun Fukue and Takushi Yokoyama, "Color Photographs of an Accretion Disk around a Black Hole," *Astronomical Society of Japan* 40, no. 1 (1988): 17, fig. 1.

RIGHT
3.23 Color photograph of an accretion disk in Fukue and Yokoyama, "Color Photographs of an Accretion Disk," 19, fig. 2a.

3.24 Fabian Oefner (Swiss, born 1984), *Black Hole, no. 4*, 2014. Inkjet print, 31½ × 47¼ in. (80 × 120 cm). Courtesy of the artist.

Oefner made this image (and that in figure 3.26) by putting paint on a drill bit and letting it spray out by centrifugal force while photographing it with a high-speed camera.

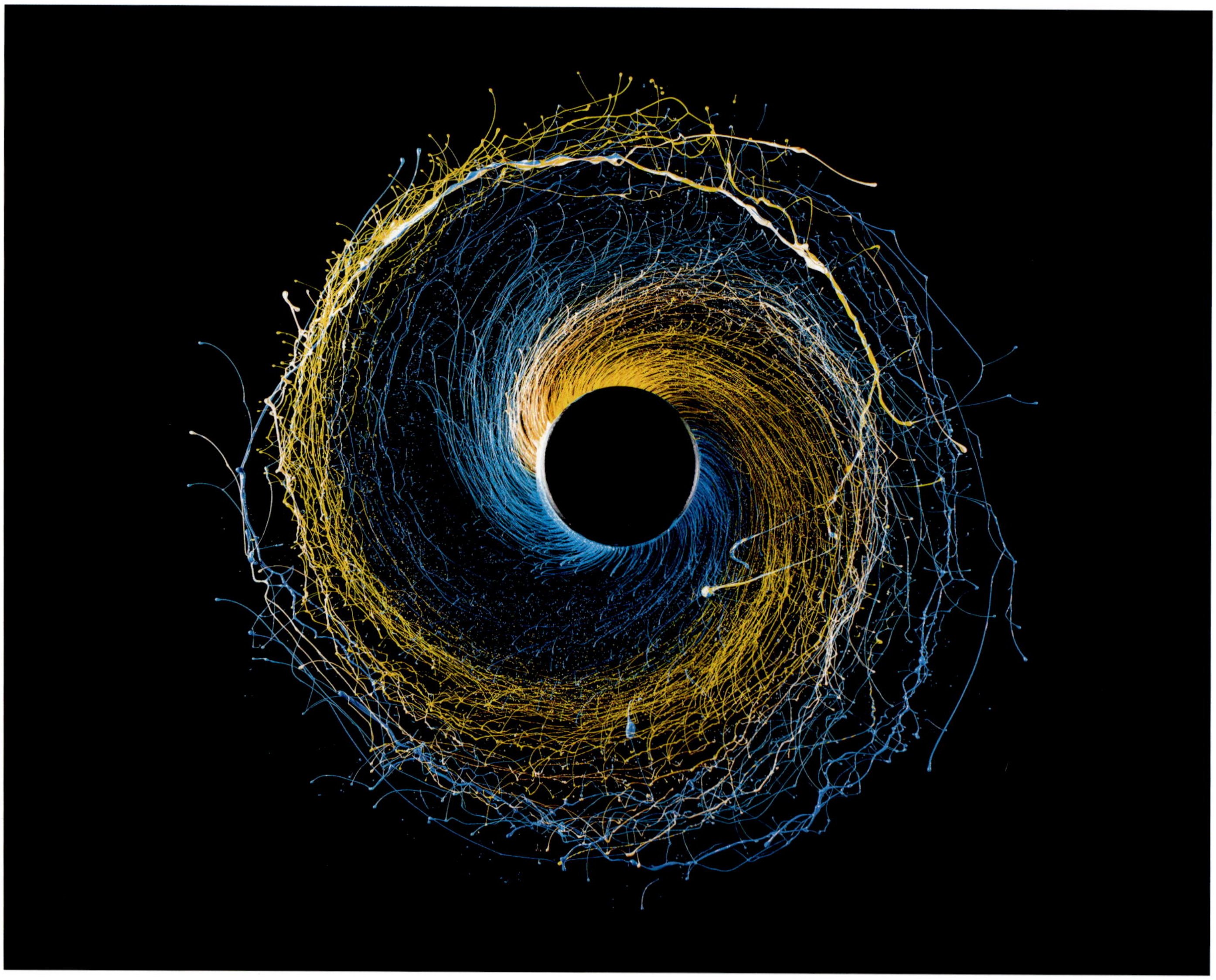

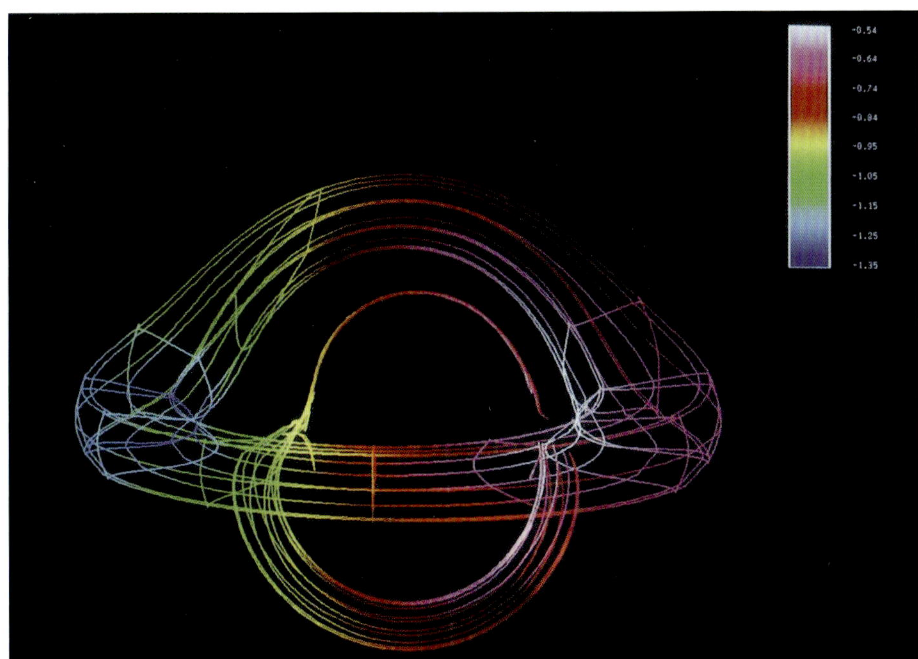

3.25 Thick accretion disk around a black hole in S. U. Viergutz, "Image Generation in Kerr Geometry: I. Analytical Investigations on the Stationary Emitter-Observer Problem," *Astronomy and Astrophysics* 272 (1993): 369, fig 10.

3.26 Fabian Oefner (Swiss, born 1984), *Black Hole, no. 2*, 2014. Inkjet print, 31½ × 47¼ in. (80 × 120 cm). Courtesy of the artist.

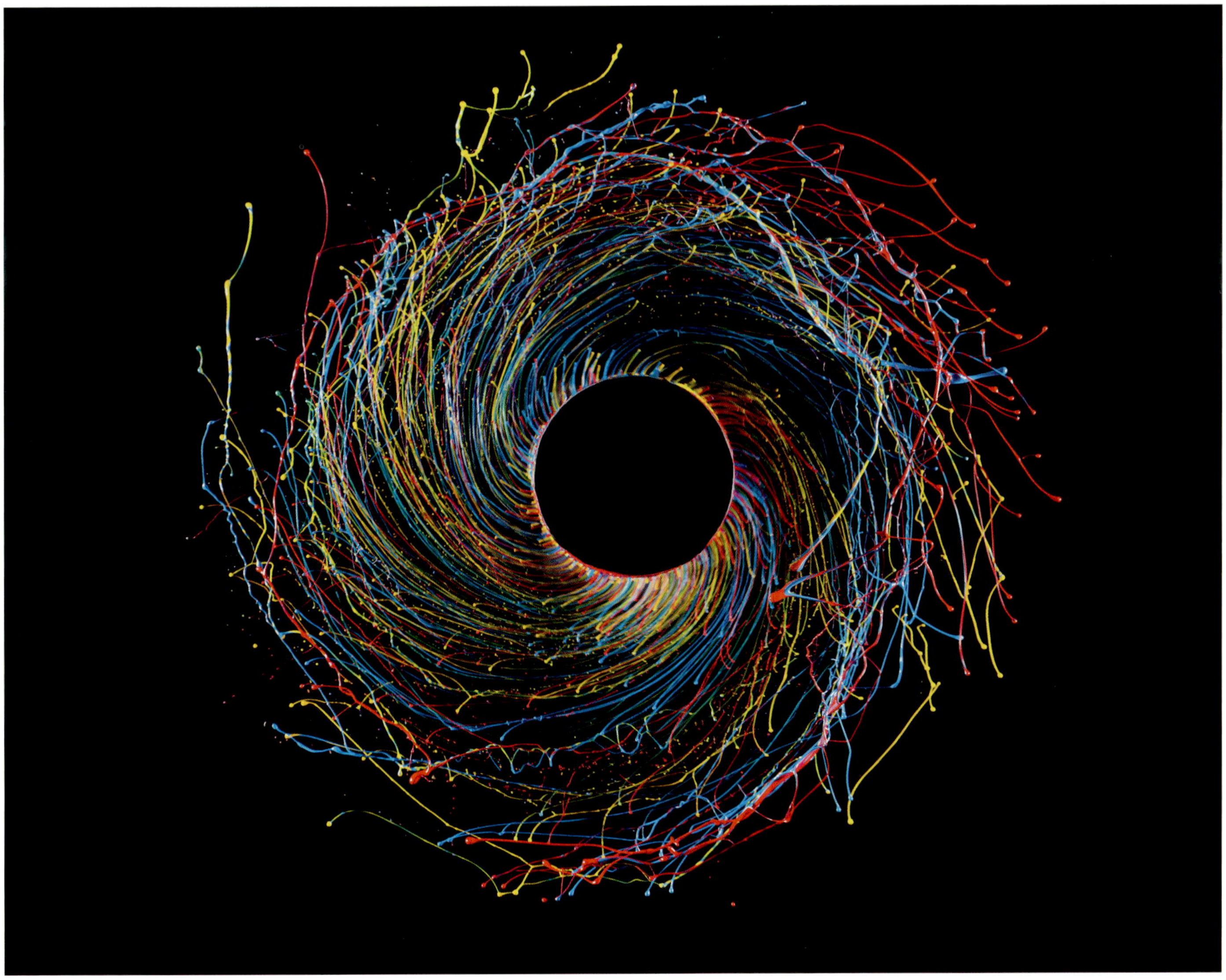

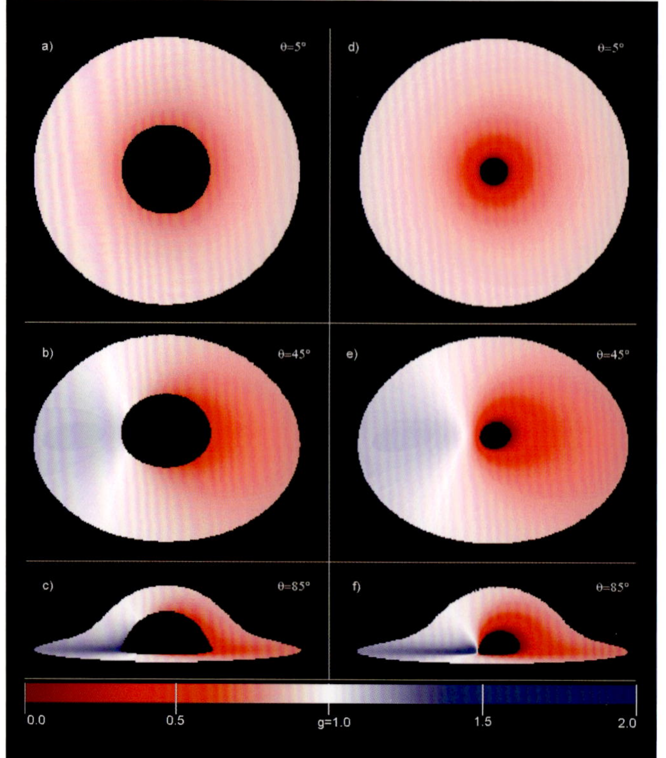

3.27 The diameter of the innermost stable circular orbit as a function of whether or not the black hole is rotating in C. Fanton et al., "Detecting Accretion Disks in Active Galactic Nuclei," *Astronomical Society of Japan* 49, no. 2 (April 1997): pl. 1, fig. 5.

In the left column, is a Schwarzschild stationary black hole: top left is seen from perpendicular to the accretion disk, bottom left is seen from edge on, and middle left is seen from between perpendicular and edge on. The innermost stable circular orbit has a large diameter. The right column shows a Kerr rotating black hole seen from the same three viewpoints. Because it is rotating, the innermost stable circular orbit has a small diameter. Both accretion disks are rotating toward an observer on the left, so they are blueshifted there.

While Luminet's monochrome picture depicted the total radiation in all wavelengths, astronomers Jun Fukue and Takushi Yokoyama imagined a visible-light photograph of an accretion disk (figure 3.22). In Luminet's drawing (figure 3.15), which is like a bolometric photograph, the left side is moving toward the observer, so it appears brighter. In Fukue and Yokoyama's visible-light photograph, the left side is likewise moving toward the observer, so it is blueshifted (figure 3.23).

Luminet, Fukue, and Yokoyama visualized thin accretion disks around Schwarzschild (stationary) black holes. Figure 3.25 is a visualization of a thick accretion disk around a Kerr (rotating) black hole from an almost edge-on viewpoint. The multicolored image is in visible light; the left side is blueshifted, as that side is rotating toward the observer. The artworks in figures 3.24 and 3.26 are metaphors for a multicolored accretion disk.

If a black hole is rotating, the speed at which it spins affects the diameter of the innermost stable circular orbit; the faster it spins, the smaller its diameter (figure 3.27). If a Kerr black hole spins extremely fast, it will distort spacetime at the inner edge of the accretion disk. Figure 3.28 shows a thin accretion disk around a maximally rotating Kerr black hole from an elevated viewpoint. The asymmetry of the disk's inner edge is the result of frame-dragging; the rotating black hole "drags" spacetime along.

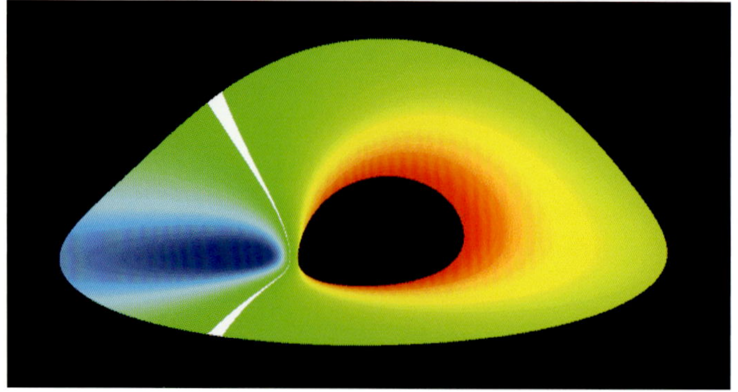

3.28 Asymmetry of the accretion disk's inner edge as the result of frame-dragging in B. Bromley et al., "Line Emission from an Accretion Disk around a Rotating Black Hole: Toward a Measurement of Frame Dragging," *Astrophysical Journal* 475, no. 1 (1997): pl. 4, fig. 1.

Seen in visible light, the accretion disk is rotating toward the observer on its left side, so it is blueshifted there and red-shifted on the right side; the white strip divides the two regions.

THE POPULARIZATION OF BLACK HOLES

In the late 1970s popular science books about black holes began appearing, including Isaac Asimov's *The Collapsing Universe: The Story of Black Holes* (1977). Having earned a PhD in chemistry, Asimov drew on a deep knowledge of science and was a skilled storyteller. Another title that contributed to the popular fascination with black holes was Stephen Hawking's *A Brief History of Time: From the Big Bang to Black Holes* (1988) and the 1991 film based on it. Inspired by the words of Hawking, the Italian art collective Opiemme painted letterforms surrounding a long shape that symbolizes an event horizon (figures 3.29 and 3.30). Carl Sagan's book *Cosmos* (1980) sold five million copies internationally. The related TV series, *Cosmos: A Personal Voyage* (1980), was hosted by Sagan and shown in sixty countries to 400 million viewers. A sequel, *Cosmos: A Spacetime Odyssey* (2014), hosted by Neil deGrasse Tyson, was shown in 125 countries to 135 million viewers. Sagan and Tyson described many scientific topics, including black holes, which were brought to life by animators (figure 3.31).

The impact of these popularizations was felt around the world, and artists in Asia mixed Western science with Eastern philosophy and history. Cai Guo-Qiang was in his twenties when he began experimenting with gunpowder as an artistic medium. When you explode a small amount of gunpowder on paper, it leaves a mark. Cai called these works "gunpowder drawings." In 1986, at age twenty-nine, he moved from his native China to Japan and became enthralled by popular books about astrophysics, especially *A Brief History of Time* and *Cosmos*, which he read in translation.[46] Cai said: "When I came to Japan, my encounters with the theories of twentieth-century astrophysics were very significant to me. The concepts of the Big Bang, black holes, the birth of stars, what is beyond the universe, time tunnels, how to leap over great distances of time and space and dialogue with something infinitely far away—these ideas were still not commonly in circulation in China at the time. They were an eye-opener for me. At the same time, many of these ideas have similarities with traditional Chinese views, with which I was familiar, of metaphysics and the universe."[47]

In 1991 Cai created large gunpowder drawings on paper mounted on wood panels, such as *The Vague Border at the Edge of Time/Space Project* (figure 3.32). Then he joined the wooden panels together, transforming them into traditional Chinese folding screens. He called the series *Primeval Fireball: The Project for Projects* because his drawings, like the cosmos, exploded into existence (figure 3.33).

3.29 Opiemme (Italian art collective), *To Stephen Hawking* (detail), 2018. Painted mural at the Museo di Arte Urbana, Turin, Italy, directed by Edoardo Di Mauro (1960–2024). Courtesy of the artists.

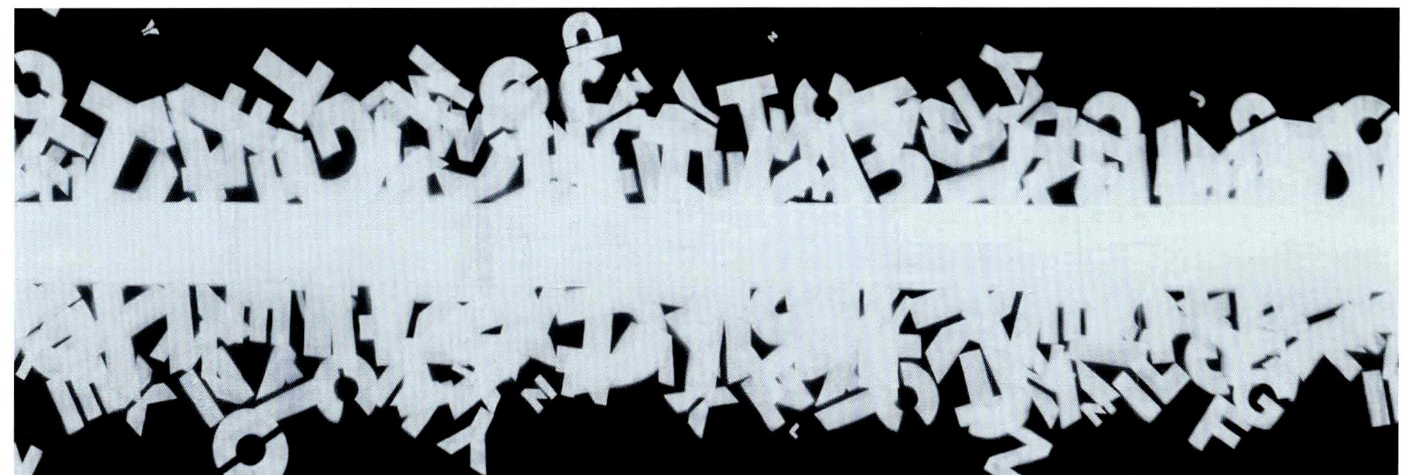

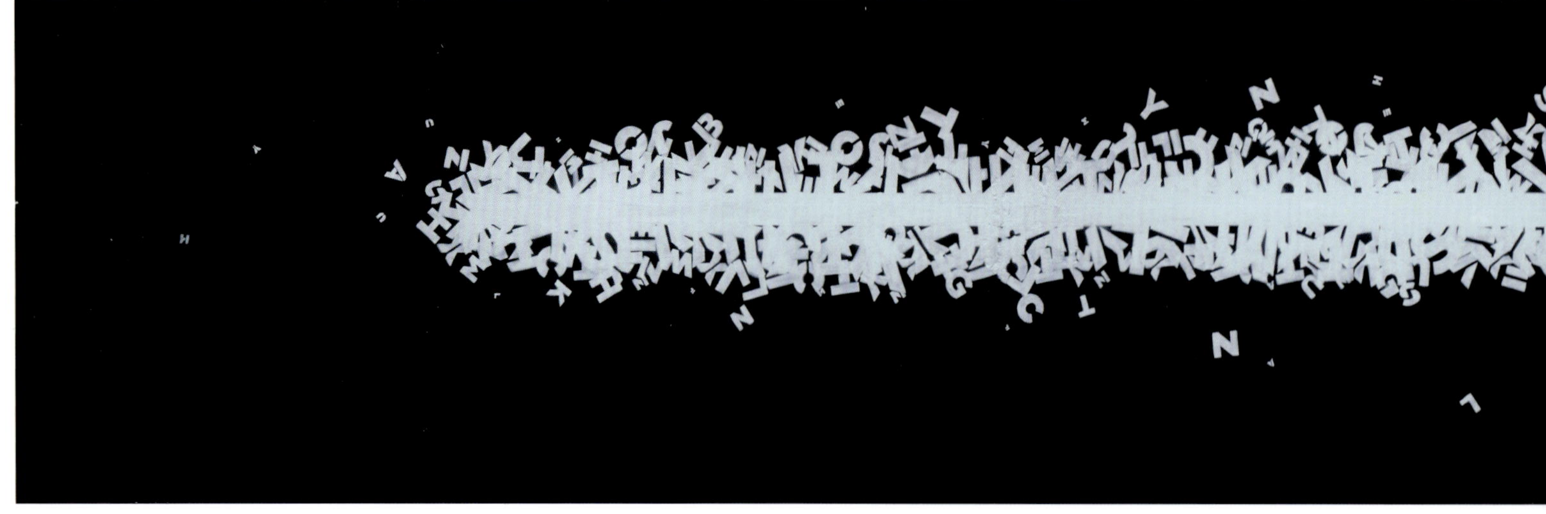

3.30 Opiemme (Italian art collective), *To Stephen Hawking*, 2018. Painted mural at the Museo di Arte Urbana, Turin, Italy, directed by Edoardo Di Mauro (1960–2024). Courtesy of the artists.

3.31 Stills from *Cosmos: A Spacetime Odyssey*, season 1, episode 4, "A Sky Full of Ghosts", written by Ann Druyan and Steven Soter, directed by Brannon Braga, hosted by Neil deGrasse Tyson, aired March 30, 2014, on the National Geographic Channel and Fox.

In *Cosmos: A Spacetime Odyssey*, the astrophysicist Neil deGrasse Tyson pilots the Ship of the Imagination, which can go anywhere, at any scale. In figure 3.31 *left* the ship approaches a black hole. In figure 3.31 *right* it explores inside the black hole, where it magically remains unharmed.

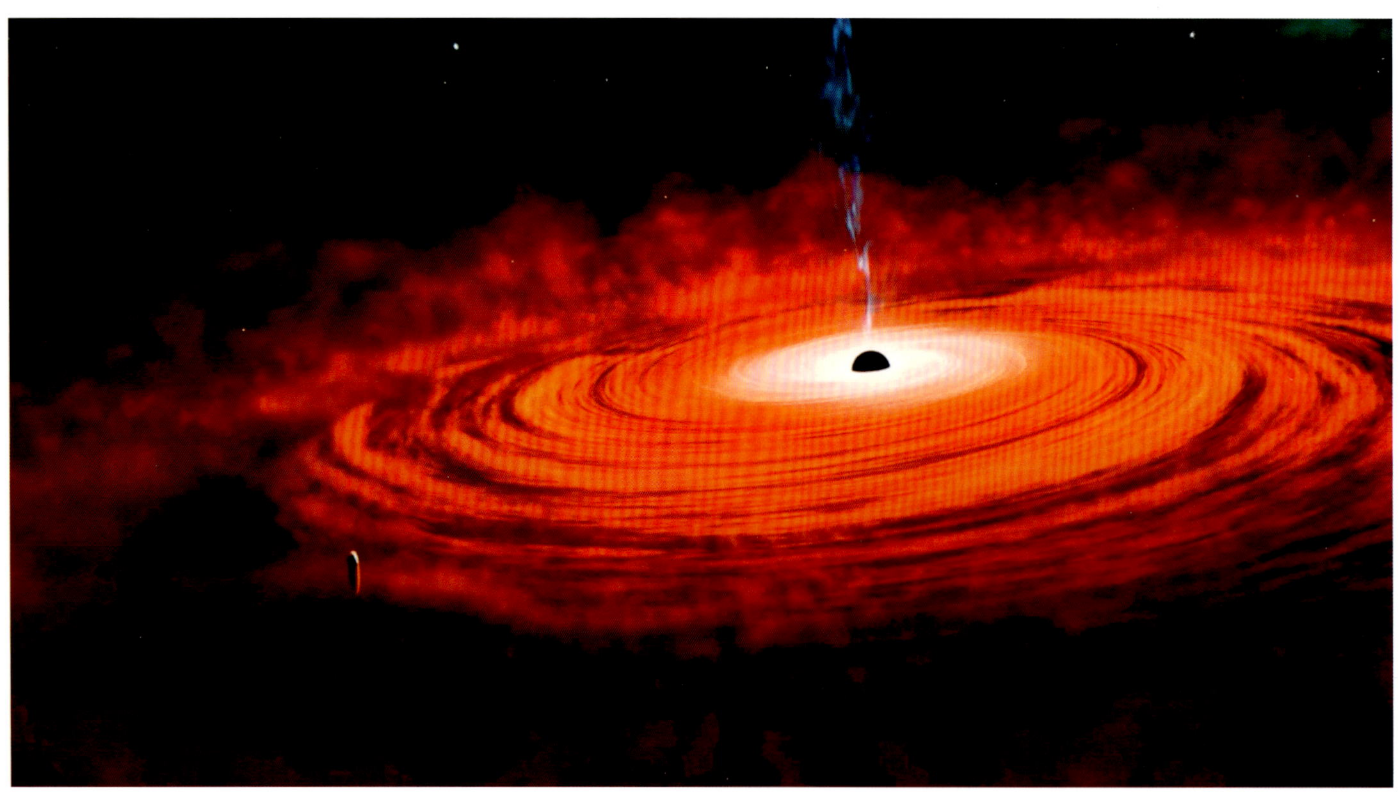

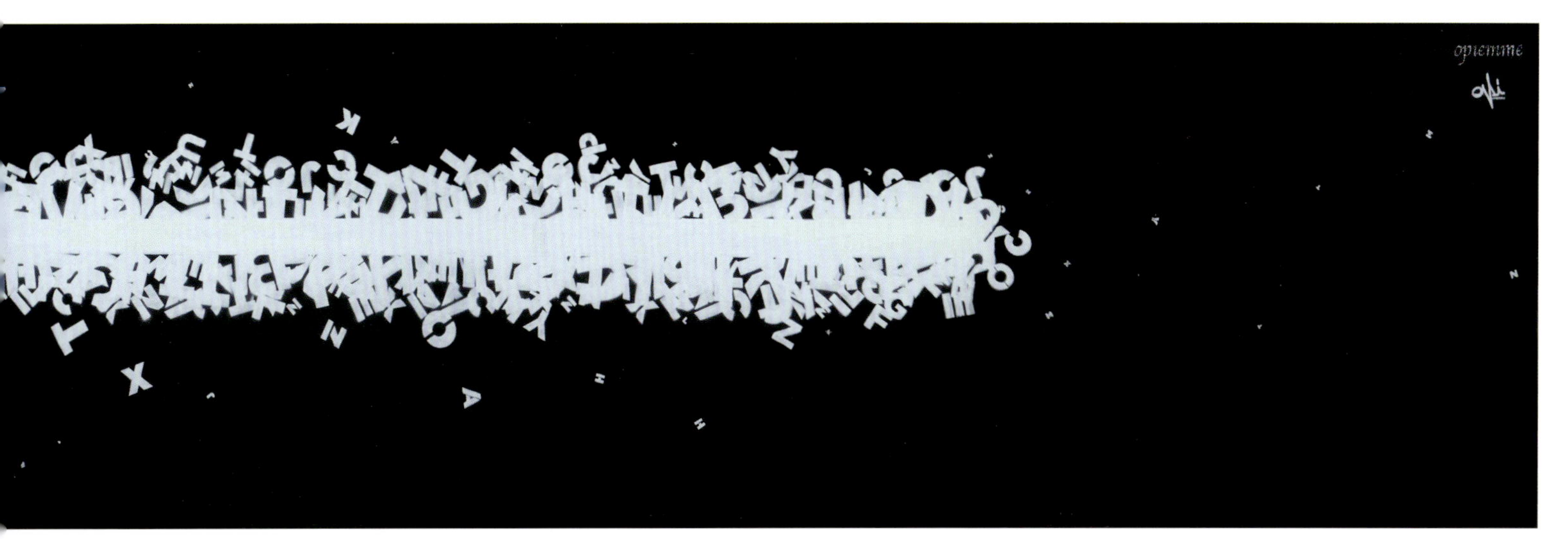

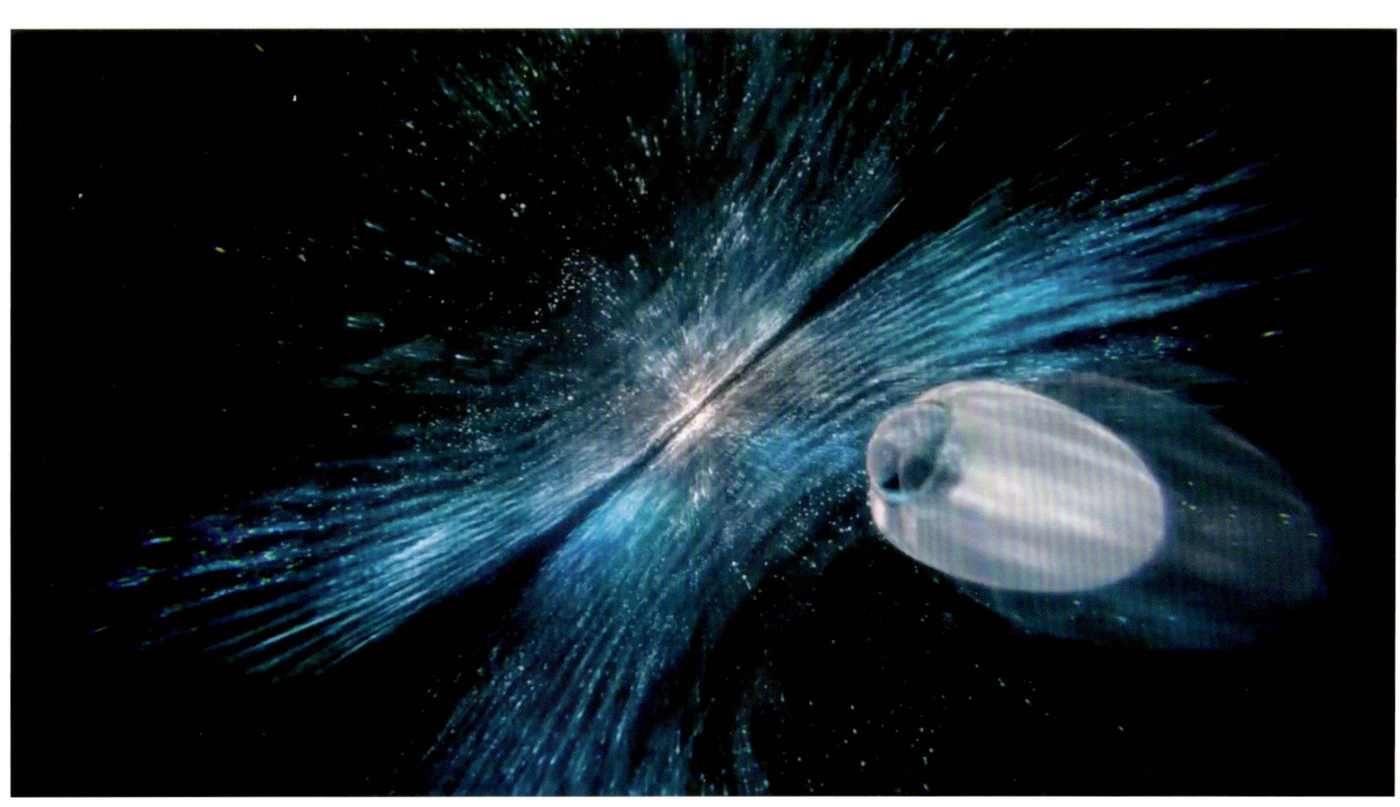

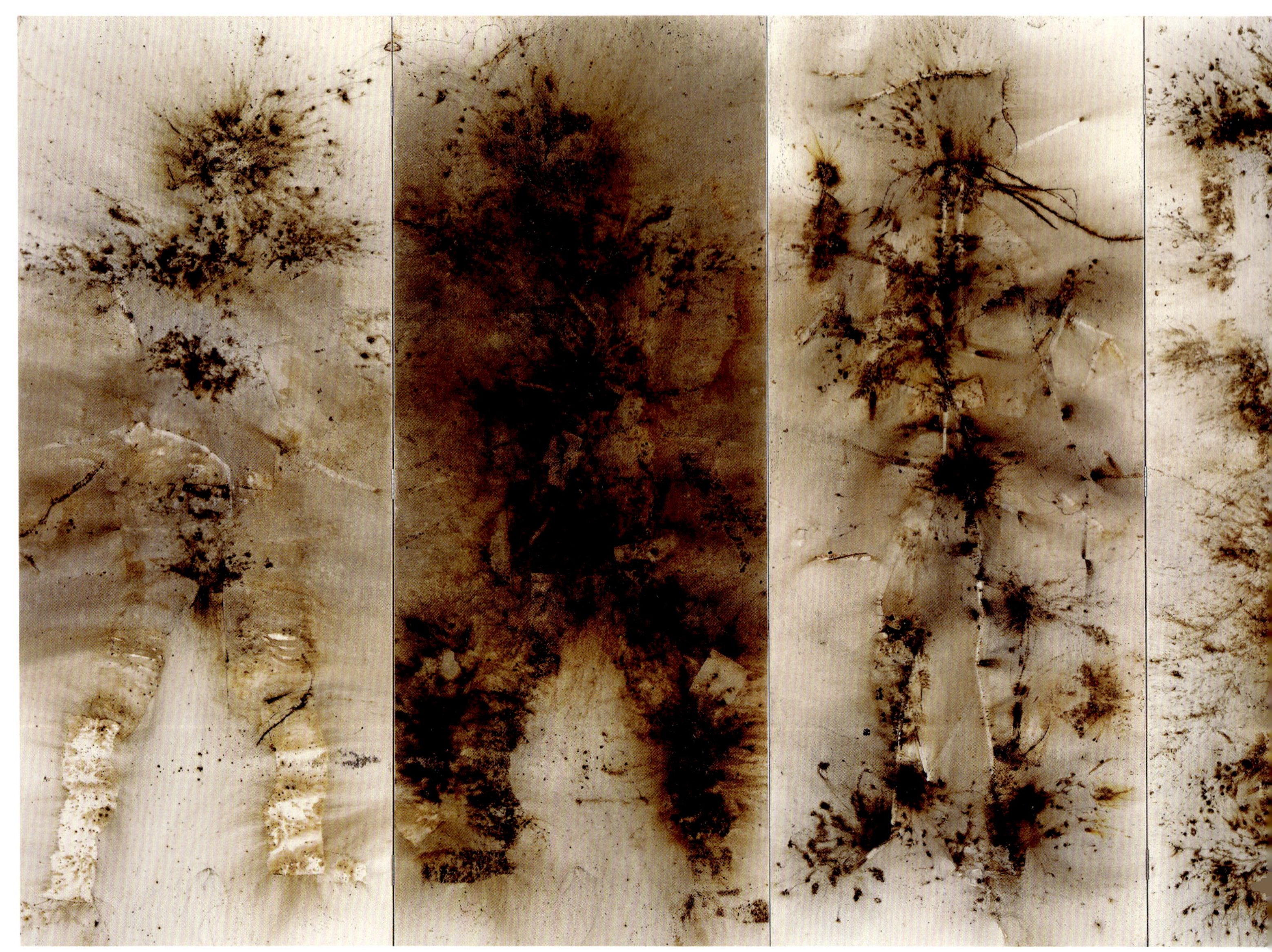

ABOVE
3.32 Cai Guo-Qiang (Chinese, born 1957), *The Vague Border at the Edge of Time/Space Project*, 1991. Gunpowder on paper mounted on wood as a seven-panel folding screen, 6 ft. 6¾ in. × 21 ft. (2 × 6.4 m) overall. Courtesy Fondation Cartier pour l'art contemporain.

OPPOSITE
3.33 Cai Guo-Qiang (Chinese, born 1957), *Primeval Fireball: The Project for Projects*, 1991. Seven gunpowder drawings; gunpowder on paper mounted on wood as folding screens. Dimensions variable. Courtesy Cai Studio.

The screen in figure 3.32 is on the right in this installation.

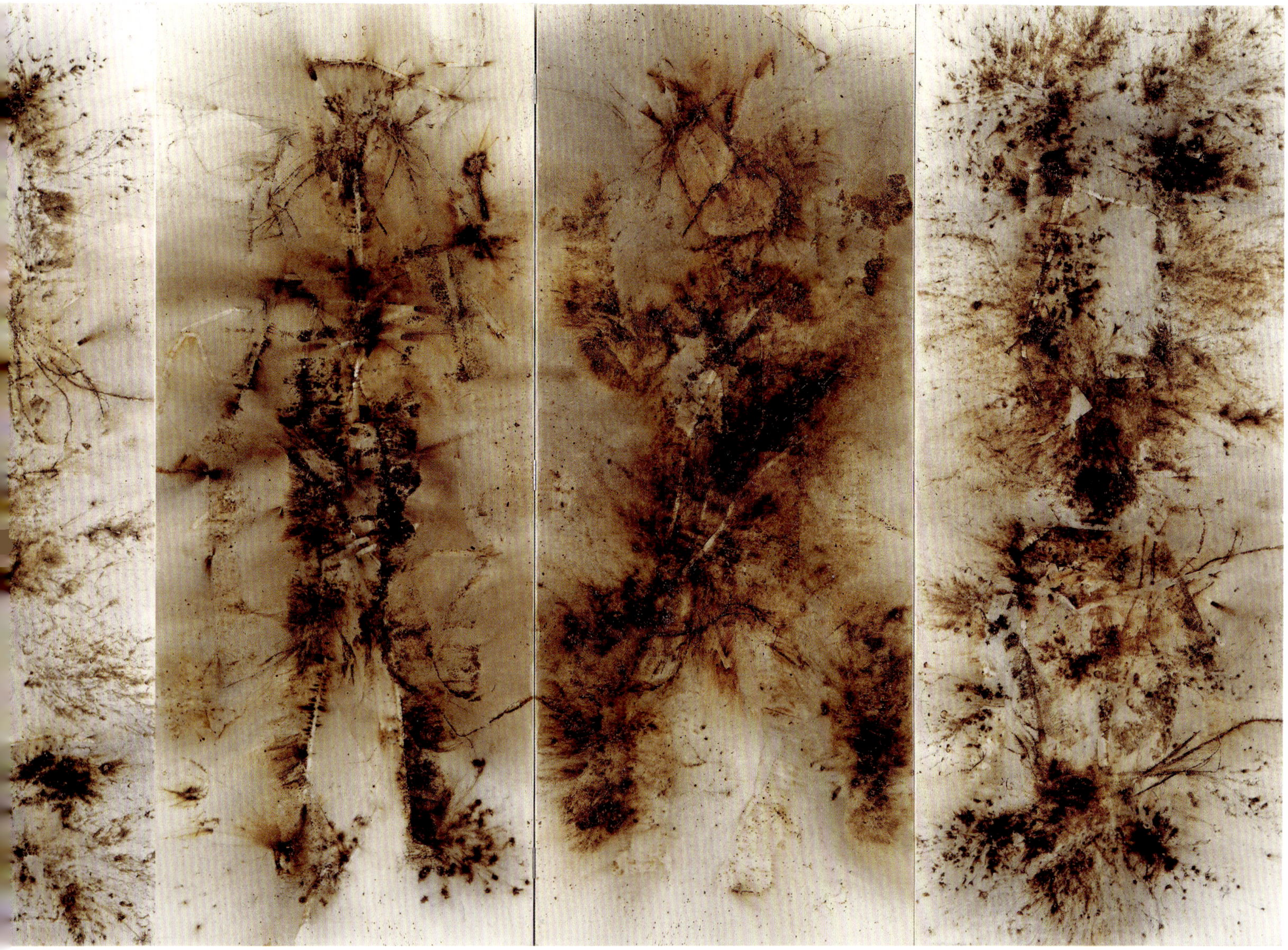

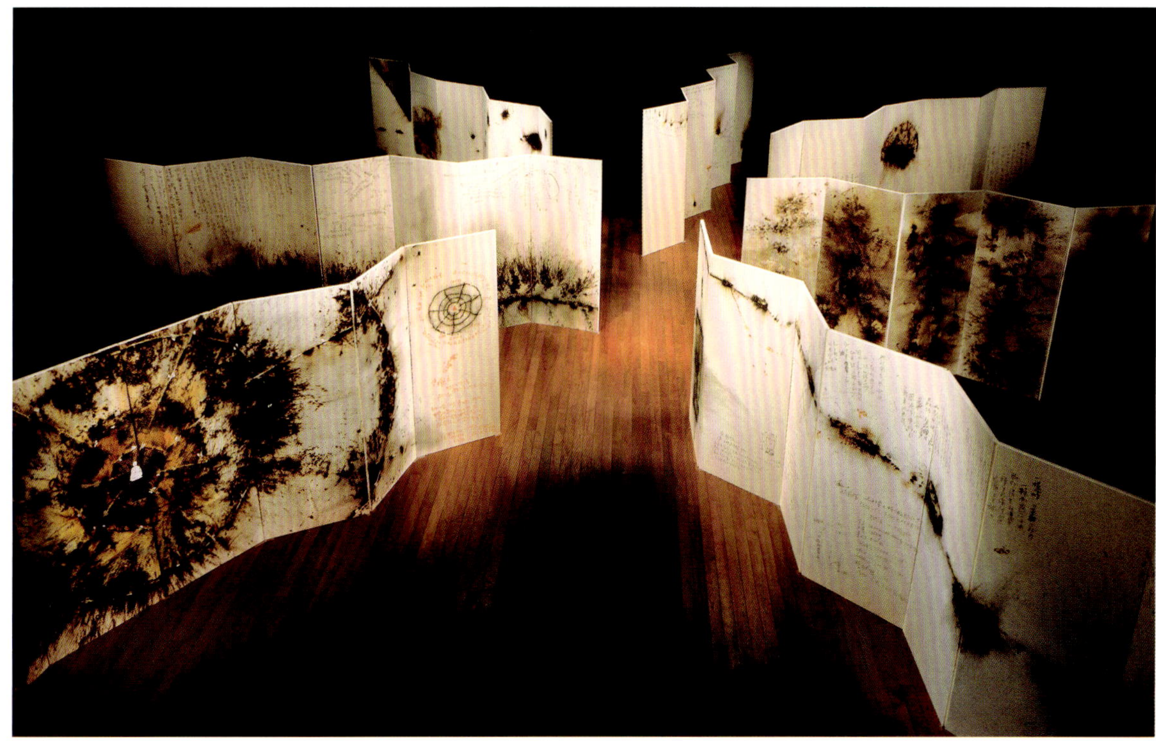

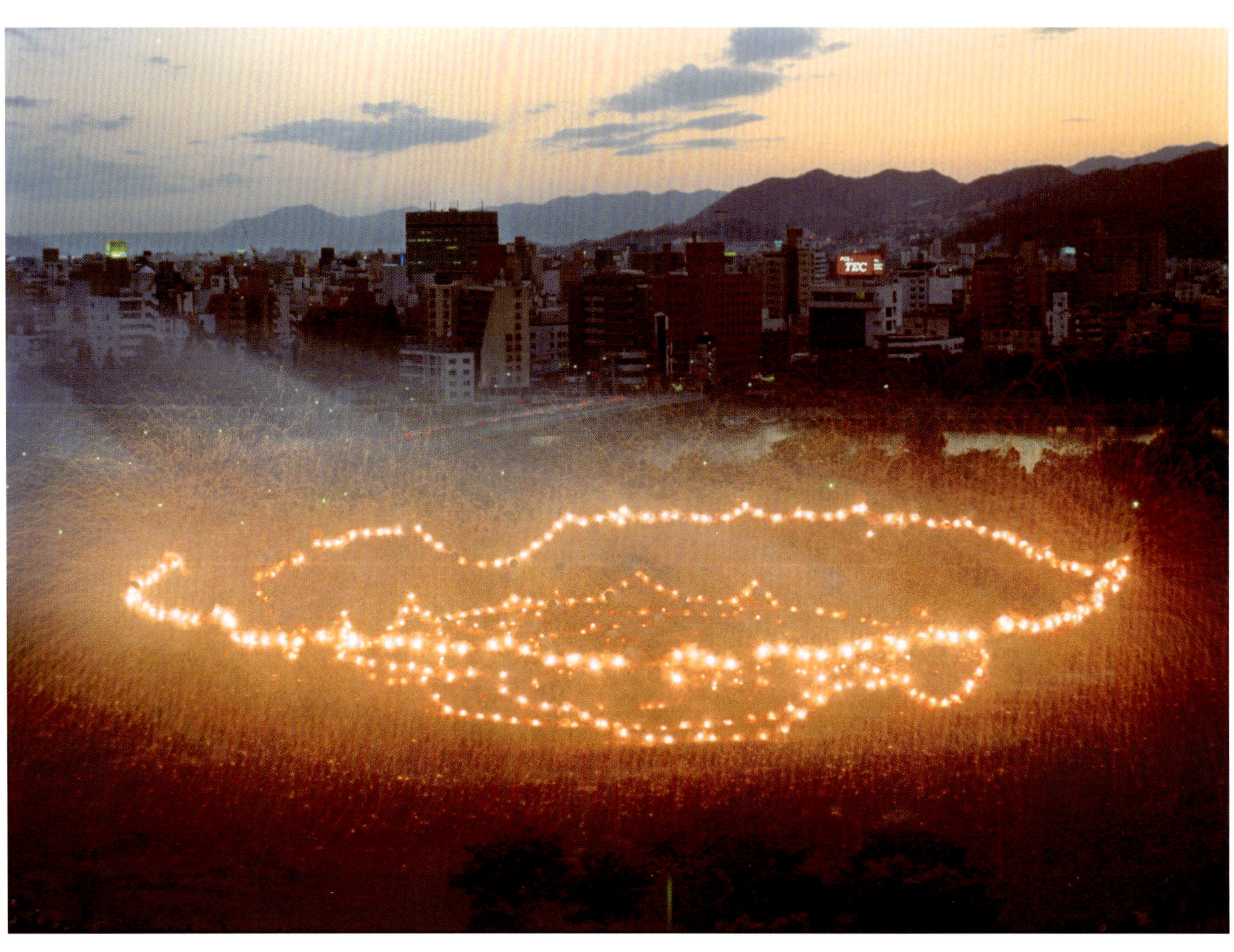

3.34 Cai Guo-Qiang (Chinese, born 1957), *The Earth Has Its Black Hole Too: Project for Extraterrestrials No. 16*, 1994. Explosion event in Hiroshima, Japan, 1994. Courtesy Cai Studio.

After reading books about the possibility of life in outer space, Cai began a series of explosion events. He calls the series *Projects for Extraterrestrials* and says of these works: "I wanted my explosions to take place in a vast open space, as if designed to be seen from well above the Earth."[48] One event in this series drew a parallel between the extreme power of a black hole and the energy unleashed by the atomic bombs in 1945. Cai titled this artwork *The Earth Has Its Black Hole Too* (1994; figure 3.34):

> This explosion project was realized at Hiroshima central park, the target of the atomic bomb. I dug a deep hole in the ground at the center of the park, and then I used 114 helium balloons at various heights to hold aloft 2,000 meters of fuse and three kilograms of gunpowder, which together formed a spiral with a 100-meter diameter, to mimic the orbits of heavenly stars. The ignition kicked off then from the highest and outermost point to the spiral, burning inward and downward in concentric circles, and disappeared into the "black hole" in the center of the park. The sound from the explosion was extremely violent; the bang echoed and rocked the entire city. My intention was to suggest that in harnessing nuclear energy, humanity has generated its own black hole in the Earth that mirrors those in space.[49]

After nine years in Japan, Cai moved permanently to New York in 1995. There he produced several explosion events, including *Transient Rainbow*, which Cai realized over the East River between Manhattan and Queens in 2002 (figure 3.35).

3.35 Cai Guo-Qiang (Chinese, born 1957), *Transient Rainbow*, 2002. Explosion event in New York, 2002. Courtesy Cai Studio.

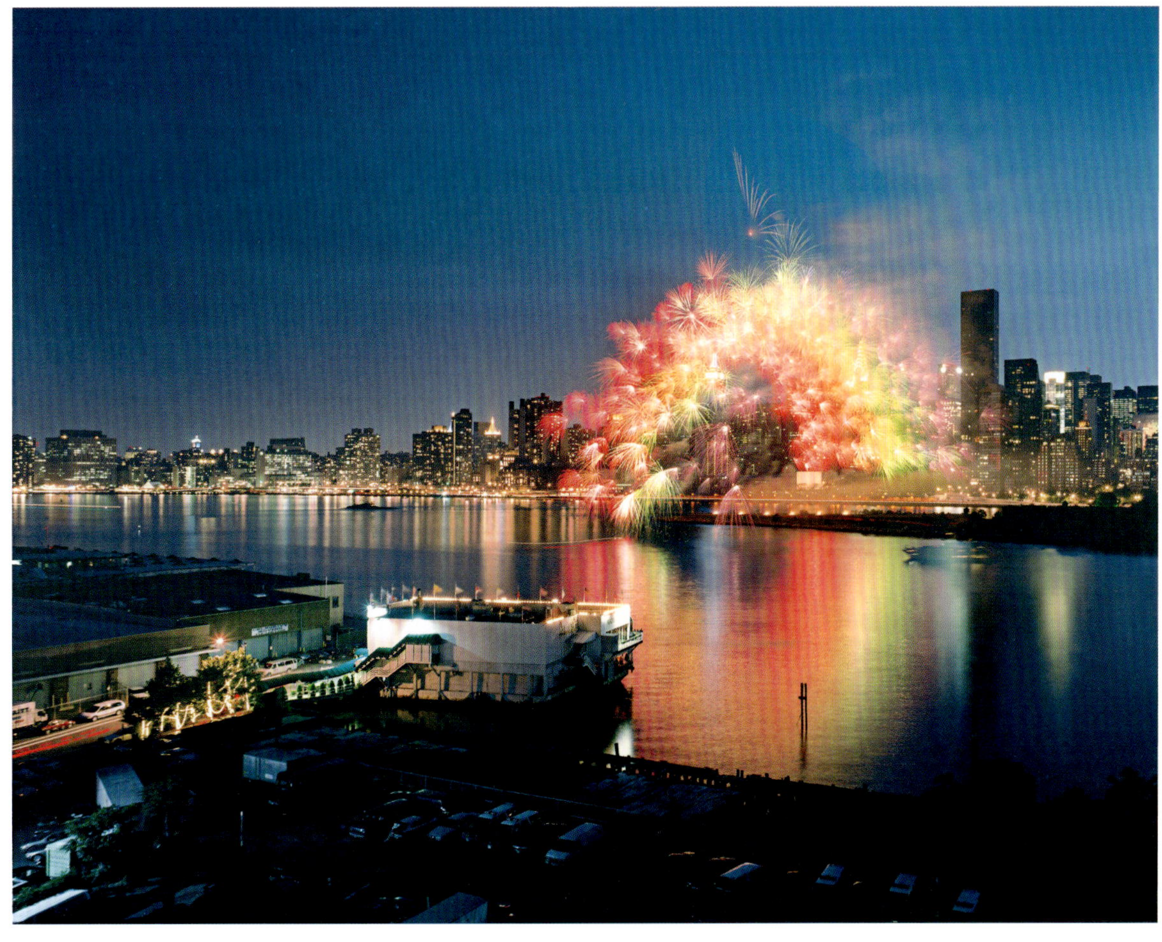

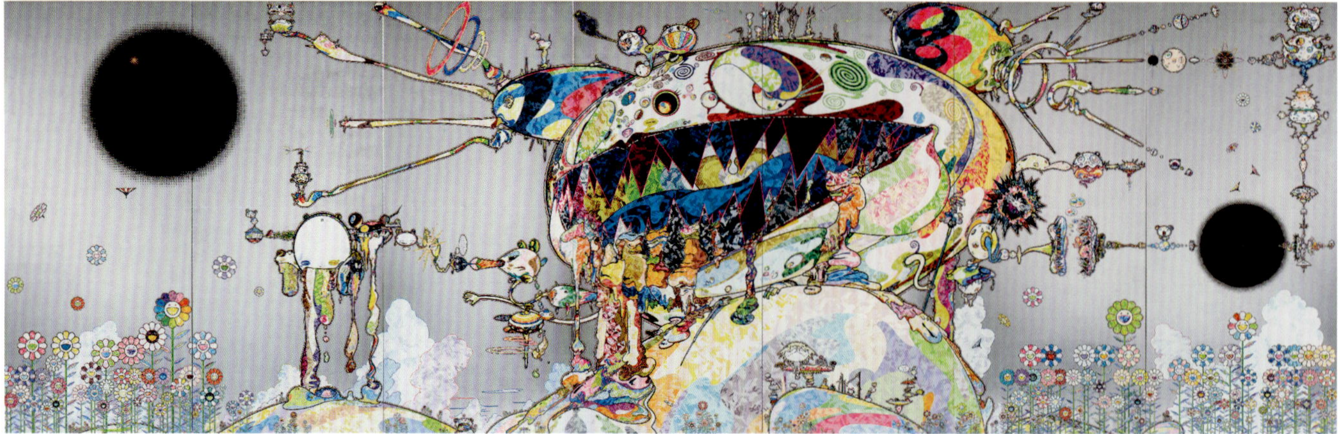

3.36 Takashi Murakami (Japanese, born 1962), *Tan Tan Bo Black Hole*, 2019. Acrylic, gold leaf, and platinum leaf on canvas mounted on aluminum frame; seven panels, 7 ft. 10½ in. × 24 ft. 1¼ in. (2.40 × 7.35 m).

The tragedy of Hiroshima and the power of atomic energy symbolized by black holes are explored in an entirely different aesthetic strategy in the work of Takashi Murakami. His art is based on the colorful characters in anime (Japanese animation) and manga (graphic novels). Murakami created a round-eared, anime-inspired alter ego, Mr. DOB (shown in its morphed form as Tan Tan Bo in figure 3.38), referring to popular gaming, manga, and anime characters such as Sonic the Hedgehog and Doraemon. The large mural *Tan Tan Bo Black Hole* (figure 3.36) shows Mr. DOB above pastel hills and valleys. The lower corners are filled with brightly colored circular flowers with smiling faces. Despite these cheerful gardens, the painting expresses the artist's apprehension about nuclear energy (notice the tiny atom in the upper part of the detail in figure 3.37). This painting was shown at the University of Chicago, where in 1942 Enrico Fermi led the group of scientists who made the world's first nuclear reactor as part of the Manhattan Project to build the atomic bomb. Recalling the victims of Hiroshima whose inner organs were liquefied by the blast, biological forms drip from Tan Tan Bo's mouth and gush out his ears.

By the 1990s science illustrators had developed a standardized image of a black hole (figure 3.39). At the center is the black hole (shown as a black sphere or a void). Jets of particles shoot out from the black hole perpendicular to the plane of a colorful accretion disk, shown as a flat disk without the bending of light due to gravity. Many artists were captivated by such images and created artworks drawing on these elements (figure 3.40).

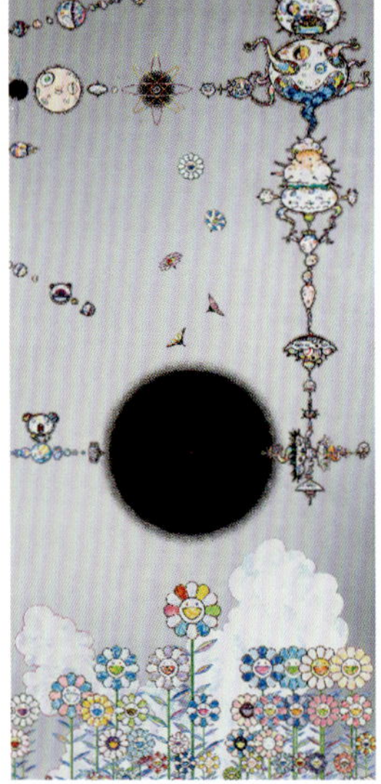

ABOVE
3.37 Takashi Murakami (Japanese, born 1962), *Tan Tan Bo Black Hole* (detail), 2019.

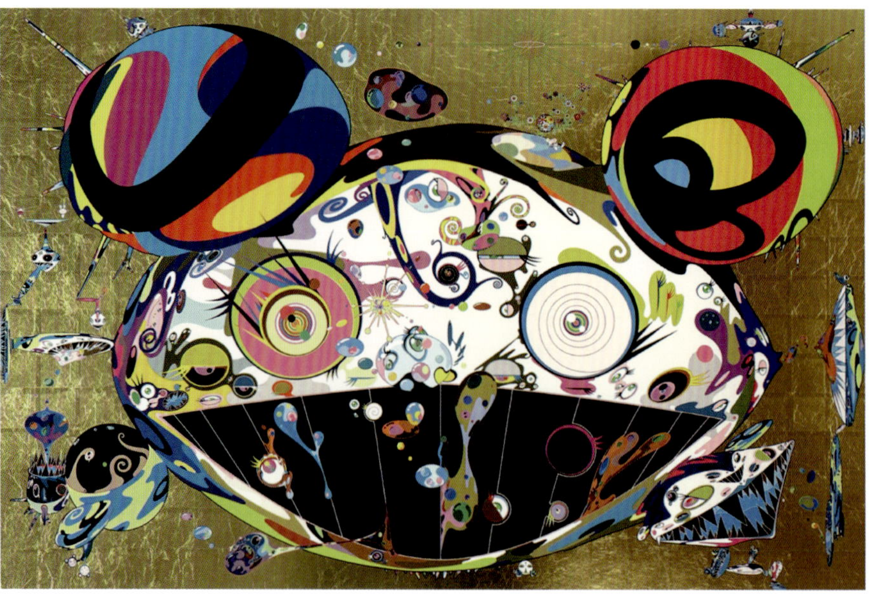

RIGHT
3.38 Takashi Murakami (Japanese, born 1962), *Tan Tan Bo*, 2016. Screen print, 31¼ × 44¼ in. (79.38 × 112.4 cm).

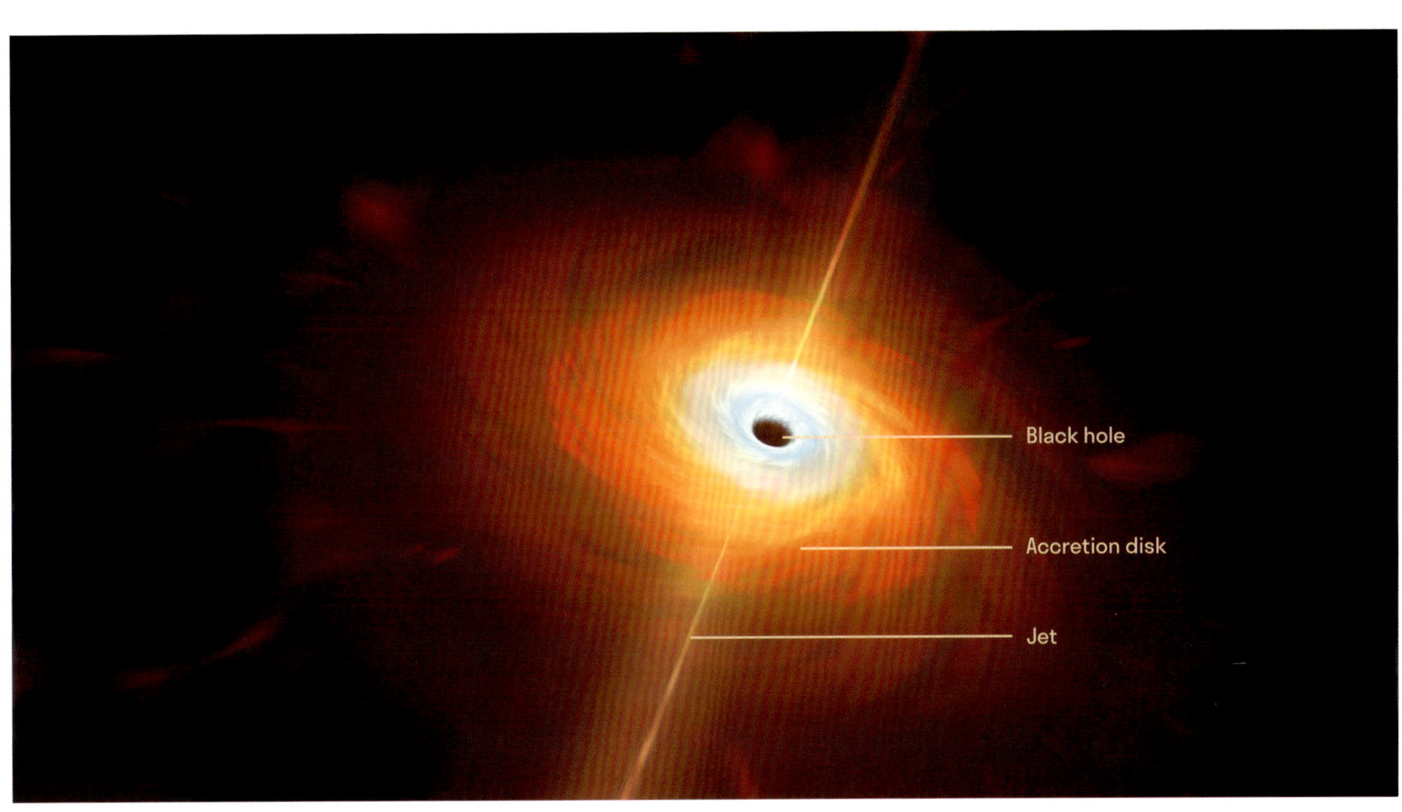

3.39 Science illustrator's standardized image of a black hole, 1990s.

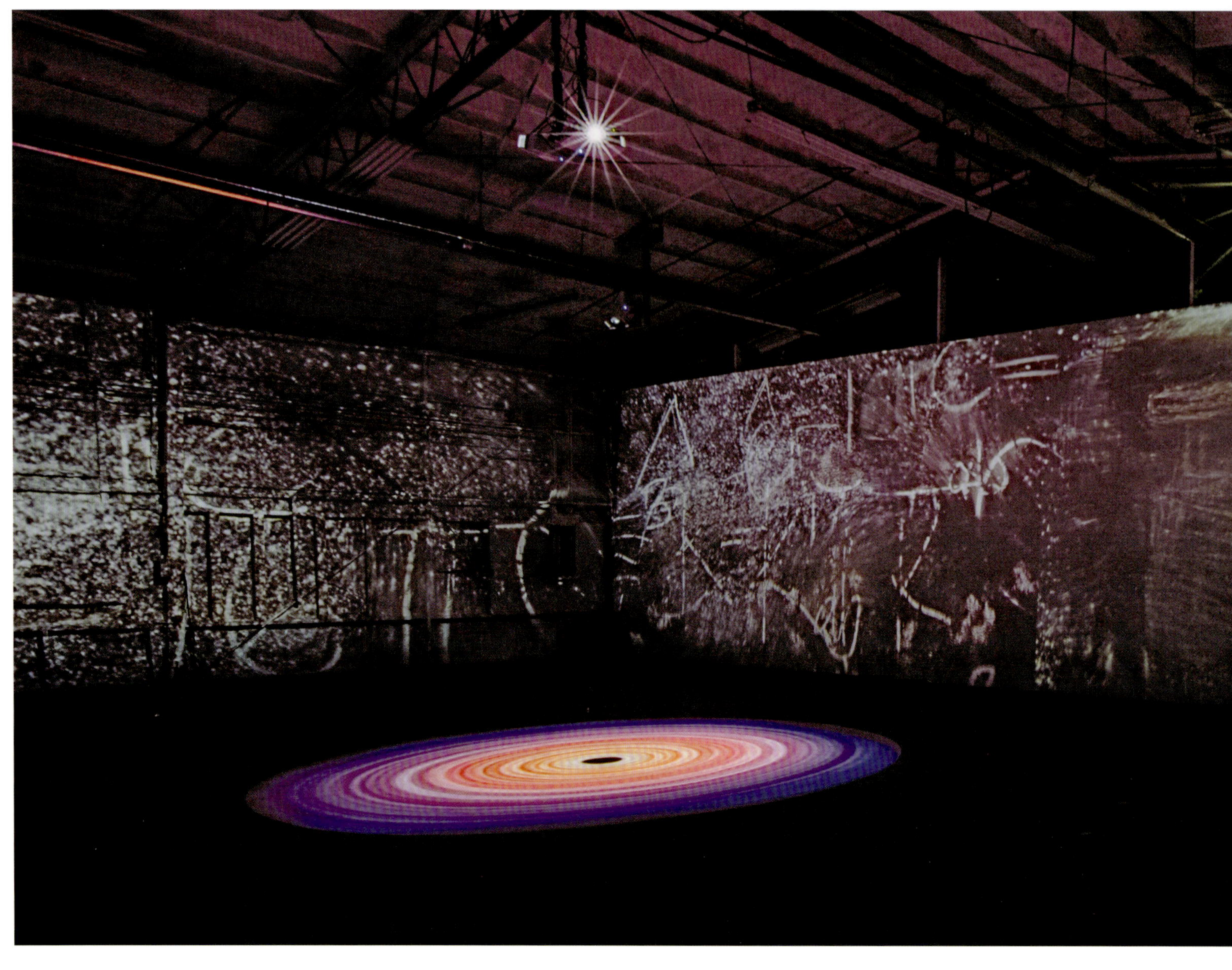

3.40 The Einstein Collective, *Black (W)hole*, 2013. Laser array, video projections. Installation view at Tinworks Art, Bozeman, Montana, 2020. Courtesy of the artists.

The Einstein Collective is a group of artists and scientists from several American universities who came together to create an installation about a black hole, which consisted of a ceiling-mounted laser array that projected a starfield across the gallery floor and two video projections. The projection on the walls was inspired by Einstein's blackboards on which he wrote equations. The projection on the floor was a black hole with a colorful accretion disk. The Einstein Collective includes Sara Mast (visual artist), Jessica Jellison (architect), Christopher O'Leary (animator), Cindy Stillwell (filmmaker), Jason Bolte (composer), Charles Kankelborg (physicist), Nico Yunes (astrophysicist), and Joey Shapiro Key (astrophysicist).

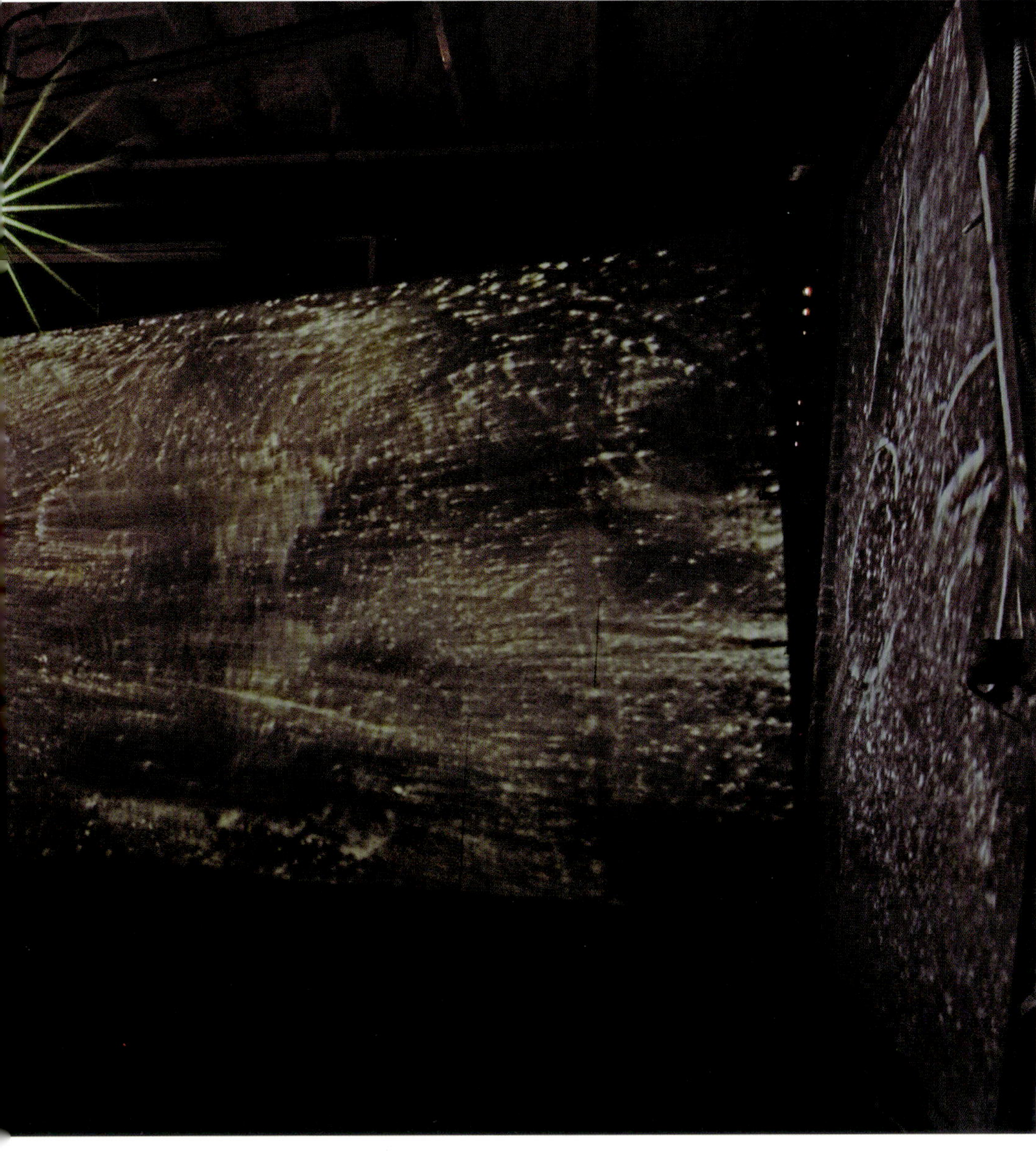

3.41 Russell Stannard, *Black Holes and Uncle Albert*, ill. John Levers (London: Faber and Faber, 1992).

This cover shows a spaceship taking off with Einstein's (fictitious) niece, Gedanken (German for "thought" as in "thought experiment"). To her uncle's relief, Gedanken returns alive from her trip to a black hole.

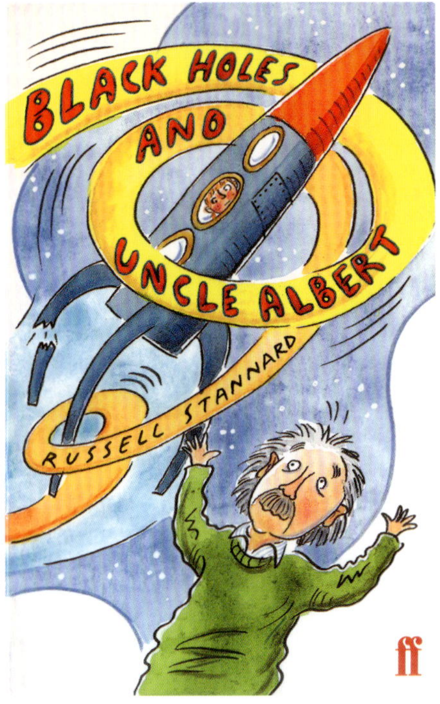

Black holes have also had starring roles in films. *The Black Hole* (1979) tells the story of an American spaceship that discovers a black hole. Nearby is the long-lost spaceship USS *Cygnus* (presumably named for Cygnus X-1) that has avoided being destroyed by the black hole's gravity because it's surrounded by inexplicable zero gravity. Despite shoddy science, the film—a Walt Disney production—demonstrated that black holes were ready for prime time. The popularization of the subject has included children's books (figure 3.41), and black holes feature in dozens of videos games.[50] Books about black holes for the general reader form a crowded field, with more than thirty titles published in English since 2020.[51] When the director Christopher Nolan made the film *Interstellar* (2014)—which is unusual for the realism of the film's black hole, named Gargantua—he enlisted physicist Kip Thorne to ensure the film's scientific accuracy (figure 3.42). Nolan stopped short of showing the Doppler effect and instead made Gargantua symmetrical. To help the public understand black holes, NASA's animators in 2019 visualized what a black hole and its accretion disk would look like from various angles (figure 3.43), including an edge-on view (figure 3.44).

3.42 Variant of the black hole accretion disk seen in the film *Interstellar* (2014) in Oliver James et al., "Gravitational Lensing by Spinning Black Holes in Astrophysics, and in the Movie *Interstellar*," *Classical and Quantum Gravity* 32, no. 6 (February 13, 2015): fig. 16.

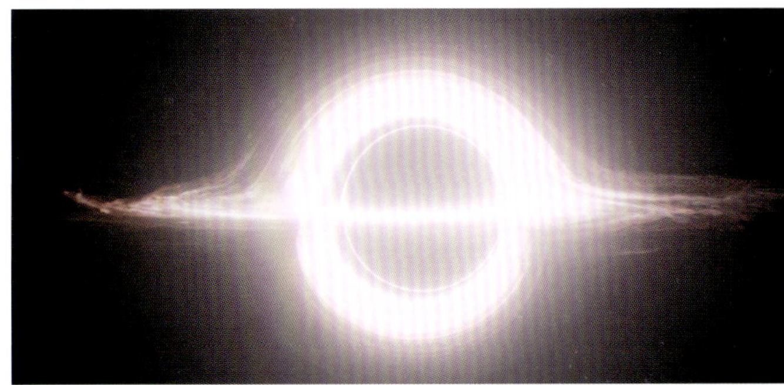

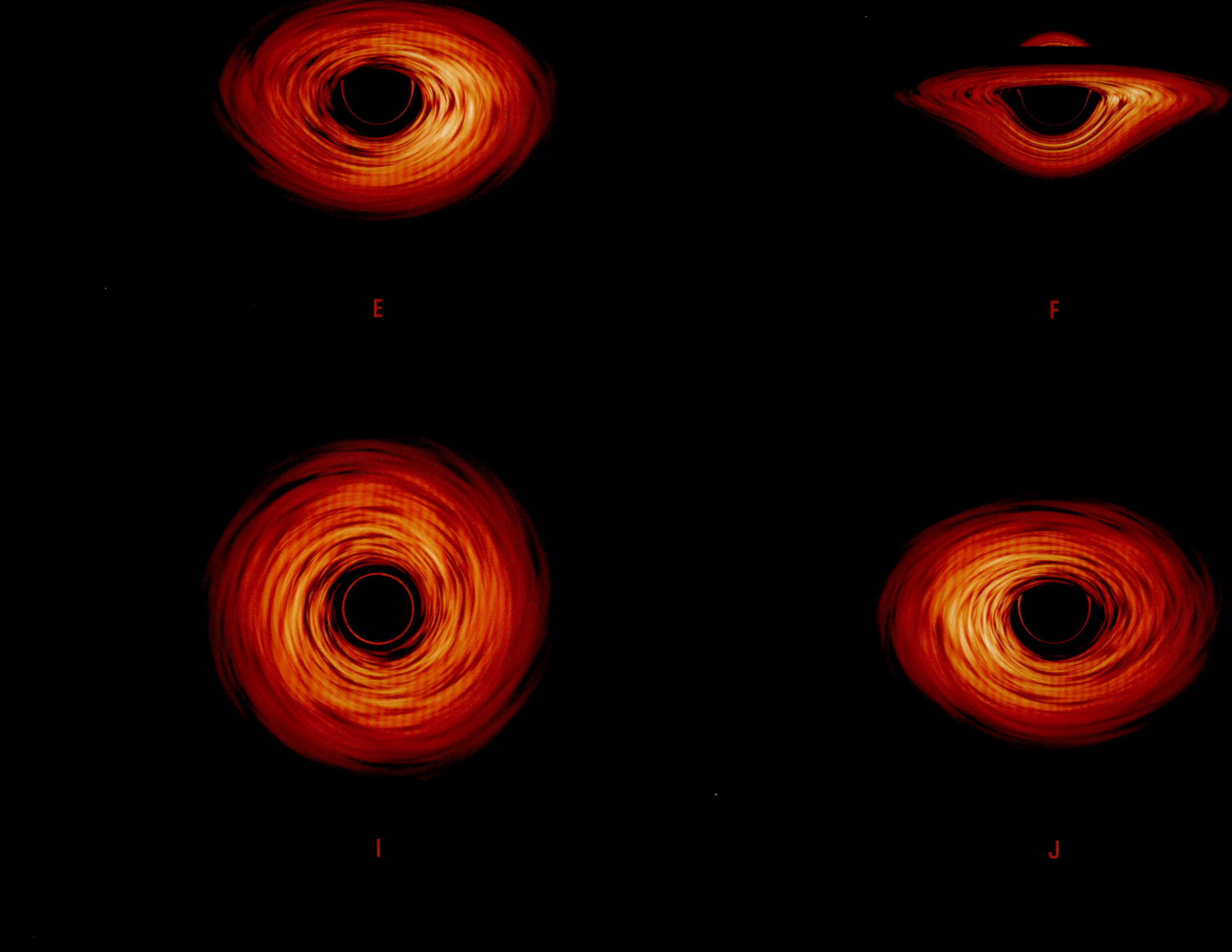

3.43 Black hole accretion disk from various angles. Visualizations by NASA's Goddard Space Flight Center and Jeremy Schnittman, 2019.

This sequence of visualizations begins at the upper left with an edge-on view (A); view D is perpendicular to the plane of the accretion disk. Note that view B is like the one Luminet showed in his 1979 drawing (figure 3.15) and C recalls the first image of a black hole, M87*, released on April 10, 2019, by the Event Horizon Telescope (figure 3.138).

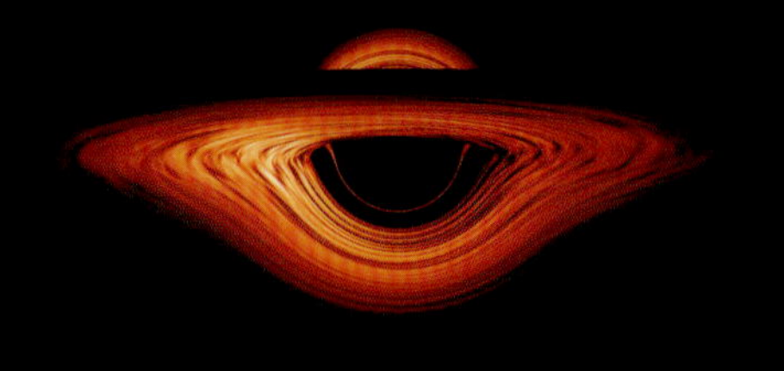

K

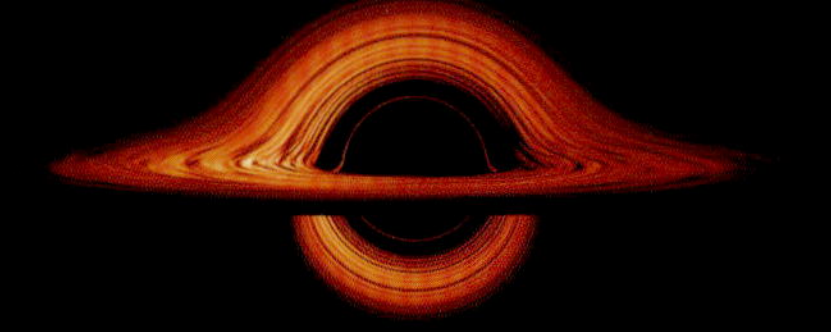

L

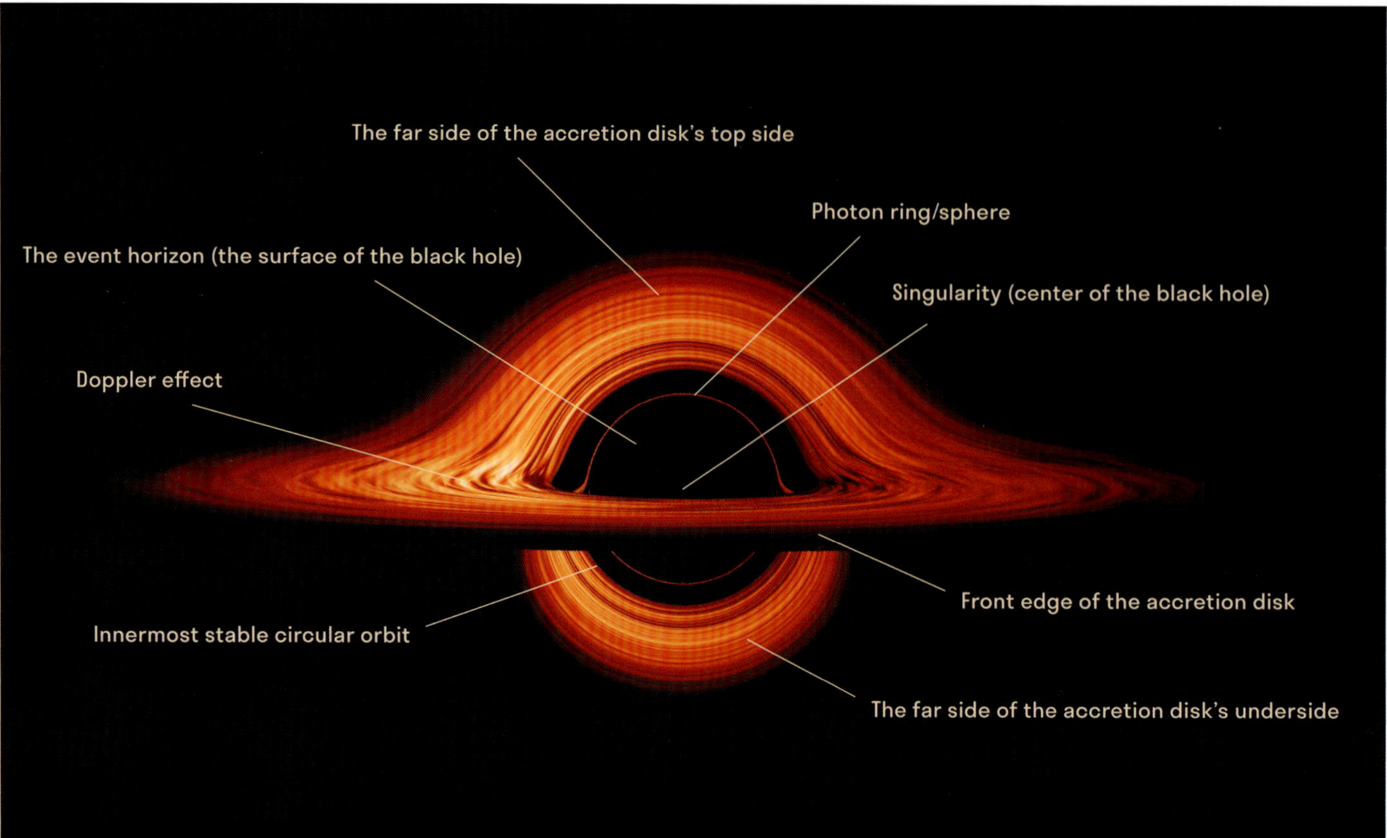

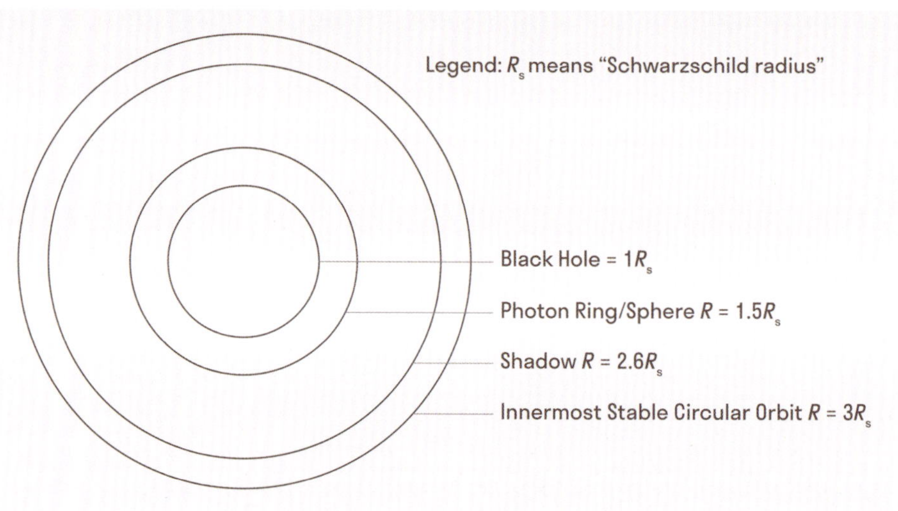

ABOVE
3.44 Edge-on view of a black hole accretion disk. Visualization by NASA's Goddard Space Flight Center and Jeremy Schnittman, 2019.

RIGHT
3.45 Dimensions in the vicinity of a black hole.

Each black hole has a unique Schwarzschild radius, which is measured from its center (the singularity) to its surface (the event horizon; see figure 1.10). Every black hole's "shadow" measures $2.6R_S$.

THE SHADOW OF A BLACK HOLE

In physics, the word "shadow" is not used in the ordinary sense of an object blocking light and casting a shadow. Astronomers name the dark area in the center of a black hole's accretion disk its *shadow*. (A black hole does not cast a shadow on its accretion disk.) The dimension of the shadow is diagrammed in figure 3.45. What we're looking at when we see the shadow is the front and back of the black hole. This magic trick happens because of the geometry of light around a black hole. In outer space, light rays go in all directions, but for simplicity's sake, imagine parallel light rays coming from the direction of an observer on Earth (figure 3.46). If light strikes the event horizon, it's gone, so the observer sees black (absence of light). If a light ray comes a little above or below the event horizon, it will follow the warped spacetime caused by the gravity of the black hole and end up crossing the

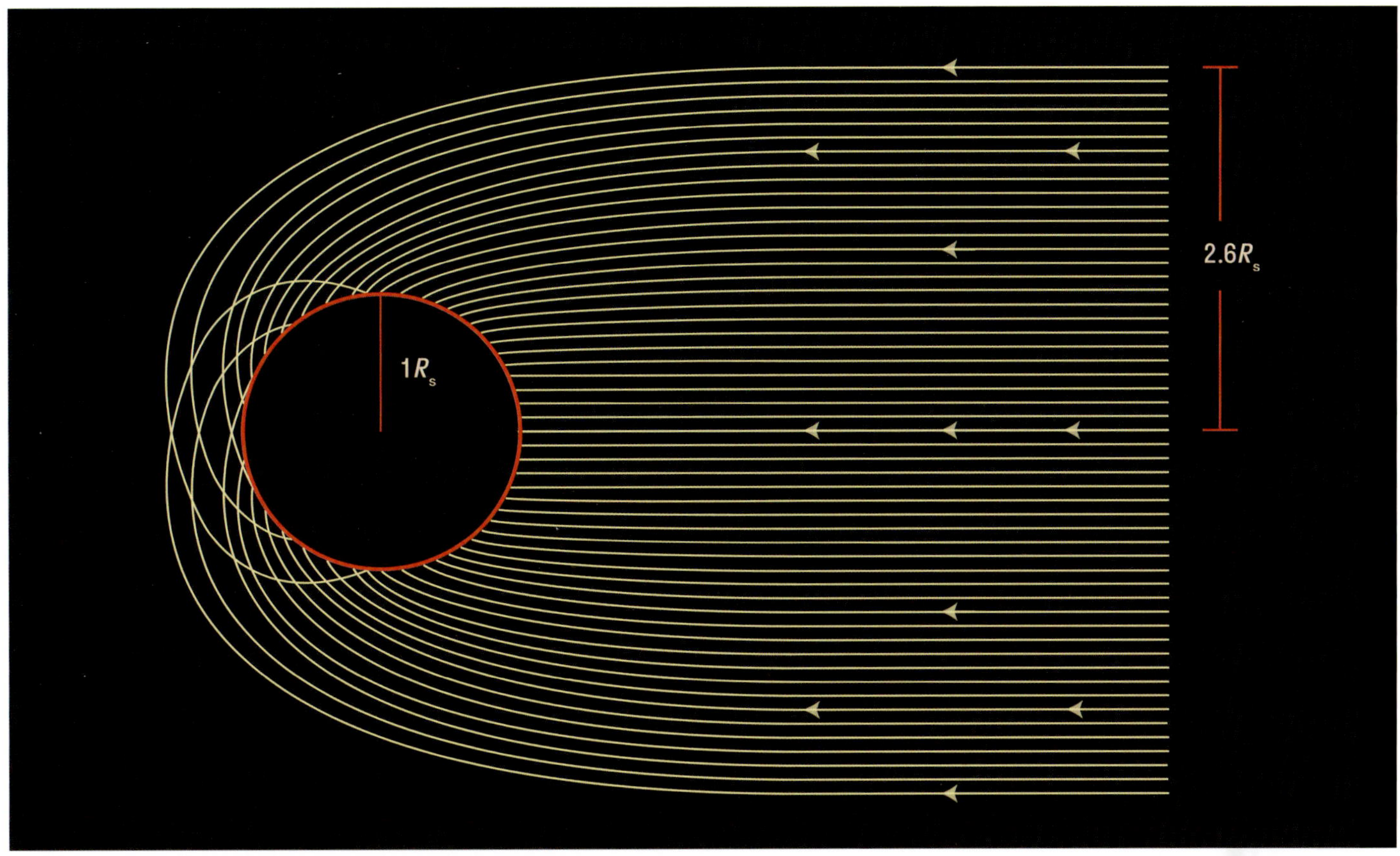

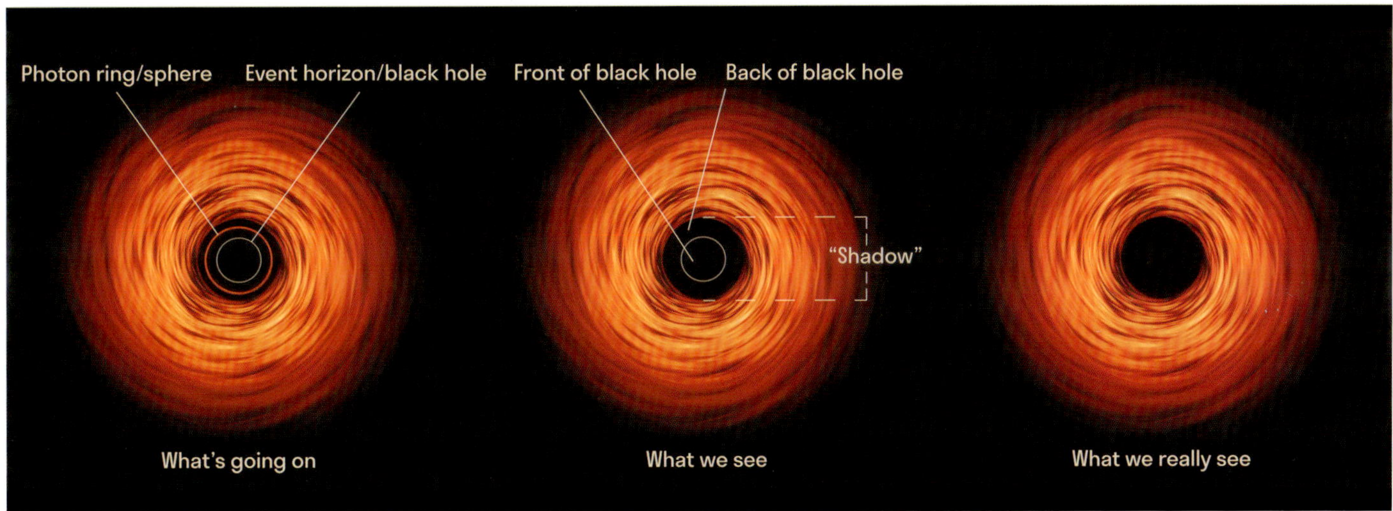

event horizon into the black hole. The same is true for the next set of light rays. To get a parallel ray that does not end up in the black hole, the light ray has to be beyond 2.6 Schwarzschild radii, so the resulting shadow is 2.6 times larger than the black hole, which measures 1 Schwarzchild radius.

Intuitively, the front of the black hole is mapped—point by point—onto the center of the circular shadow, and the back of the black hole is mapped onto a ring surrounding the front. So, from our viewpoint on Earth, we "see" the entire surface of the black hole, frontside and backside (which is what the shadow represents), as a result of the warping of spacetime by the black hole's extreme gravity. (We don't actually see anything in the dark area in the center of the accretion disk, only absence of light.) Figure 3.47 shows what we'd see if we looked at the shadow from a viewpoint perpendicular to the plane of the accretion disk. On the left, the photon ring/sphere is between the inner edge of the accretion disk and the black hole. In the

TOP
3.46 The "shadow" of a black hole.

ABOVE
3.47 What's going on, what we see, and what we really see. Visualization by NASA's Goddard Space Flight Center and Jeremy Schnittman, 2019.

This is what we'd see if we're looking at the black hole and its accretion disk from a viewpoint perpendicular to the accretion disk.

OPPOSITE
3.48 Yambe Tam (American, born 1989), *Shadow of Unknowing*, 2024. Stills from an interactive digital work patterned on a video game. Courtesy of the artist.

middle image, the front of the black hole is in the center, and the back of the black hole is around it; front and back together comprise the black hole's shadow. On the right is what we really see (without labels); since we're looking at the front and back of the black hole, we just see black.

The scientific concept of a black hole's shadow resonates with *Shadow of Unknowing* (figure 3.48), an artwork by Yambe Tam, an American of Chinese descent and a Zen Buddhist. In this interactive digital work patterned on video-game technology, viewers embark on a journey into a black hole, where they pursue a dual scientific and spiritual exploration. Along the way, viewers accumulate bits of knowledge that both affirm and challenge their concepts of self, reality, space, and time. The name of the artwork echoes a book by the Japanese author Jun'ichirō Tanizaki, *In Praise of Shadows* (1933), in which he wondered what Japanese science would be like if it had developed following Eastern traditions. Rather than Newton and Einstein asking questions about the nature of light, Japanese scientists would have explored the nature of darkness. The title of Tam's artwork also mirrors *The Cloud of Unknowing*, a fourteenth-century anonymous book of Christian mysticism in the tradition of Pseudo-Dionysius and negative theology.

In *Shadow of Unknowing*, when the viewer's avatar crosses the event horizon, it enters the shadow of the black hole (figures 3.48 *top* and 3.49). The symbols orbiting the avatar are fragments of knowledge that the viewer collects throughout their journey. The symbols, written in oracle bone script (the oldest type of Chinese writing), form a phrase from the seventh-century Buddhist text the Heart Sutra: 色即是空, 空即是色 (Form is emptiness, emptiness is form). Figure 3.48 *center* and *bottom* show views from inside the black hole. Consistent with Eastern philosophy, Tam's work asserts that only when entering the black hole—the "shadow of unknowing"—and only by embracing the mysterious darkness will viewers be able to gather knowledge; as they accumulate these granules of knowledge, they learn that darkness is not absence but infinite potentiality, an infinite density with enormous energy. *Shadow of Unknowing* is a contemporary characterization of the sublime with a distinctly Asian perspective: a sublime darkness, or, as the artist has said, "a vast, radiant void that inspires feelings that oscillate between terror and bliss . . . an attempt to establish truth in a vast, unknowable universe."[52] Immersed in darkness, unknowing becomes awareness. Tam's mysterious *Shadow of Unknowing* recalls a remark by Einstein: "The most beautiful experience we can have is the mysterious. It is the fundamental emotion that stands at the cradle of true art and science."[53]

3.49 Yambe Tam (American, born 1989), *Shadow of Unknowing* (detail), 2024.

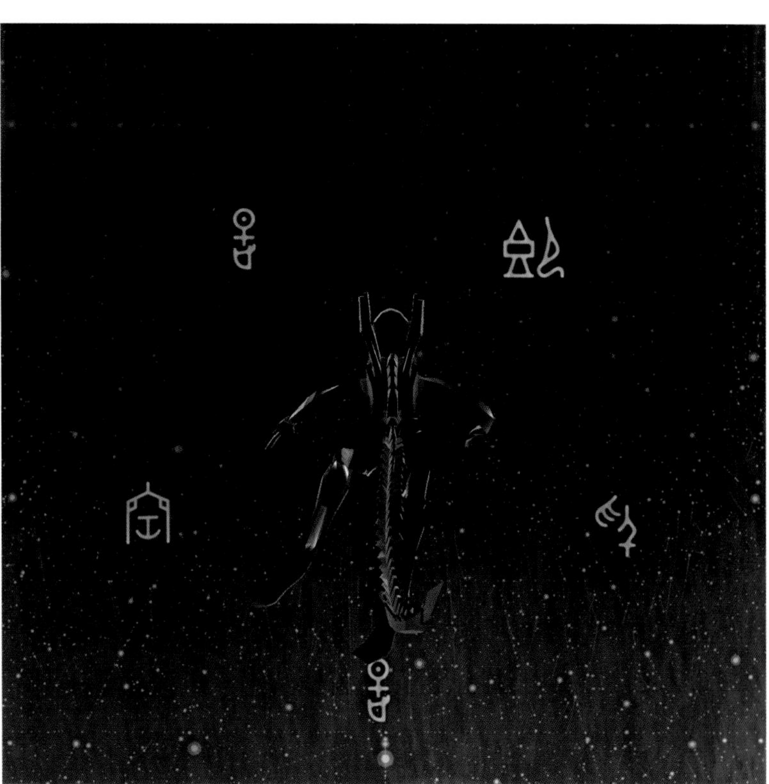

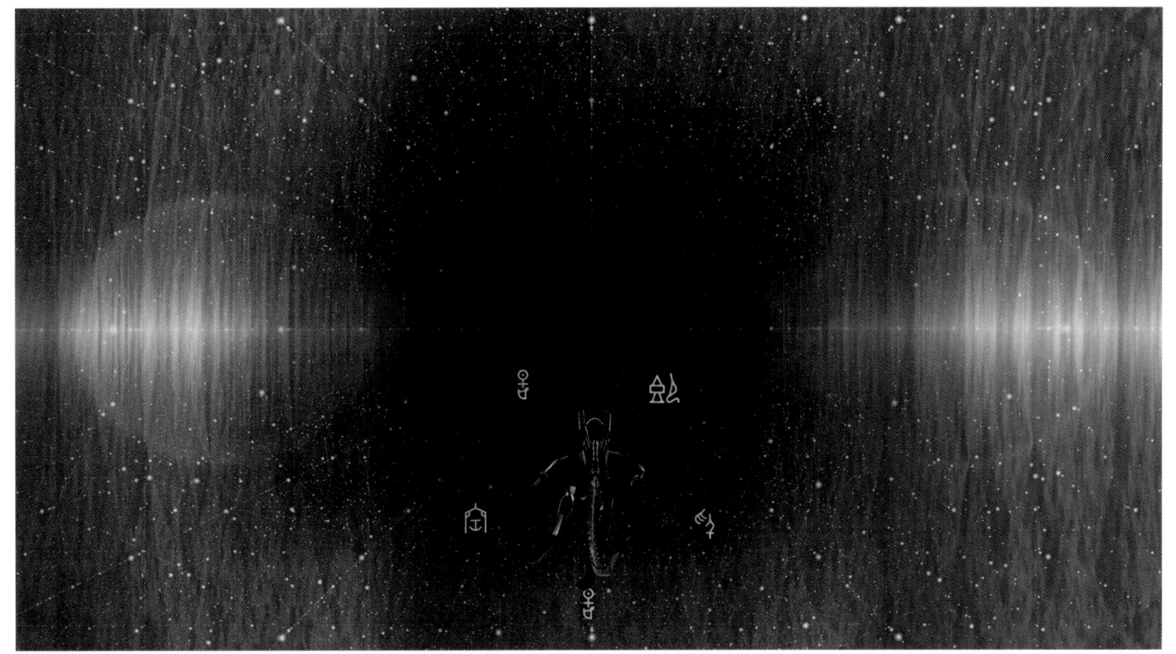

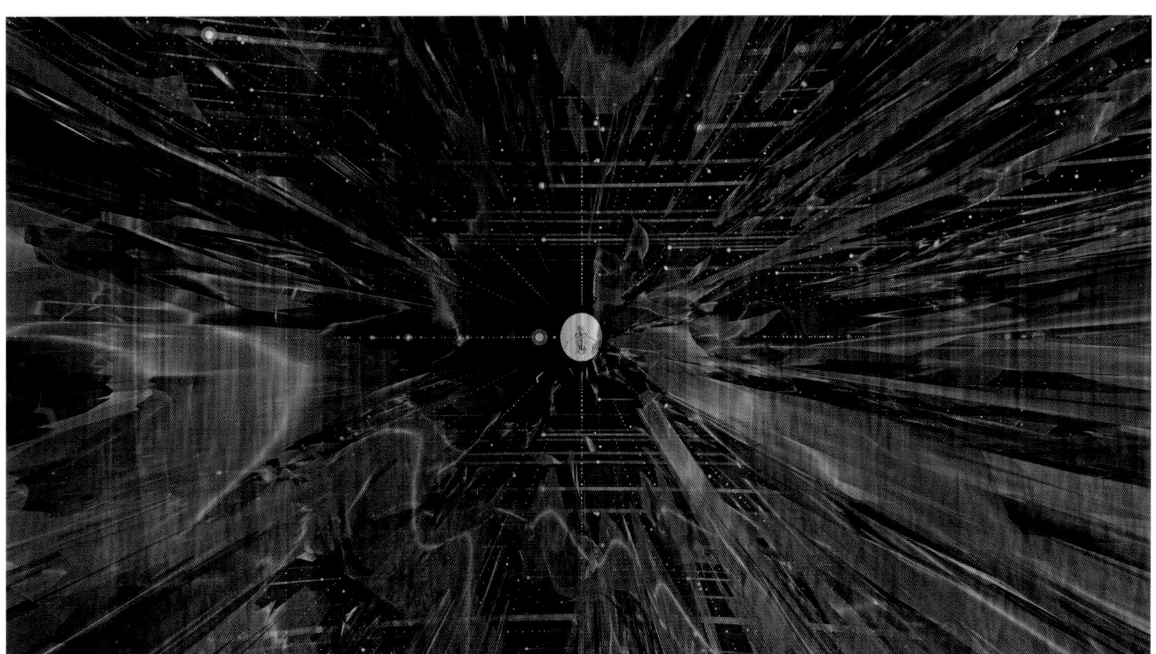

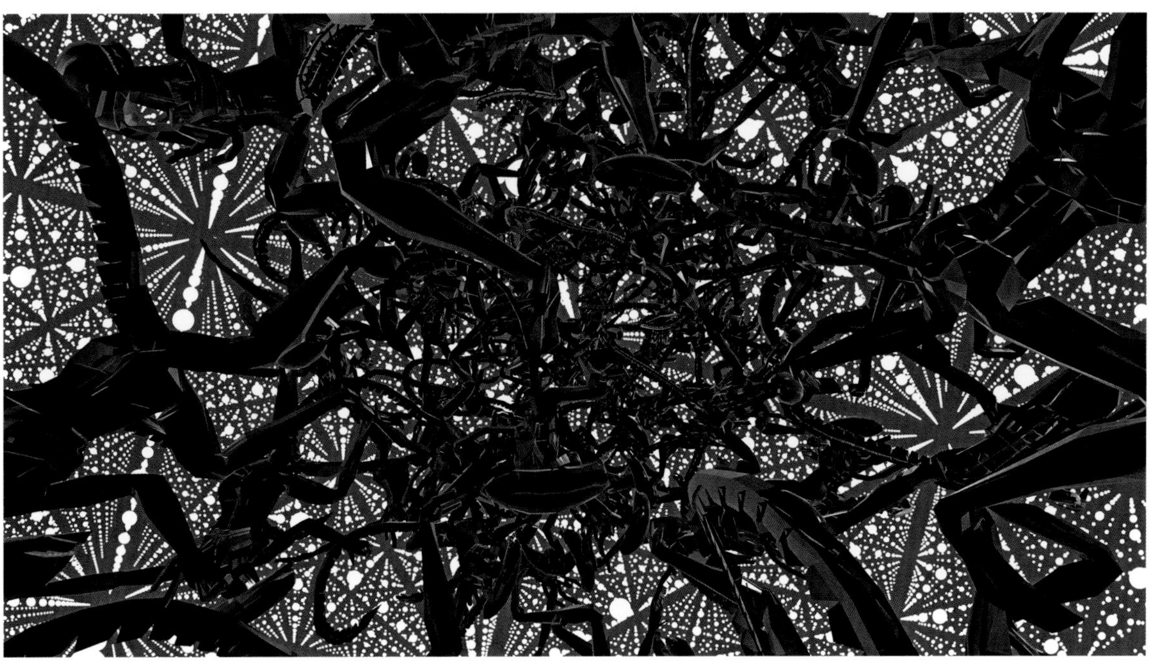

BLACK HOLES AS METAPHORS FOR DISAPPEARANCE, DEPRESSION, AND DESTRUCTION

Marco Poloni produced a constellation of artworks called *The Majorana Experiment* after Ettore Majorana, an Italian physicist who disappeared at sea in 1938 during a turbulent time in Italian politics. Majorana worked at the forefront of nuclear physics in Enrico Fermi's laboratory in Rome, firing neutrons into an atom of uranium. Majorana and Fermi were distressed by the rise of Benito Mussolini, who was increasingly allied with Adolf Hitler. After the Italian Fascists passed anti-Semitic laws in 1938, Fermi fled with his Jewish wife and their small children; he traveled to Stockholm with his family to receive the 1938 Nobel Prize in Physics but never returned to Rome, emigrating instead to the United States. Majorana withdrew all the money from his bank account and was last seen buying a ticket for the boat from Naples to Palermo. Between 2008 and 2010 Poloni produced three films, a series of photographs, and a set of historical documents about Majorana (figure 3.50). When the film laboratory accidentally made a hole in a picture of the sea (figure 3.51), Poloni kept it as part of the work—a trace of Majorana's disappearance. This photograph, called *Black Hole*, represents a void, a conspicuous absence, that draws our attention.

Disappearance takes on other grim political meanings in a work by Kamil Hassim, an artist of Indian descent who was born and raised in South Africa. In 2022 Hassim was artist-in-residence at the Conseil européen pour la recherche nucléaire (CERN), which operates the most powerful particle accelerator in the world, the Large Hadron Collider. He produced an artwork titled *Event Horizon* by positioning sources of light and prisms throughout Constitution Hill in Johannesburg. This is no ordinary exhibition space. Constitution Hill was formerly the Old Fort Prison Complex, which was completed in 1902 and remained in use until 1983. Nelson Mandela spent some of his twenty-seven years in prison there. The anti-apartheid politician Robert Sobukwe and Mahatma Gandhi were imprisoned in Number Four, the section of the segregated complex that housed dark-skinned male prisoners in inhumane conditions. Hassim lit some of the spaces at the site—including the courtyard of Number Four (figure 3.52)—with a uniform red light; other spaces featured spectra created by prisms (figure 3.53). The terms "inescapable" and "point of no return" can

3.50 Marco Poloni (Italian and Swiss, born 1962), *Majorana Eigenstates* from *The Majorana Experiment*, 2008. Still from video, 46 min. Courtesy of Galerie Campagne Première, Berlin.

Here, an actor sits in the dimly lit cabin of a boat playing chess as Ettore Majorana did. In quantum mechanics, an *eigenstate* is the state of a system such as an atom or molecule.

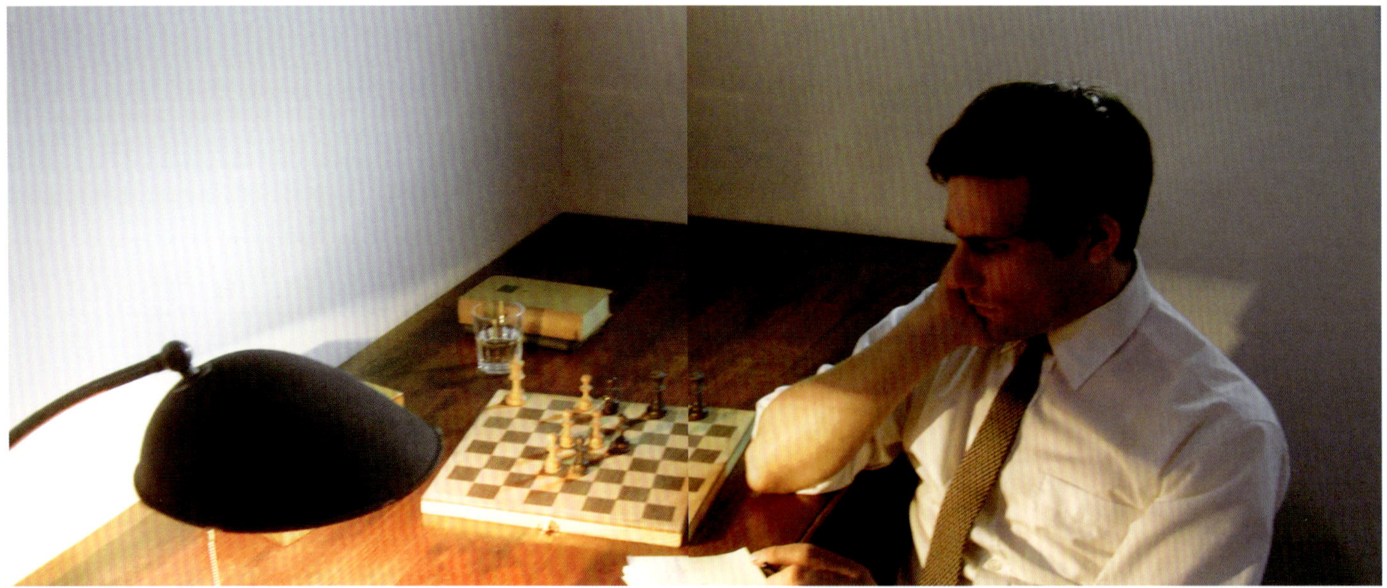

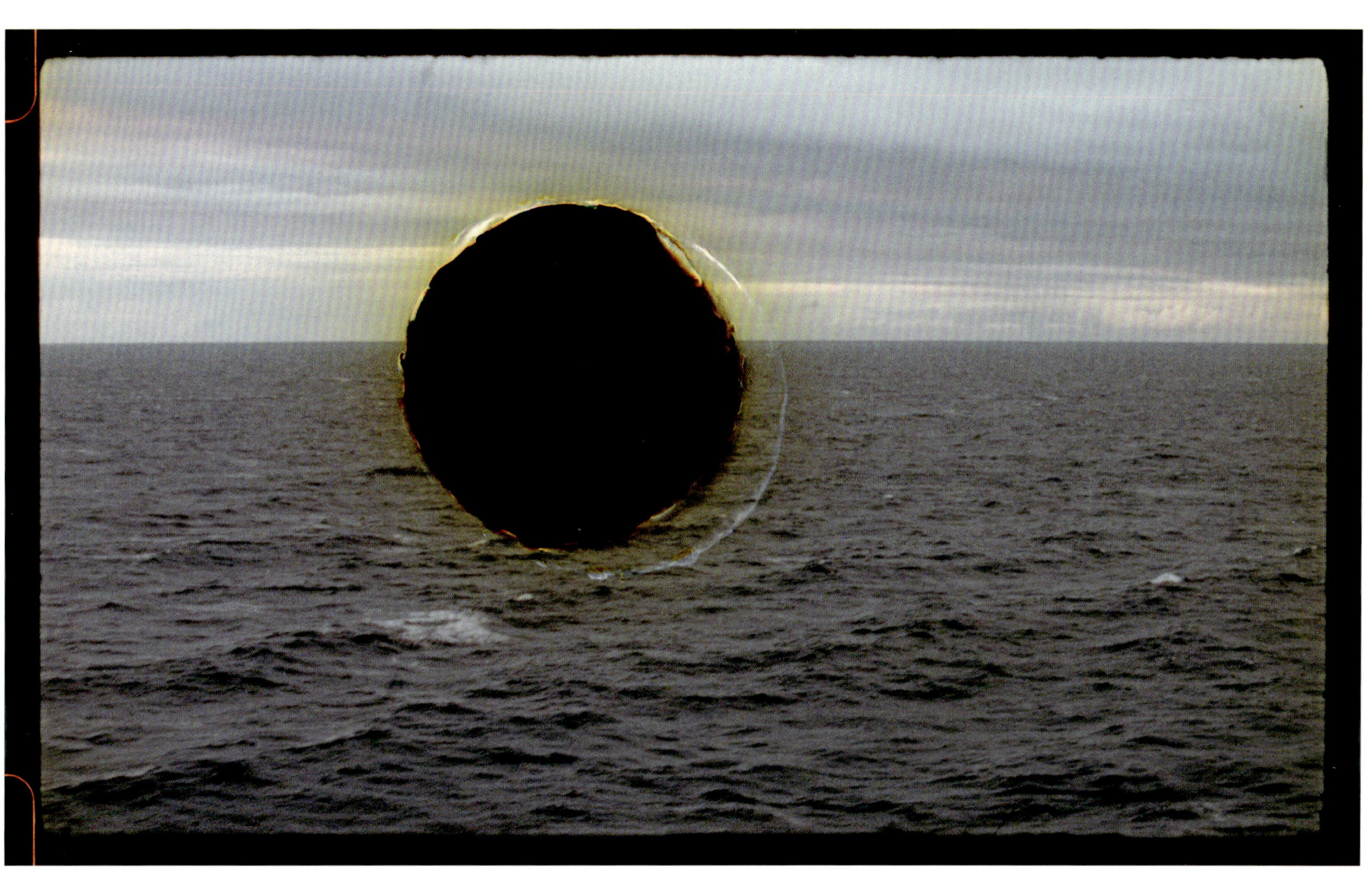

3.51 Marco Poloni (Italian and Swiss, born 1962), *Black Hole* from *The Majorana Experiment*, 2010. C-print from a Super-16mm film, 25½ × 39¼ in. (65 × 100 cm). Courtesy of Galerie Campagne Première, Berlin.

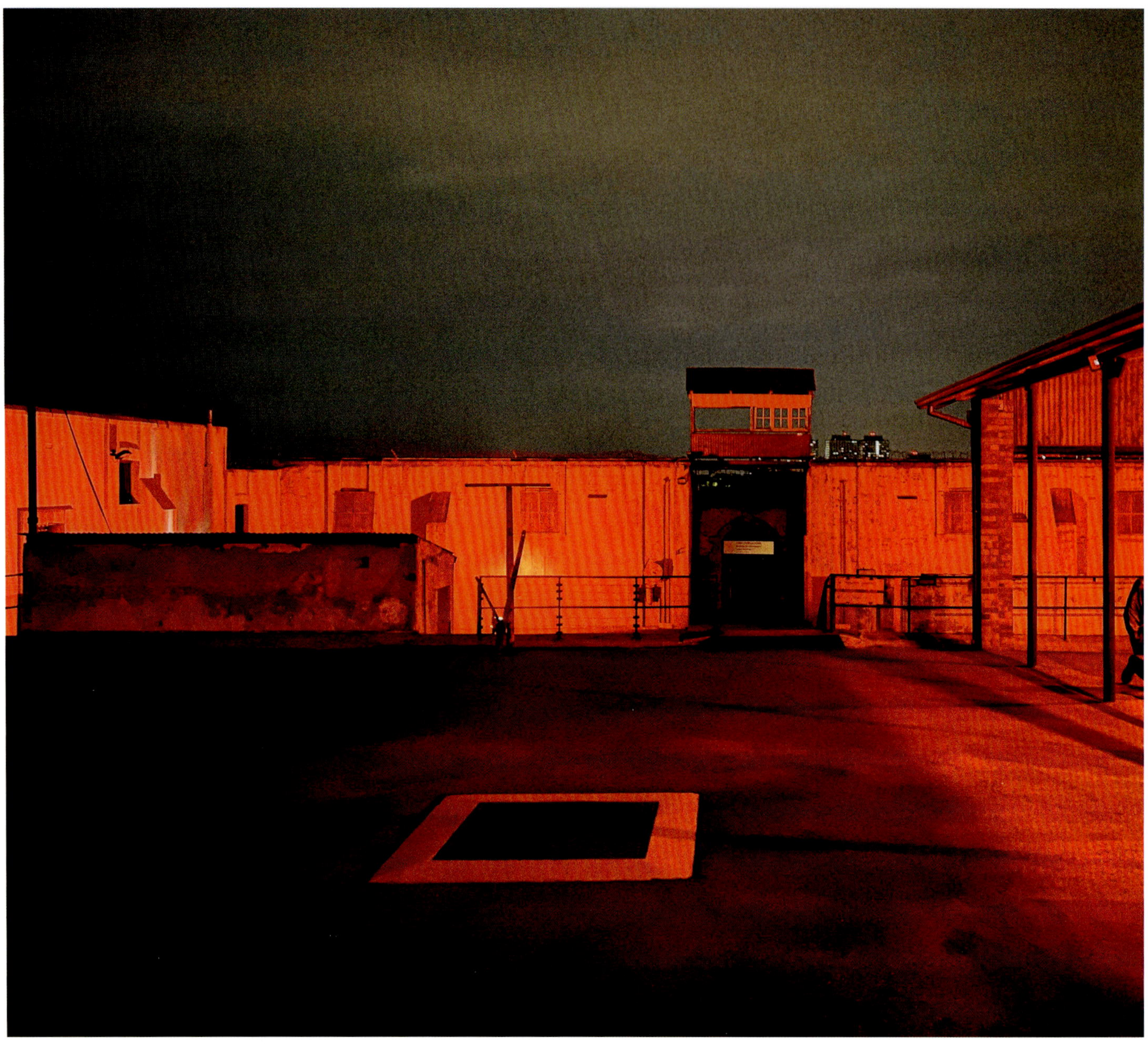

3.52 Kamil Hassim (South African, born 1997), *Event Horizon*, 2023. Installation view at Constitution Hill in Johannesburg. Courtesy of the artist.

certainly be applied to the Old Fort Prison Complex. In a prism, light separates into a colorful spectrum; once the event horizon of a black hole is crossed, all light disappears.

The Madras Atomic Power Station and the Indira Gandhi Centre for Atomic Research are located in the state of Tamil Nadu, India, where the artist Shanthi Chandrasekar was born and raised. Growing up in this community, Chandrasekar became intrigued by the nature of light and studied physics as an undergraduate. She was also trained in traditional Thanjavur-style painting (colorful relief painting of Hindu deities) and kolam (geometric line drawings). At daybreak in India, millions of women draw a kolam in rice flour at the entrance of the home to welcome visitors and bring prosperity. Throughout the day, people walk on the kolam and it gets blown away. Thus, a new kolam is drawn each morning. The tradition of kolams originated in antiquity in Tamil Nadu. In her painting *Kolam—Universe* (figure 3.54), Chandrasekar combined Western abstract art with a native Indian tradition, in effect welcoming viewers to the universe.

Chandrasekar's *Information Paradox* (figure 3.55) makes an analogy between objects disappearing into a black hole and the loss of memory. The

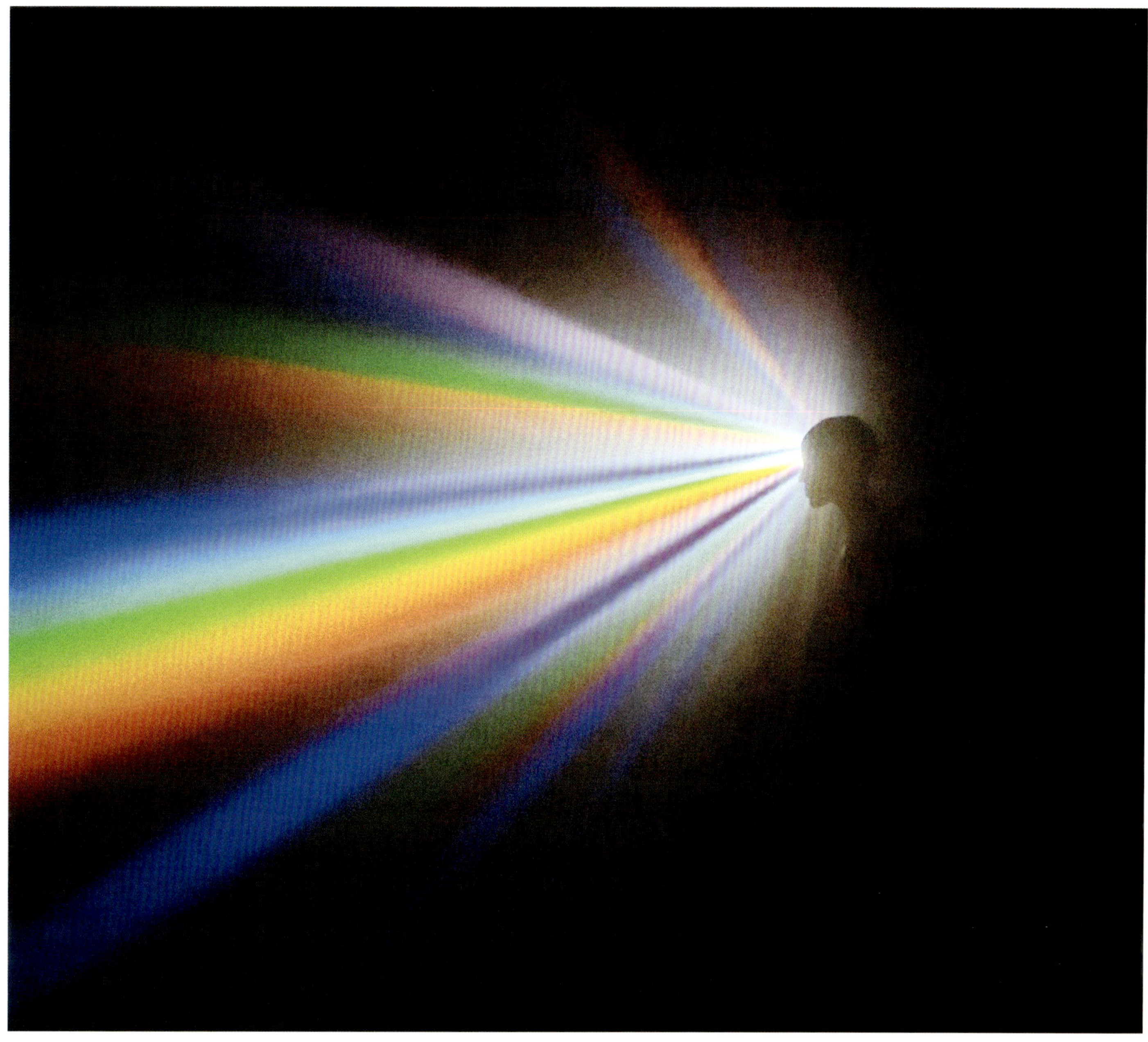

3.53 Kamil Hassim (South African, born 1997), *Event Horizon*, 2023. Installation view at Constitution Hill in Johannesburg. Courtesy of the artist.

information paradox describes what happens to objects that cross the event horizon into a black hole. If they are permanently lost, this creates a paradox since a core principle of physics is that information can never be destroyed. In this work, Chandrasekar arranged images of dendrites (parts of nerve cells) around a circle; dendrites gather electricity—the carrier of cerebral information—and bring it to cell bodies, which, in turn, transmit the charge down axons that disappear into a black hole, where memories are lost.

A character in Wong Kar-wai's film *In the Mood for Love* (2000) wants memories to disappear, and the director used a black hole as a metaphor for disappearance of information. Chow Mo-wan, a journalist, and Su Lizhen, a secretary, are attracted to each other, but they never become lovers. Midway through the film, Chow gets a job in Singapore, so he doesn't see Su anymore, but he keeps thinking about her. He doesn't want anyone to know his feelings, so he travels to Angkor Wat in Cambodia. While a Buddhist monk looks on, he finds a deep hole in the ancient temple and whispers his secret into the hole, where the information disappears forever (figure 3.56). Therefore, Chow's secret is safe. No one, not even himself, will ever know his deepest feelings.

3.54 Shanthi Chandrasekar (American, born India 1967), *Kolam—Universe*, 2021. Acrylic on canvas, 10 × 10 in. (25.4 × 25.4 cm). Courtesy of the artist.

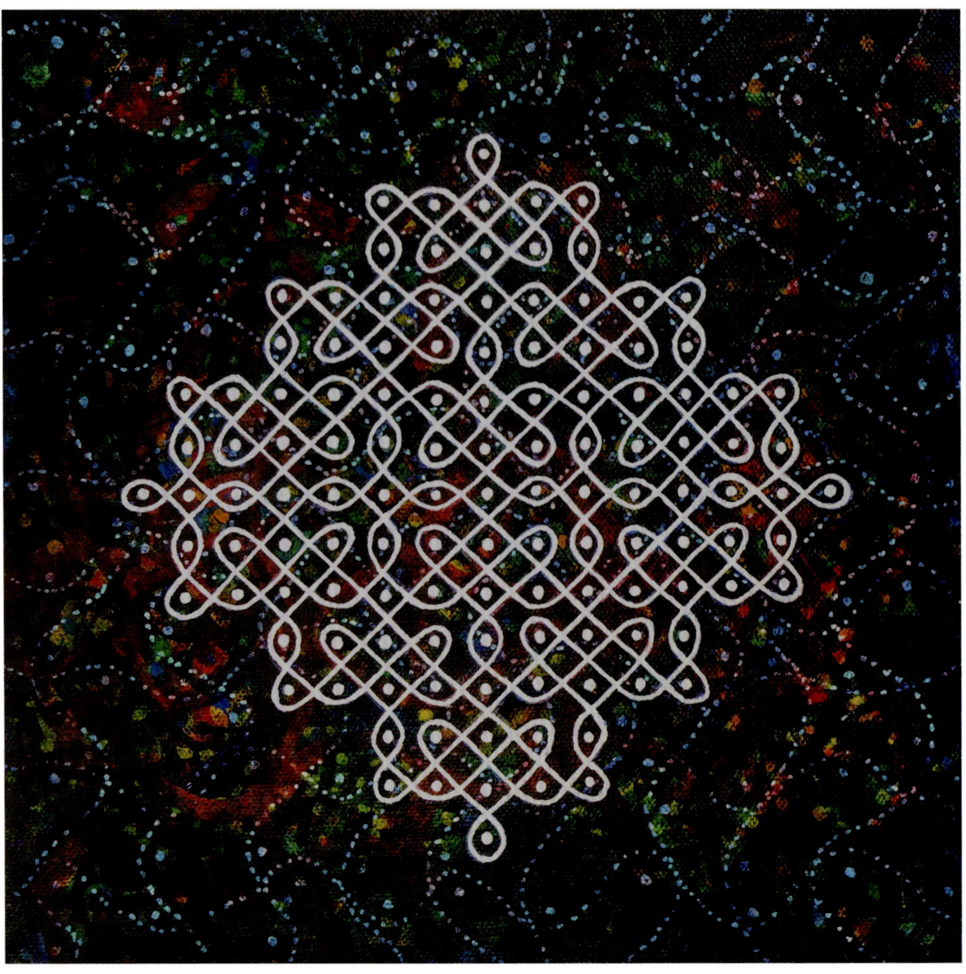

3.55 Shanthi Chandrasekar (American, born India 1967), *Information Paradox*, 2018. Pen and ink on paper, 30 × 22 in. (76 × 55.8 cm). Courtesy of the artist.

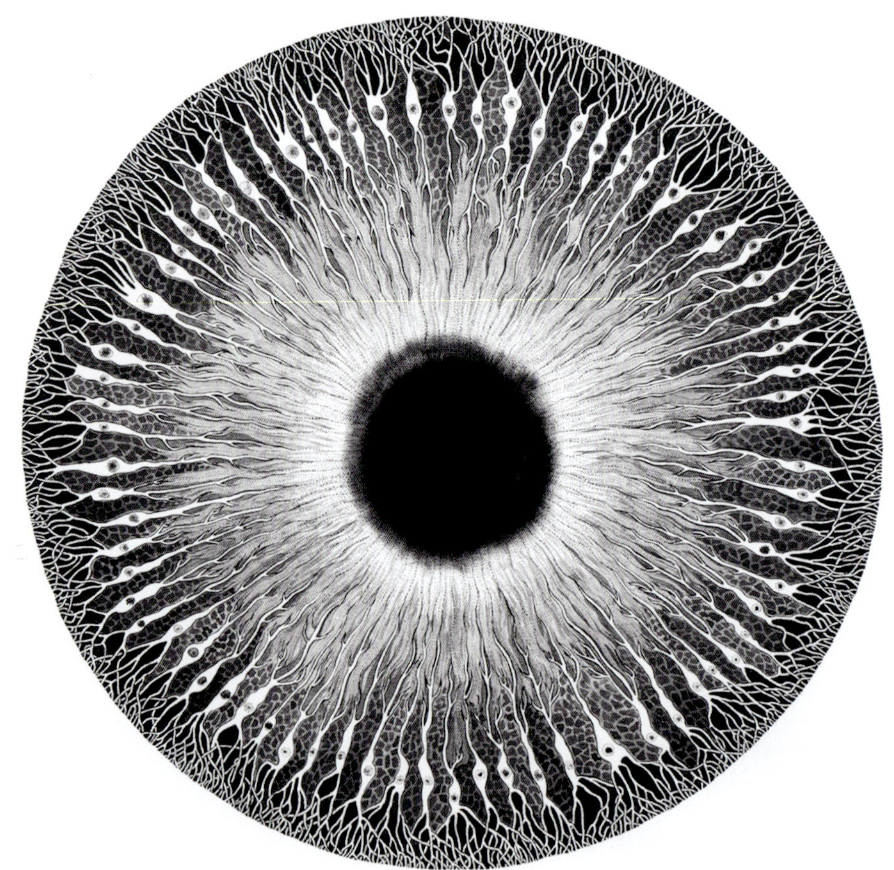

OPPOSITE
3.56 Still from *In the Mood for Love*, 2000, directed by Wong Kar-wai.

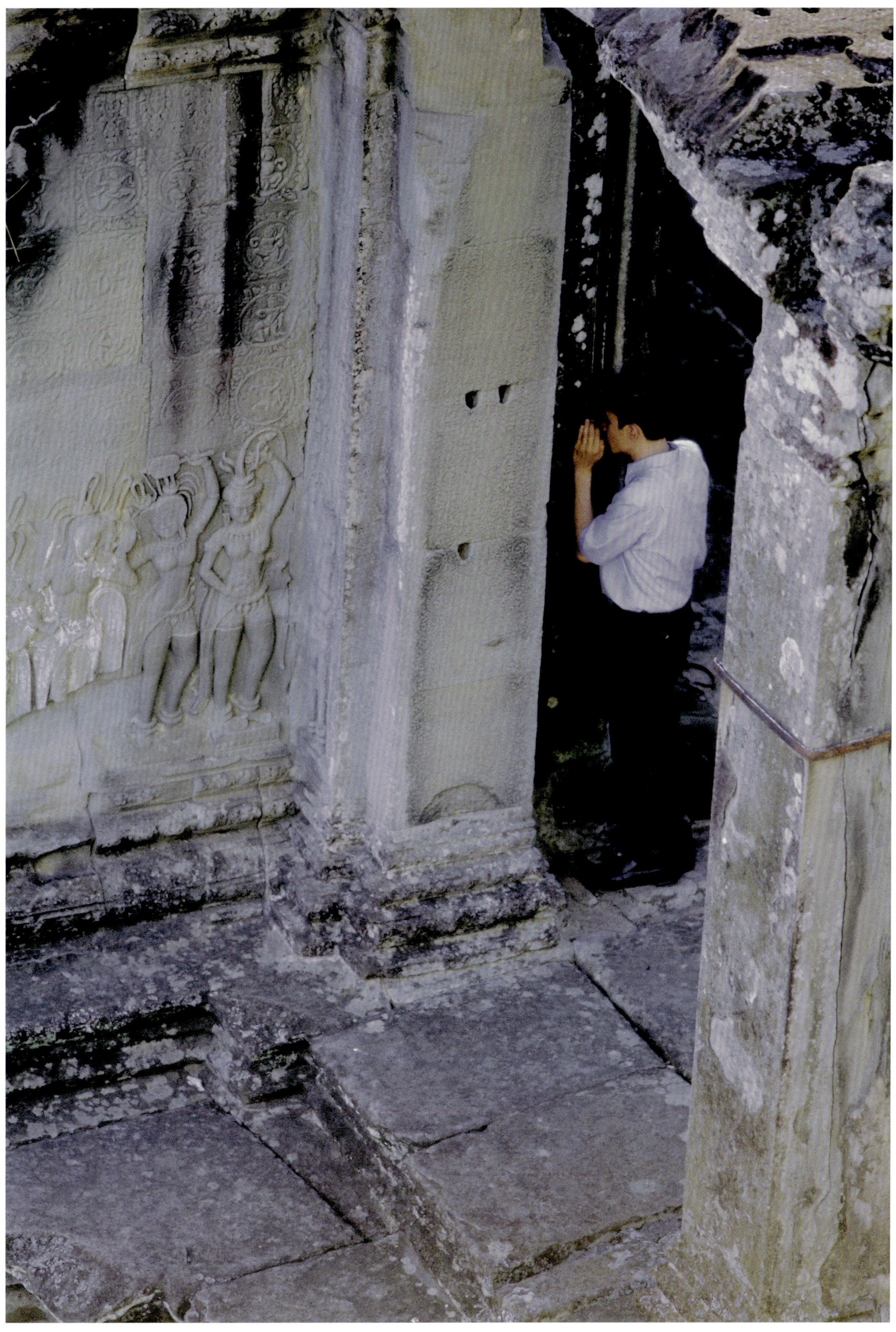

3.57 Moonassi (Daehyun Kim; Korean, born 1980), *Black Hole*, 2018. Ink on paper, 26¾ × 16½ in. (68 × 42 cm). Courtesy of the artist.

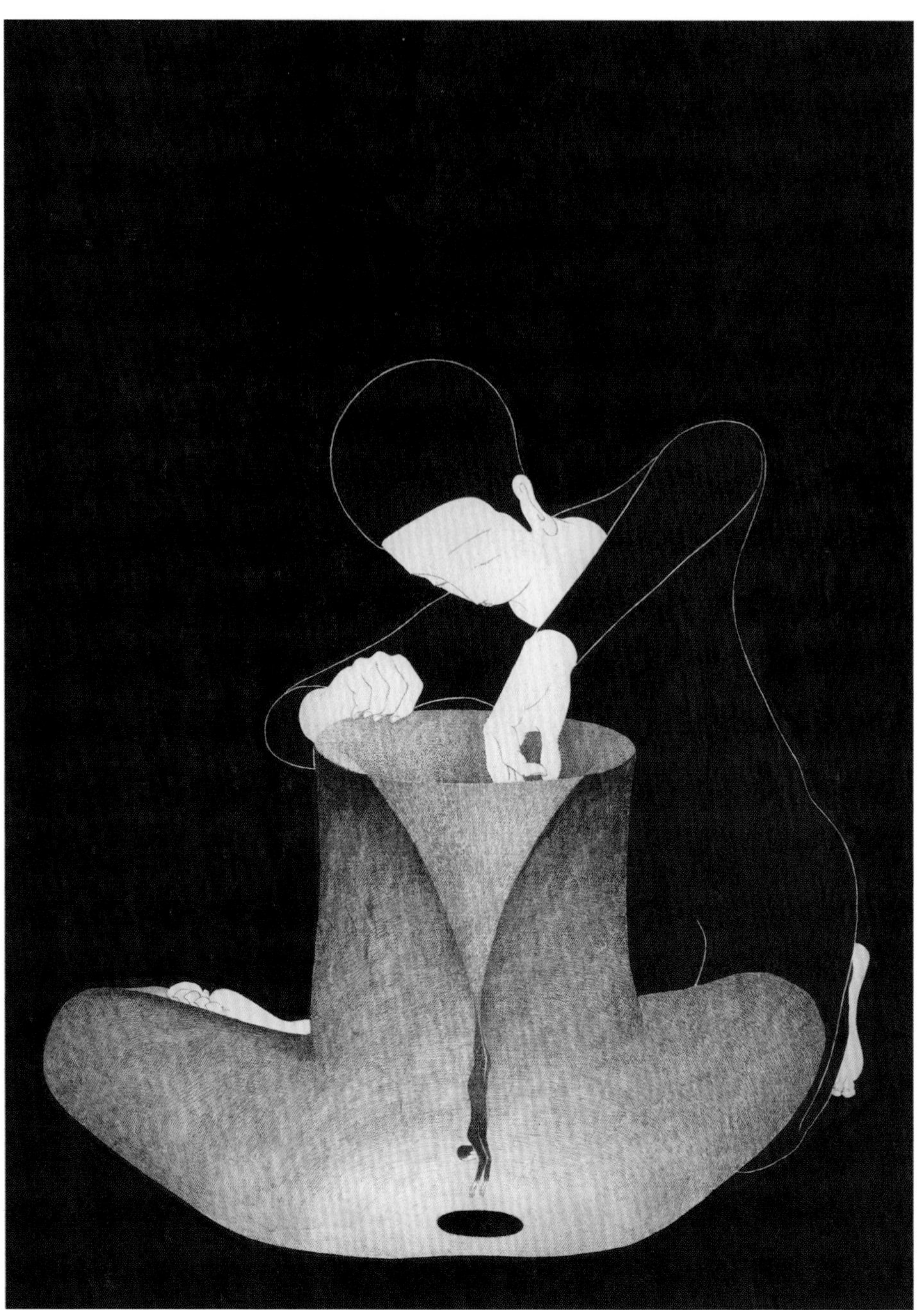

The Korean artist Moonassi's name is a pseudonym derived from a Korean word for anatta, the Buddhist concept of non-self. In his ink painting titled *Black Hole* (figure 3.57), a person watches their self (symbolized by a miniature figure) disappear into the singularity of a black hole during meditation (symbolized by the bottom half of a person sitting in the lotus position). In 2018 in South Korea, where suicide is the leading cause of death for men and women between fifteen and thirty-five years of age, *Vogue Korea* chose this artwork to illustrate an article on depression.[54] While the magazine was perhaps suggesting that depression feels like falling into a black hole, Buddhist doctrine suggests that an embrace of non-self may be a path out of depression.

Another form of trauma is represented in a painting by the American artist David Huffman. In *The Black Hole and a Traumanaut's Uncertain Journey* (figure 3.58), a black hole hovers above colorful clouds as a figure below enters a spaceship. As the title makes clear, that figure is not an astronaut (*astro* is Latin for "star"; *naut* is from the Greek for "voyager") but a *traumanaut* (*trauma* is from the Greek for "wound"). The figure

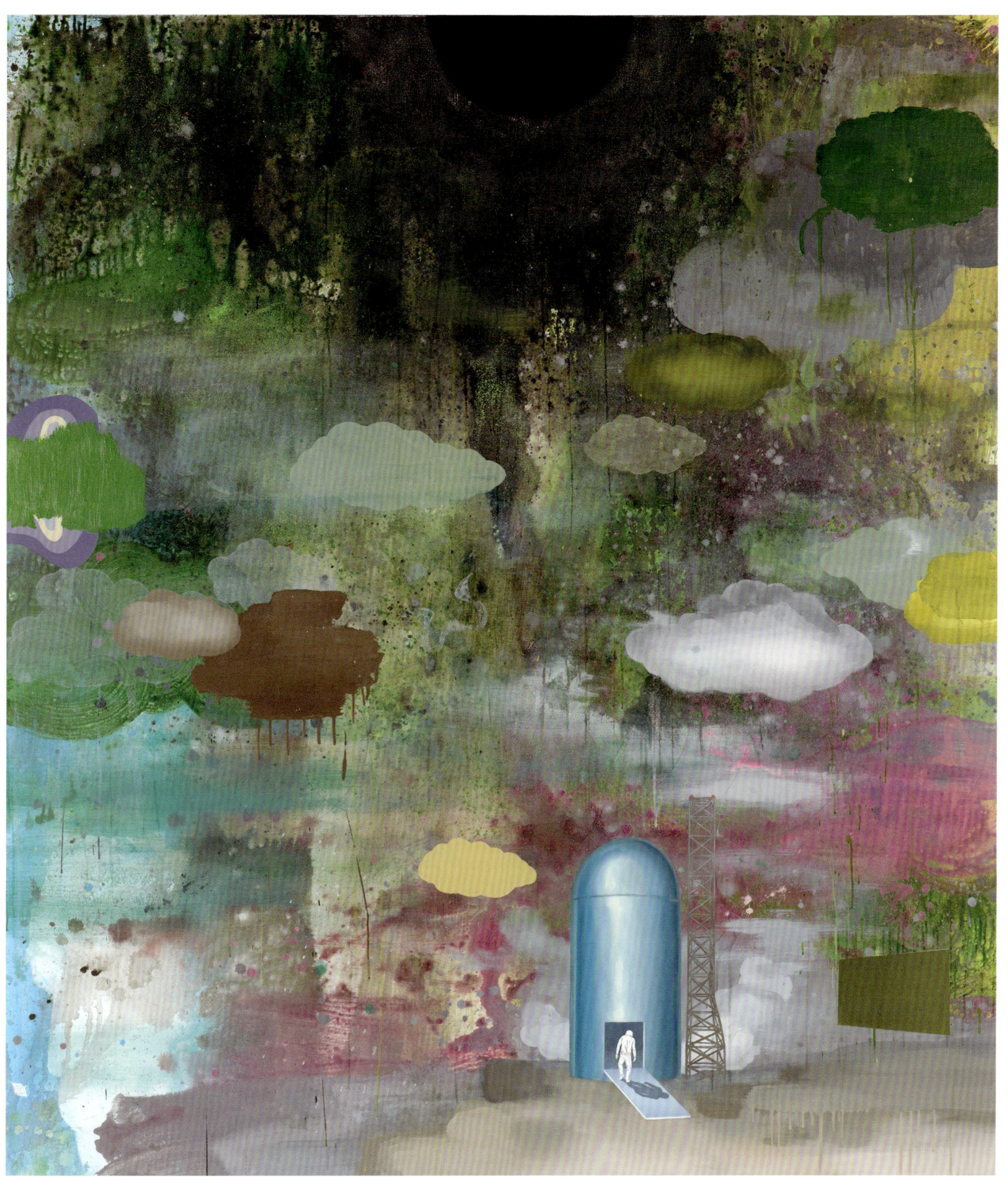

3.58 David Huffman (American, born 1963), *The Black Hole and a Traumanaut's Uncertain Journey*, 2008. Acrylic, glitter, oil and spray paint on canvas, 82 × 72 in. (208.3 × 182.9 cm). San Francisco Museum of Modern Art, Accessions Committee Fund purchase. Courtesy of Jessica Silverman Gallery, San Francisco.

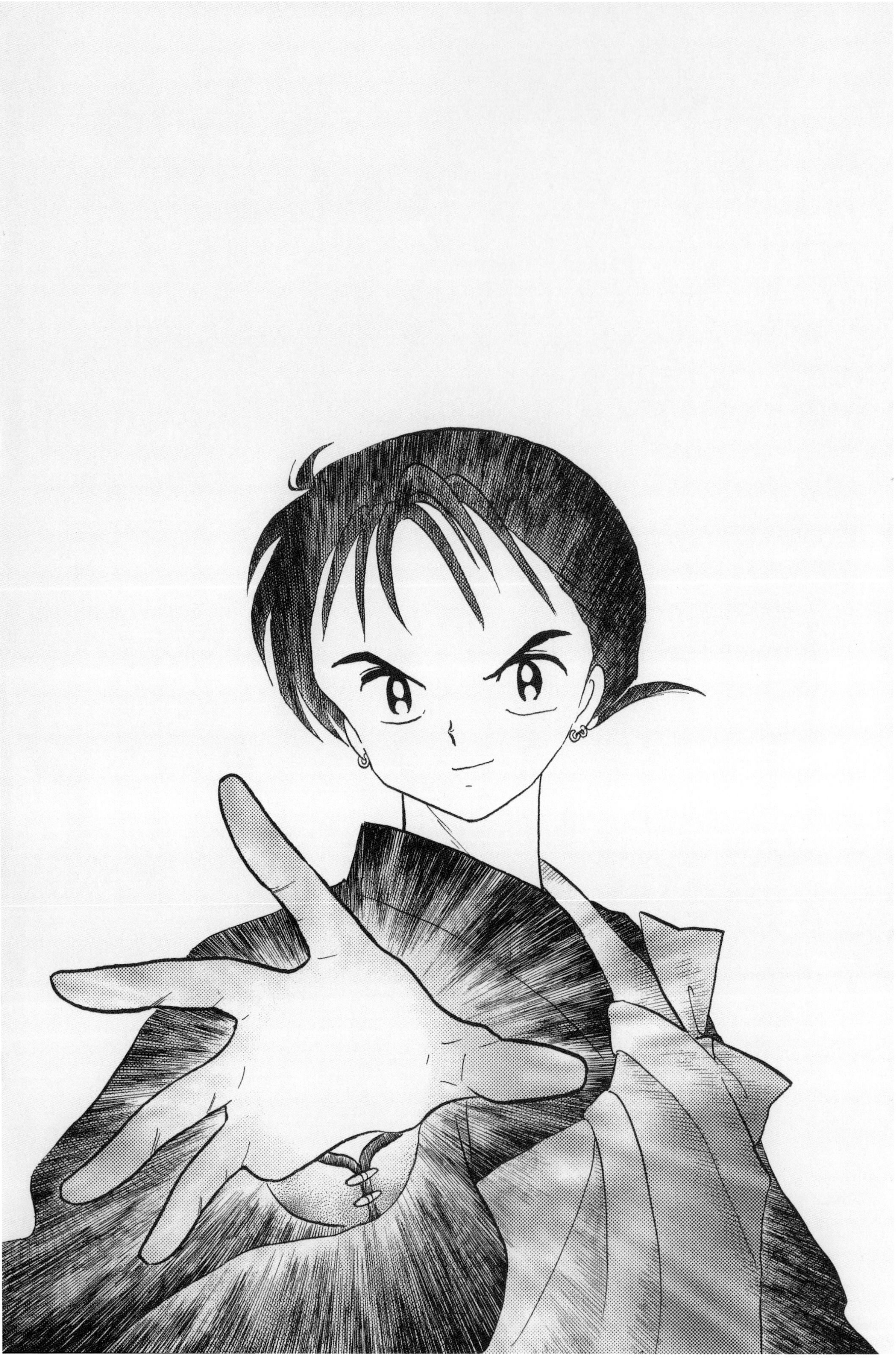

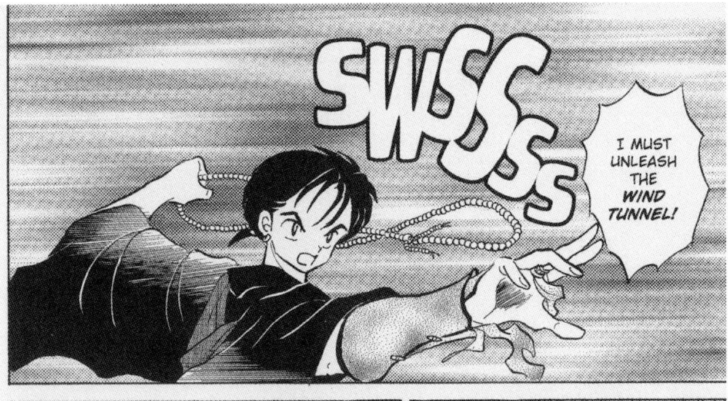

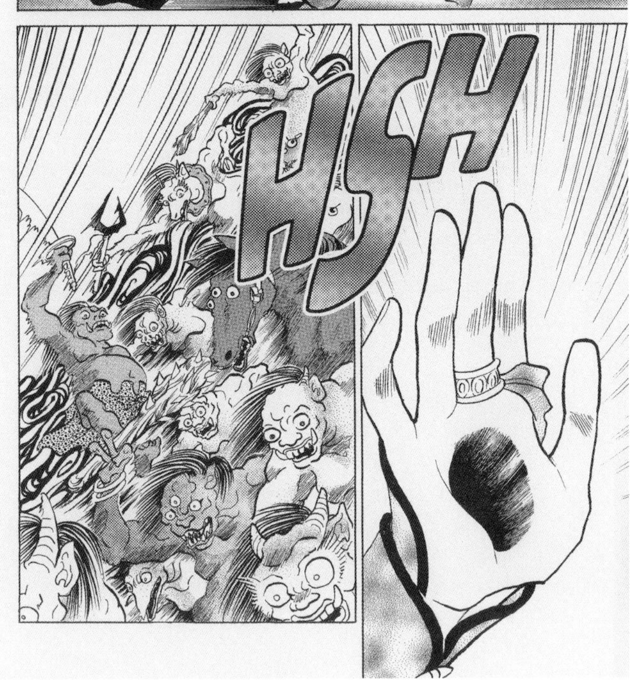

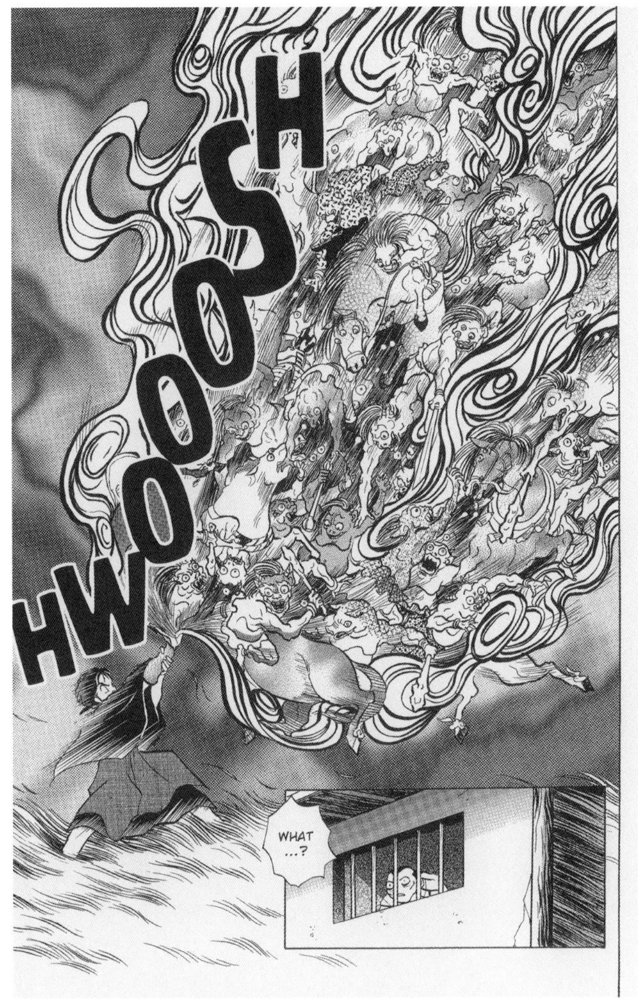

represents an enslaved person who was brought to America from Africa. Huffman's traumanaut is embarking on an uncertain journey through a dangerous universe full of black holes, always looking for home.

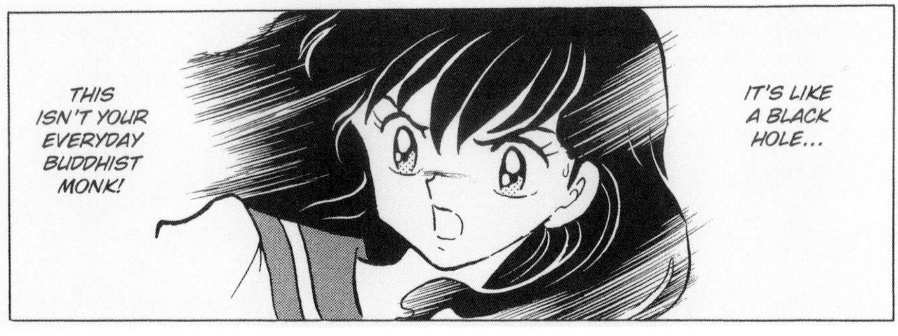

Destruction is a theme in Rumiko Takahashi's manga, *Inuyasha*, featuring Miroku, an eighteen-year-old Buddhist monk. Miroku's grandfather was cursed by an evil villain to have a "Wind Tunnel" (black hole) embedded in the palm of his right hand, and all his male descendants retain the curse (figure 3.59). The Wind Tunnel becomes stronger each year, until it threatens to swallow the hand and then the whole person. Miroku manages to turn his Wind Tunnel from an evil curse into a powerful weapon, but he must avoid being destroyed by it (figure 3.60).

A different kind of fantasy world was created by the Korean illustrator Sangho Bang. What's inside the myriad black holes in *Spaceship* (figure 3.61)? The artist said: "Our imagination lets us enter the black hole and probe its furthest reaches.... When I peer into those holes, I see other worlds inside. And the worlds in those holes have holes of their own. This chain of interlinked black holes means we can never know the exact scale and location of these worlds."[55] Exploring the black holes, the artist discovers magical creatures in psychedelic landscapes (figure 3.62).

OPPOSITE

3.59 Miroku showing his Wind Tunnel in Rumiko Takahashi, *Inuyasha*, trans. Mari Morimoto (San Francisco: VIZ Media, 2010), 2:457; originally published in *Weekly Shōnen Sunday*, episode 53, January 2, 1998.

ABOVE

3.60 Scenes from Rumiko Takahashi, *Inuyasha*, 469, 536–537.

In the scenes above, an artist has painted monsters that come to life and attack Miroku, who unleashes his Wind Tunnel and sucks them all in while the startled artist looks on from below. In the lower right, the heroine of *Inuyasha* watches with astonishment.

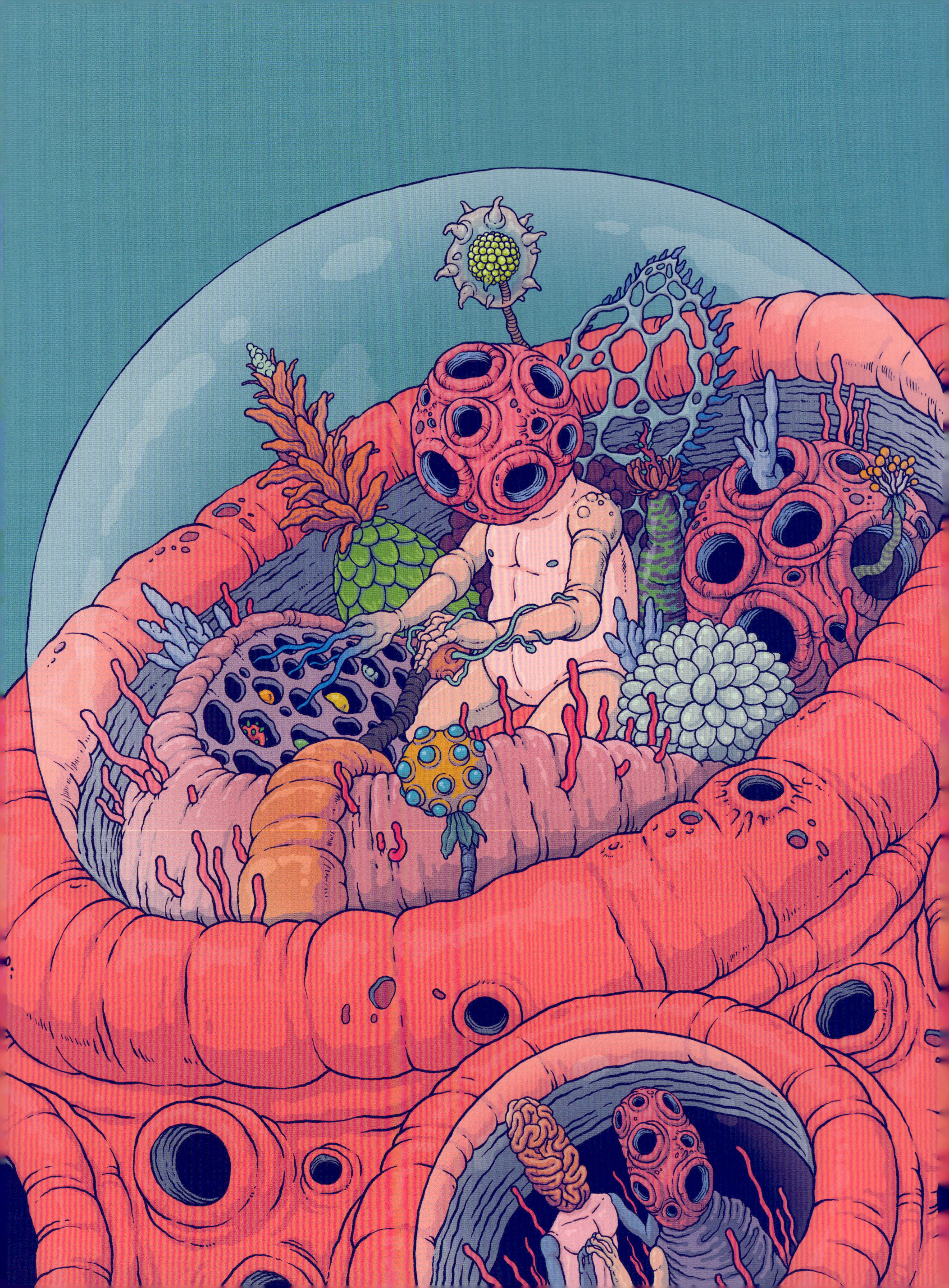

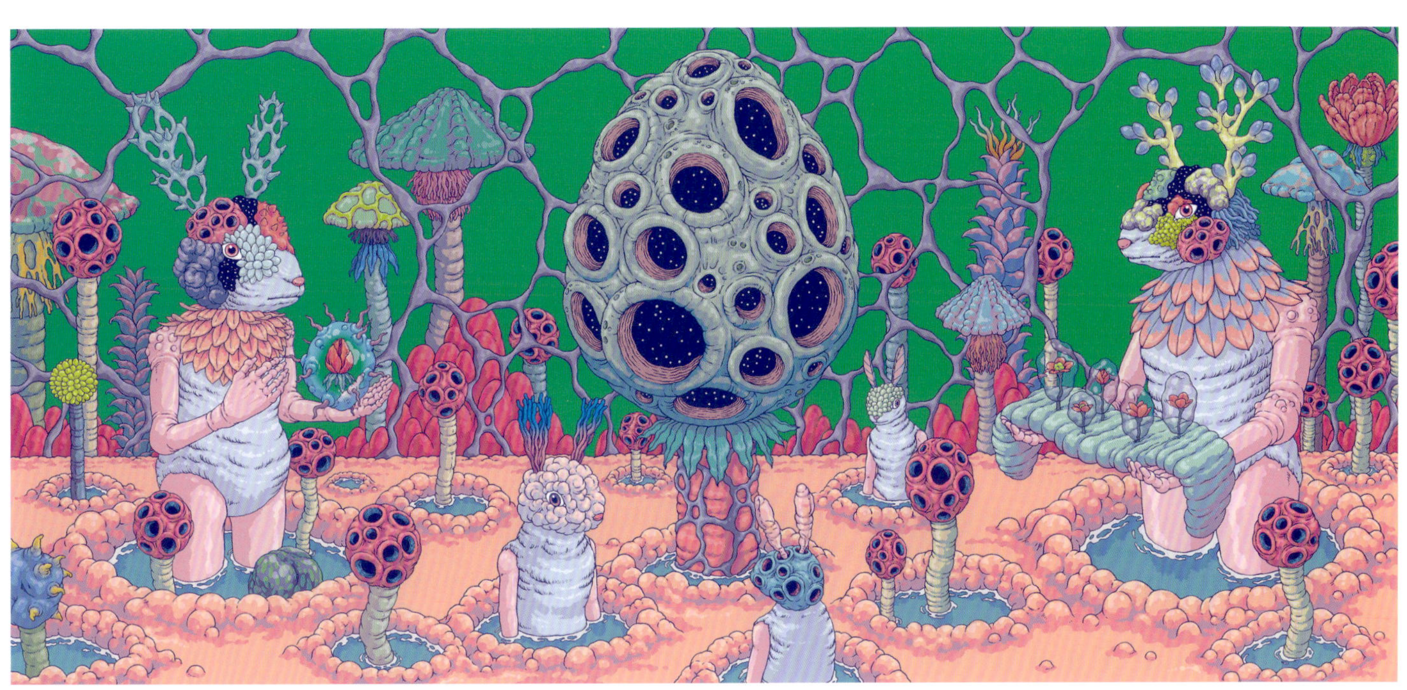

OPPOSITE
3.61 Sangho Bang (Korean, born 1991), *Spaceship* (detail), 2018. Digital print, 21½ × 23½ in. (55 × 60 cm). Courtesy of the artist.

ABOVE
3.62 Sangho Bang (Korean, born 1991), *On the Moon*, 2018. Digital print, 16½ × 35 in. (42 × 89 cm). Courtesy of the artist.

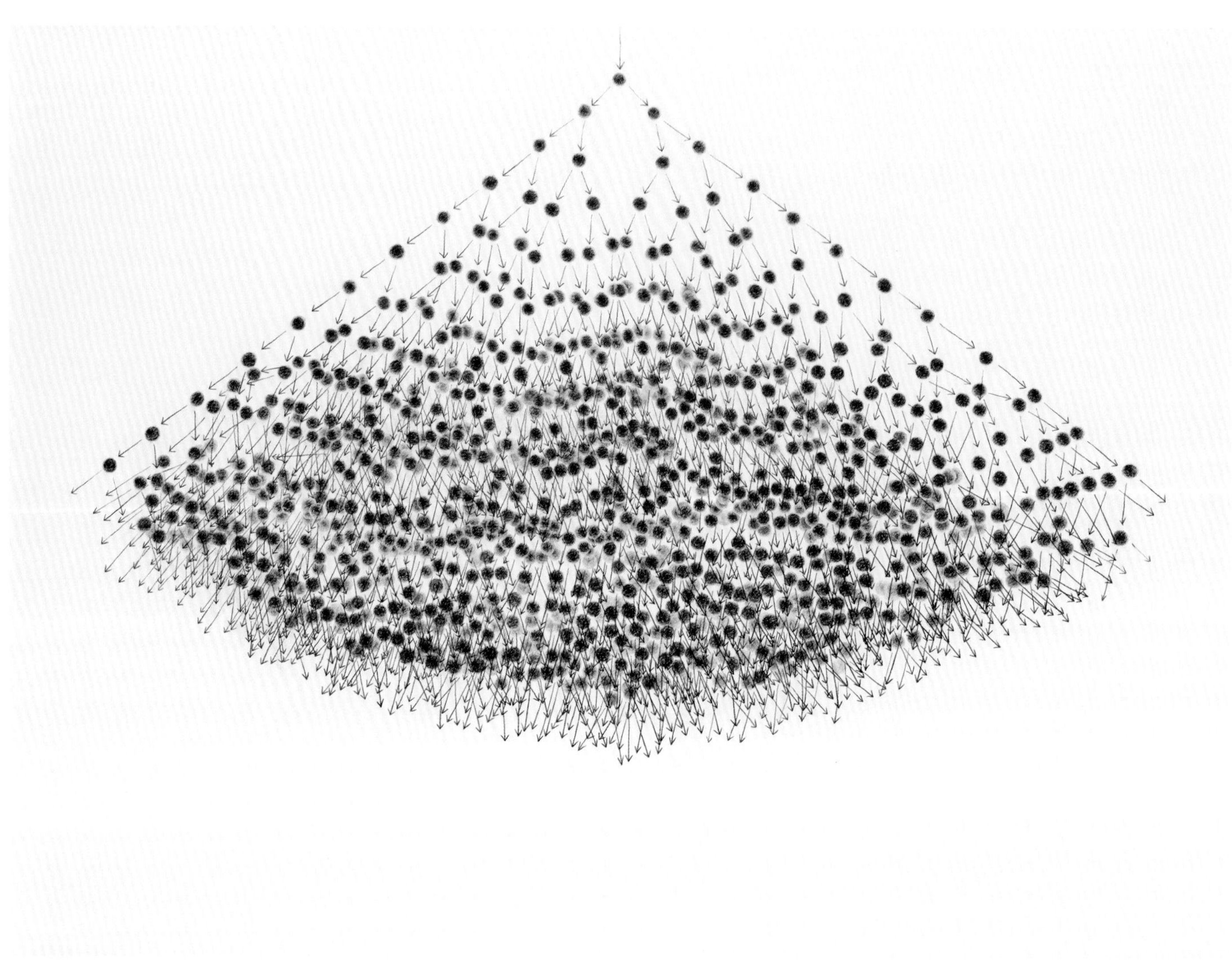

3.63 Ingrid Koenig (Canadian, born 1956), *Chain Reaction*, 2009, from the series *Navigating the Uncertainty Principle*. Graphite on paper, 38 × 50 in. (96.5 × 127 cm). Courtesy of the artist.

The Canadian artist Ingrid Koenig explores the everyday implications of physics, as the title of her series *Navigating the Uncertainty Principle* indicates. In one work (figure 3.63), she created a drawing based on a scientific diagram of a chain reaction: for example, a neutron splits an atom of uranium, which releases neutrons that split other uranium atoms, beginning a chain reaction. At the top of the drawing, the process is clear and orderly, but by the time the viewer gets to the bottom there have been twelve iterations and—as in daily life—the process looks messy and far less certain. In *Black Hole* (figure 3.64), matter goes into a black hole and toward the singularity and so—again, as in life—all matter will eventually be destroyed.

The American artist James Rosenquist is best known for his colorful painted montages of advertising imagery that helped define the Pop Art movement in the 1960s. He also occasionally created paintings in a gray monochrome, giving his work a more modernist aesthetic, as in *Time Dust-Black Hole* (figure 3.65).

When Jean-Pierre Luminet described the infinite properties of black holes in his 2010 book *Le destin de l'univers: Trous noirs et énergie sombre* (The fate of the universe: Black holes and dark energy), it inspired the Spanish composer Hèctor Parra to write *Caressant l'horizon* (Caressing the horizon; figures 3.66 and 3.67). In this work, Parra aimed to create the musical space of a black hole:

> *Caressing the Horizon* becomes a musical journey of an almost dreamlike character which takes us to the limit of known physical forces, where matter is so concentrated that the curvature of spacetime becomes infinite and space itself, according to Einstein's general relativity, is torn apart. Thus, while trying to "caress," to "acoustically palpate" black holes and their event horizons, *Caressing the Horizon* takes the form of a "thought experiment" which crystallizes, as a musical architecture, into a listening experience. But, given that we are prisoners of time—like everything that happens in the universe—this journey is an experience to live, to experience sensually, and not to think in the abstract.[56]

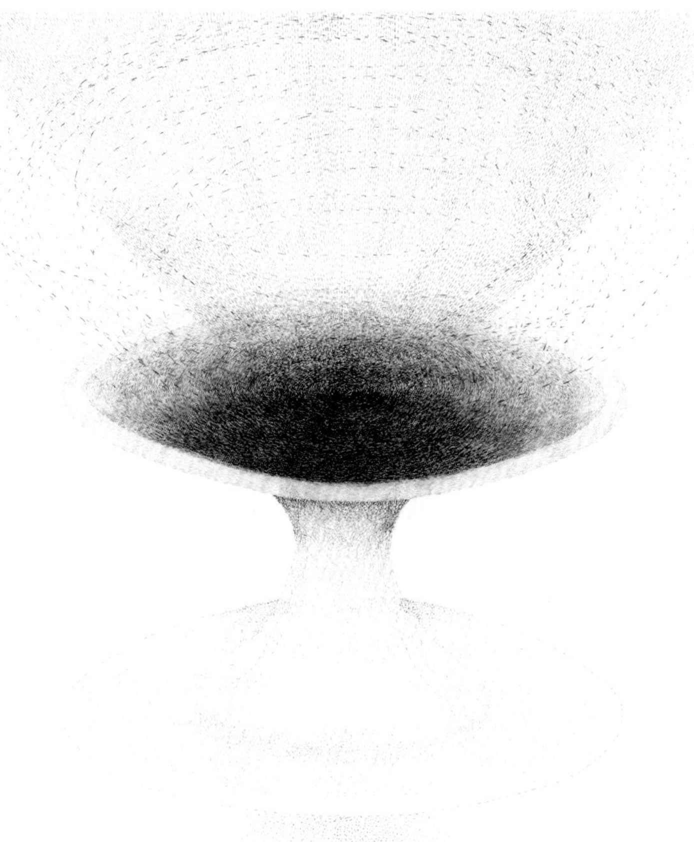

3.64 Ingrid Koenig (Canadian, born 1956), *Black Hole*, 2009, from the series *Navigating the Uncertainty Principle*. Graphite on paper, 50 × 38 in. (127 × 96.5 cm). Courtesy of the artist.

3.65 James Rosenquist (American, 1933–2017), *Time Dust-Black Hole*, 1992. Oil and acrylic on canvas, 7 × 35 ft. (2.13 × 10.67 m). Courtesy of Kasmin Gallery, New York.

A black hole is at the center of this enormous painting, surrounded by a collage of disconnected images, resulting in a surreal quality. From left to right, we see a tiny sailboat on its side, pencils forming an X, a tin can, asteroids, rolled $100 bills, an origami dragonfly, the outline of a penny with the inscription IN GOD WE TRUST, more pencils, a French horn, tests for color blindness, and a punctured space-food bag.

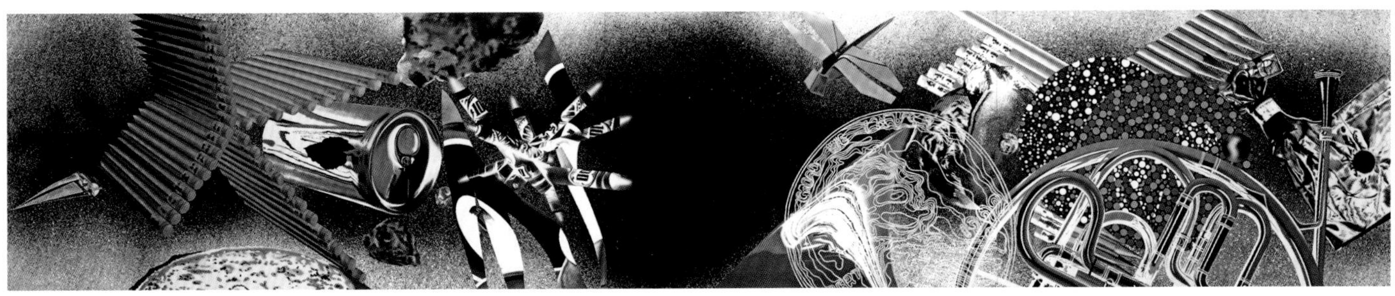

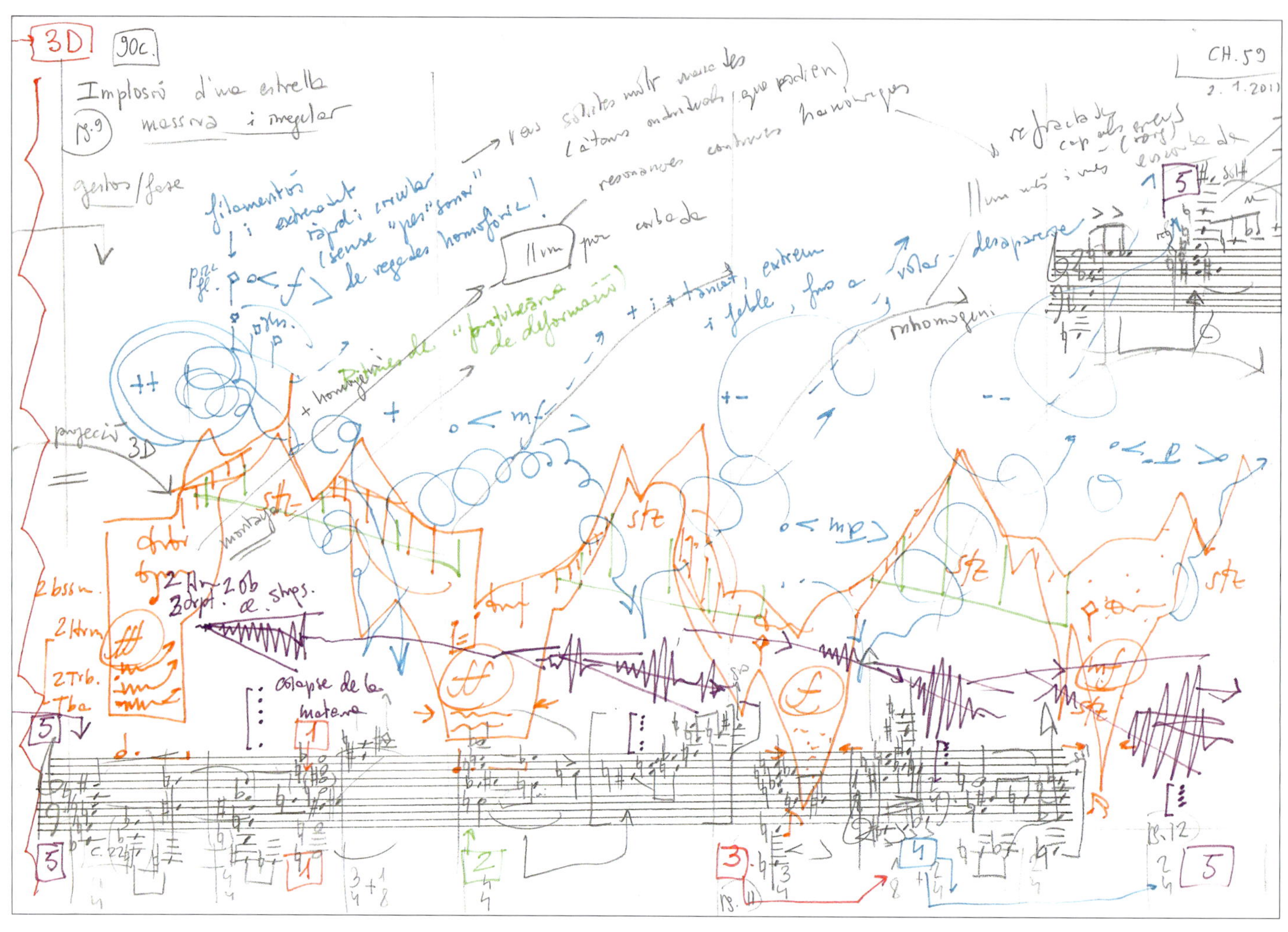

3.66 Hèctor Parra (Spanish, born 1976), *Caressant l'horizon* (Caressing the horizon), compositional sketch, 2011. Pencil on paper. Courtesy of the artist.

Parra lives in Paris, and this piece was commissioned by the French foundation Mécénat Musical Société Générale, so he titled the work in French. This diagram outlines the structure of the musical composition, including harmony and rhythm. This is a sketch for the first section of *Caressant l'horizon*; he wrote it in Catalan, his native language. In the upper left, Parra tells us what he intends to communicate: *Implosió d'una estrella massiva*—Catalan for "Implosion of a massive star."

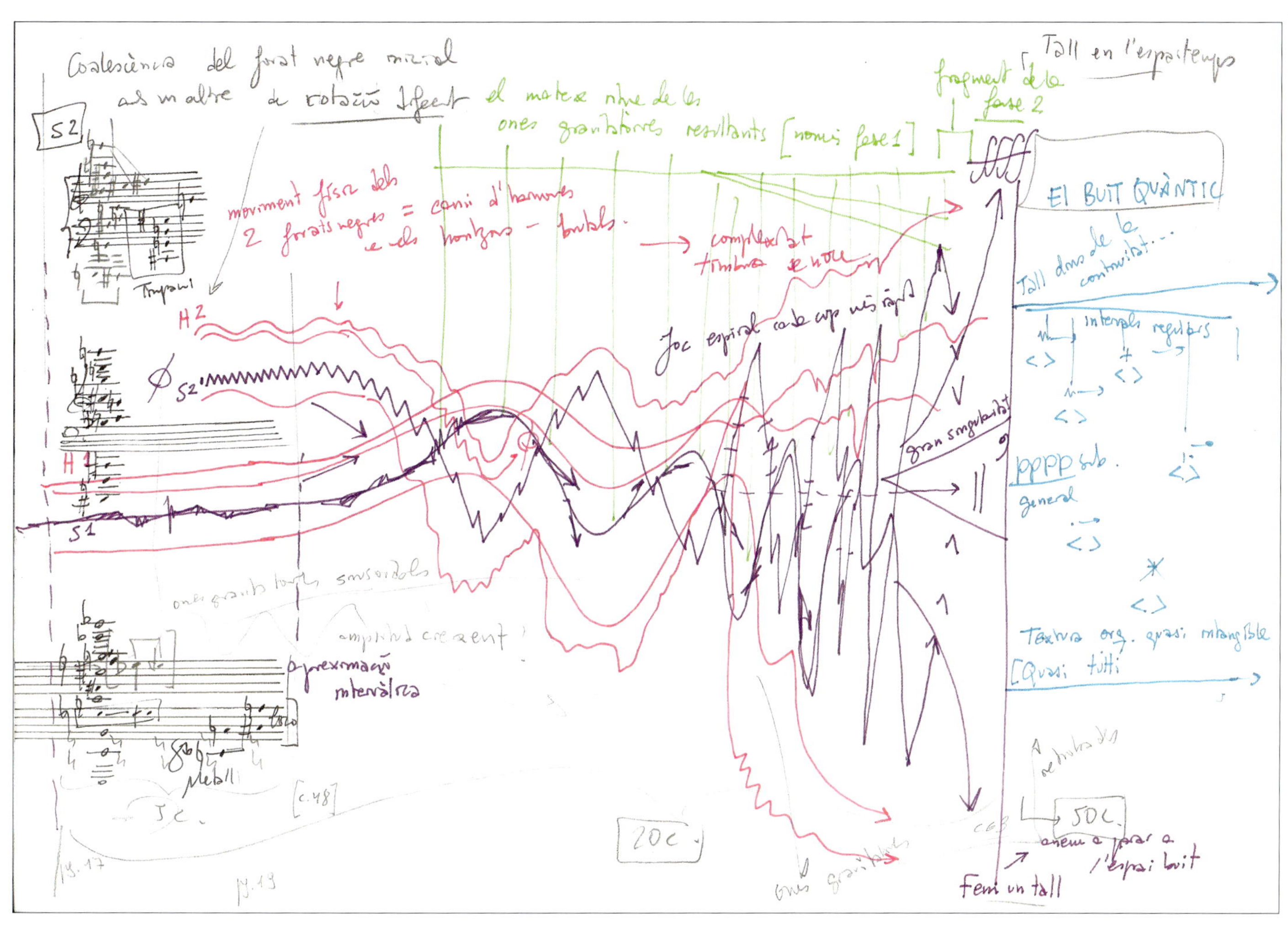

3.67 Hèctor Parra (Spanish, born 1976), *Caressant l'horizon* (Caressing the horizon), compositional sketch, 2011. Pencil on paper. Courtesy of the artist.

In the upper left, Parra has written *Coalescència del forats negres*, Catalan for "Merger of black holes."

SUPERMASSIVE BLACK HOLES

Martin Schwarzschild (the son of Karl Schwarzschild) earned a PhD in astrophysics from the University of Göttingen in 1935. The following year, he fled Germany to the United States, where he joined the faculty at Princeton University and developed a method for building galaxy models by simulating stellar orbits (figure 3.68). It's poignant that during World War I, Karl Schwarzschild was in a trench defending Germany, and two decades later Germany ousted his son because he was a Jew.

Using the Schwarzschild method, the astronomer Tod Lauer analyzed images of galaxies taken by the Hubble Space Telescope (figure 3.69). Lauer found that only when he added a black hole at the center of a galaxy did observation match model. Thus, scientists began to understand the role of black holes in the structure of the universe, affirming philosophical traditions that put a void at the heart of reality.

In 2003 NASA detected a waveform produced by the black hole at the center of the Perseus galaxy cluster. Pressure waves sent out by the black hole caused ripples in the cluster's hot gas, which NASA recorded (figure 3.71) and translated into a musical note fifty-seven octaves below middle C. This inspired the Syrian-born American artist Diana Al-Hadid to create *Portal to a Black Hole* (figure 3.70), a large sculpture made of musical components, including a dome resting on organ pipes and a spiral staircase composed of piano keys. In 2022 NASA released a new sonification of the same black hole, which inspired another artist, the English photographer John White. He painted the bottom of a petri dish black, filled it with water, and set it on top of a speaker. As he played the sound of the black hole through the speaker, the water began to vibrate. Shooting directly down at the petri dish with a macro lens and a halo light in a darkened room, he captured the vibration in a photograph titled *Black Echo* (figure 3.72).

3.68 Schwarzschild method for building galaxy models, in Martin Schwarzschild, "A Numerical Model for a Triaxial Stellar System in Dynamical Equilibrium," *Astrophysical Journal* 232 (1979): 242, fig. 2; 244, fig. 3.

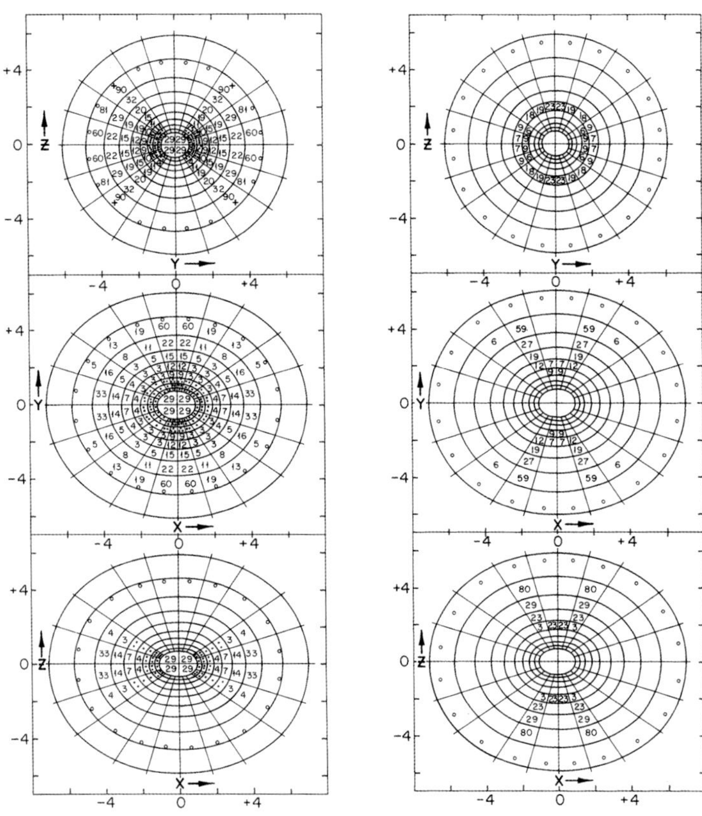

The center of our own galaxy is in the direction of the constellation Sagittarius. In the 1960s astronomers found a source of strong radio waves there. By the 1990s telescopes had been developed with mirrors large enough to see stars through the interstellar clouds that hide the center of the Milky Way, and the astronomers Reinhard Genzel and Andrea Ghez began tracing the paths of stars near the source of radio waves in Sagittarius. Genzel worked from Cerro Paranal, a mountain in Chile, using the Very Large Telescope, which has the world's largest monolithic mirror, with a diameter of more than 26 feet (8 m). Ghez used the Keck Observatory located on Mauna Kea, the highest point in Hawaii. Each of its mirrors—which are 33 feet (10 m) in diameter—consists of thirty-six hexagonal segments that can be controlled separately for better focus. An obstacle for both astronomers was atmospheric water vapor composed of large bubbles of air that expand and contract as the temperature changes and act like tiny lenses that refract starlight, distorting images. To correct this, both telescopes employ adaptive optics and have been fitted with additional thin mirrors that compensate for

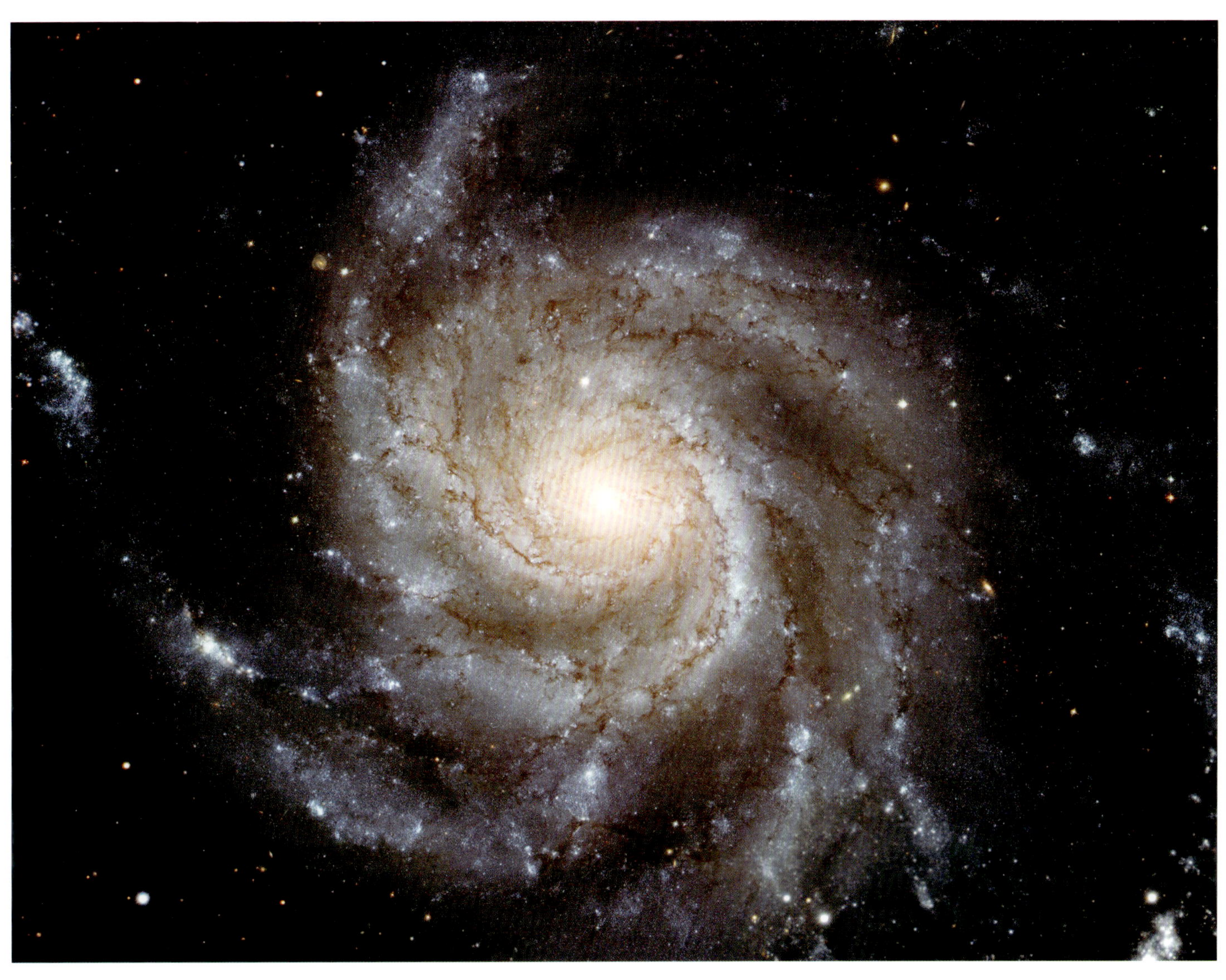

3.69 The Pinwheel Galaxy. NASA and ESA.
The Pinwheel Galaxy (Messier 101) is twenty-three million light-years away in the constellation Ursa Major.

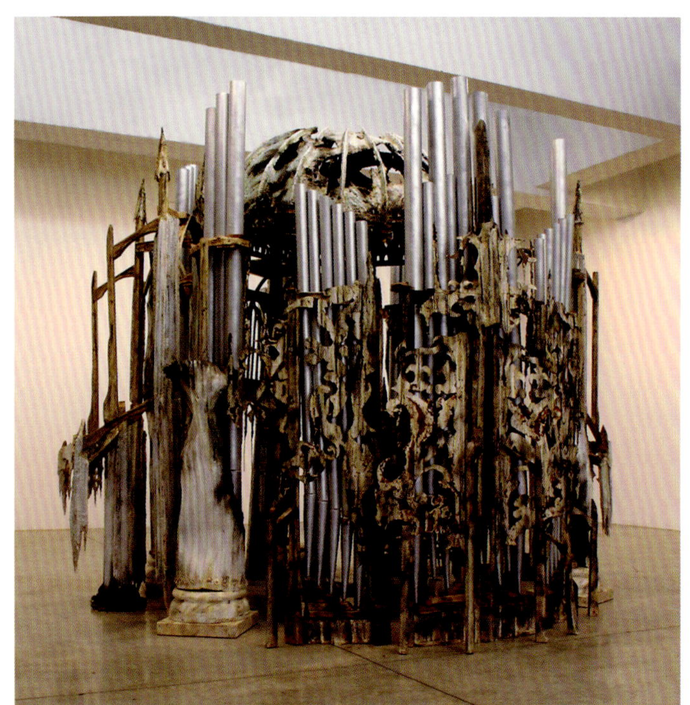

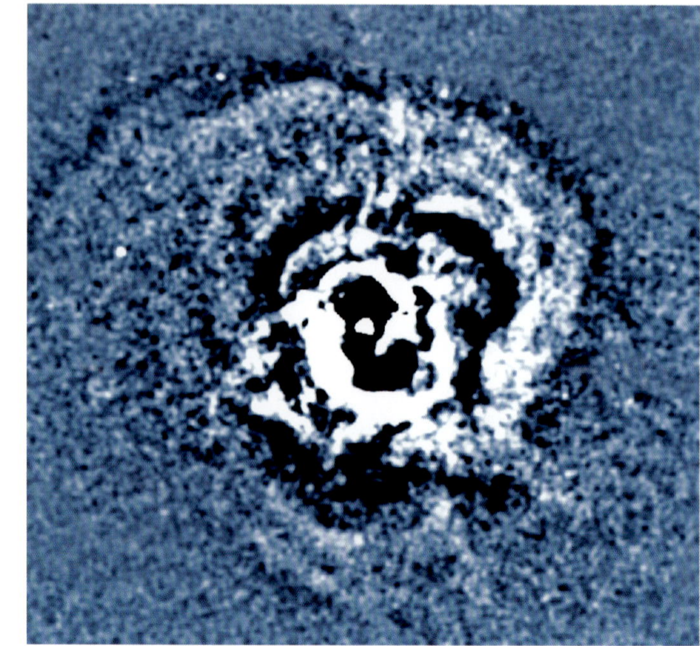

LEFT
3.70 Diana Al-Hadid (American, born Syria 1981), *Portal to a Black Hole*, 2007. Wood, plaster, fiberglass, polystyrene, plastic, cardboard, pigment, 10 ft. × 13 ft. 1 in. × 14 ft. (3.04 × 3.98 × 4.26 m). Courtesy of Kasmin Gallery, New York.

RIGHT
3.71 Ripples caused by pressure waves from the black hole at the center of the Perseus galaxy cluster. NASA/CXC/IoA/A. Fabian et al.

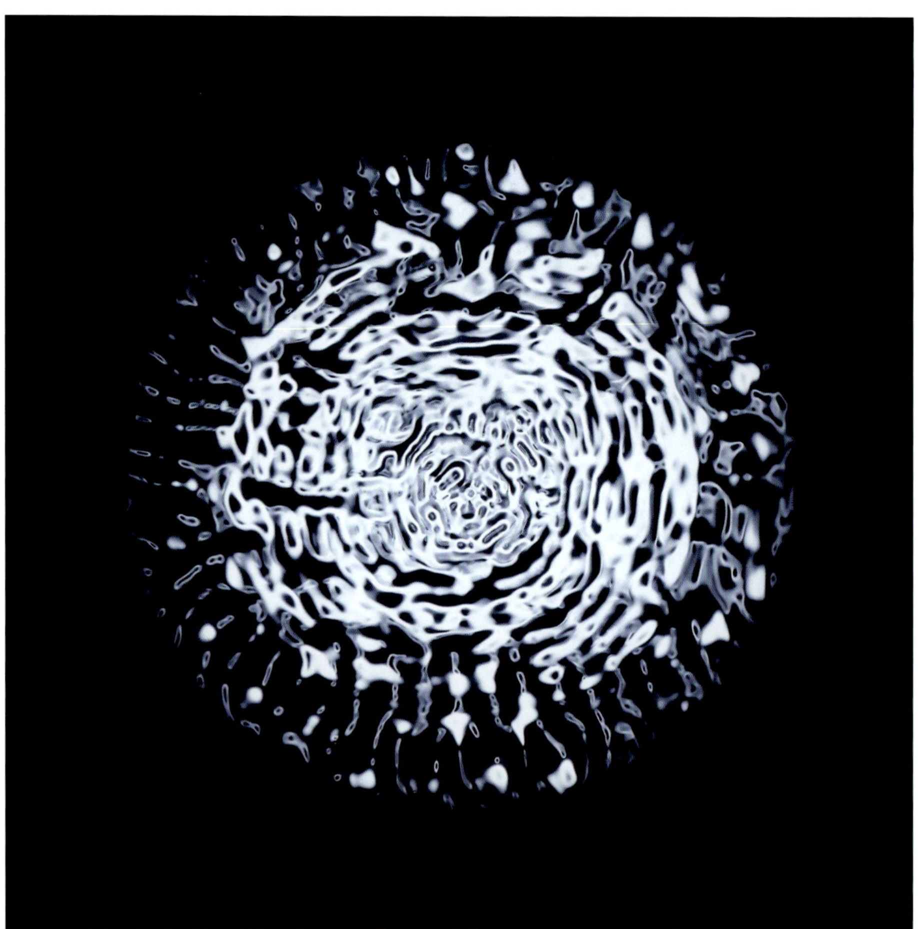

3.72 John White (English, born 1978), *Black Echo*, 2023. Digital photograph. Courtesy of the artist.

the water vapor. While Genzel and Ghez see nothing using these telescopes to view the galactic center, they have successfully charted the paths of stars near it. Their decades of observations agree that stars orbit the unseen center of the Milky Way in elliptical paths, speeding up when they get close, slowing down when they're far away (figure 3.73). Nothing but a black hole the size of four million solar masses can account for Genzel's and Ghez's observations. The Milky Way's black hole is named Sagittarius A*—which, confusingly, is pronounced "Sagittarius A star," although it is, of course, not a star.

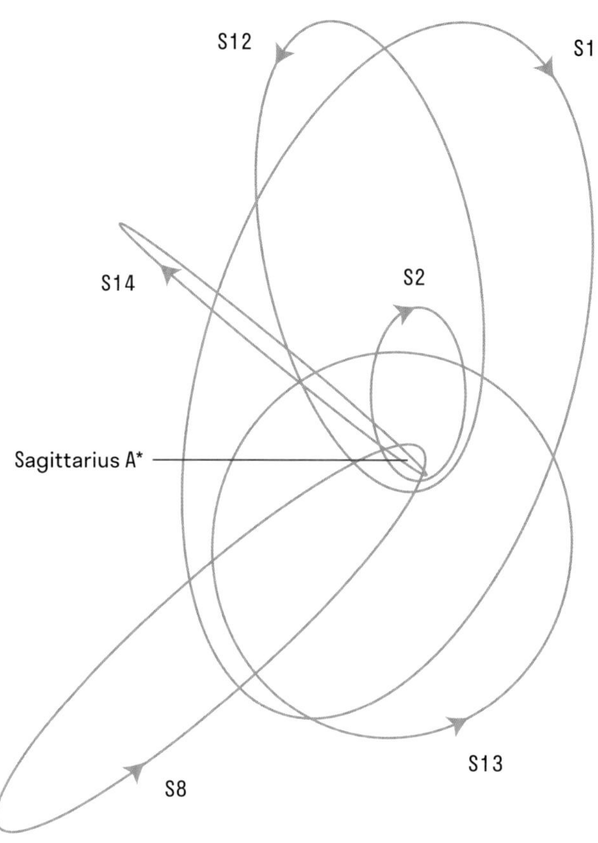

3.73 Stars orbiting the black hole at the center of the Milky Way.

Astronomers have mapped many stars (S) orbiting in elliptical paths around the center of the Milky Way. One star (S2) made a complete orbit in slightly more than sixteen years.

The black hole at the center of the Milky Way has inspired Danh Võ to make an analogy between the Milky Way and the United States in his artwork titled *Massive Black Hole in the Dark Heart of Our Milky Way* (figure 3.74). During the Vietnam War, Võ was an infant during the fall of Saigon, after which a vast number of Vietnamese civilians were sent to prison camps. When Võ was four years old, his family fled in a small boat, hand-built by his father, on the treacherous South China Sea. They were picked up by a Danish merchant ship and granted asylum in Denmark.

In Võ's artwork, American flags with thirteen stars (the flag adopted in 1777) and contemporary tourist shopping bags with silhouettes of the Statue of Liberty, both covered in gold leaf, hang from the ceiling. Other panels contain text from the nineteenth-century Grimm Brothers' tale "Cinderella" written in German Gothic script by Võ's father, an expert calligrapher. Like the Milky Way with its twinkling stars, emblems of the United States glitter as in a fairy tale, but both the galaxy and the country conceal dark hearts: Sagittarius A* lurks at the center of the Milky Way, devouring anything in its vicinity, and the United States fought a proxy war in Vietnam, where as many as two million civilians died.

Lucas J. Rougeux was also inspired by "our galaxy's void heart," as he calls it.[57] In 2014 astronomers watched as what appeared to be a cloud of dust (G2) approached Sagittarius A*. They expected the space cloud to be sucked into the black hole, but it survived the encounter. (Astronomers now believe that G2 was a binary star system that orbited the black hole in tandem, eventually merging into an extremely large star.) After learning about G2, Rougeux created a series of artworks about black holes (figures 0.3 on page 14 and 3.75) that were shown in a 2022 exhibition titled *The Soul Gravity—Guided to Black*. On the occasion of the exhibition, the artist said, "The delicacy and amorphous nature of a space cloud is directly connected to my own sense of queer identity."[58] He added, "I am a cloud of space dust. I am a collection of particles dealing with depression. I am weaving through waves of spacetime and isolation. My work is the product of this existentialism, loneliness, and search for a connection to the sublime. . . . This exhibition displays the balance and ever-presence of life and death through the overlapping lenses of religion and astrophysics. Symbolic through lines in this show include the amorphous space cloud, the soul as recycled energy, the mysterious finality of death, and the void of black holes."[59]

Cygnus X-1 formed from the collapse of a star, so it's called a *stellar-mass black hole* (figures 3.76 and 3.77). It has approximately twenty-one solar

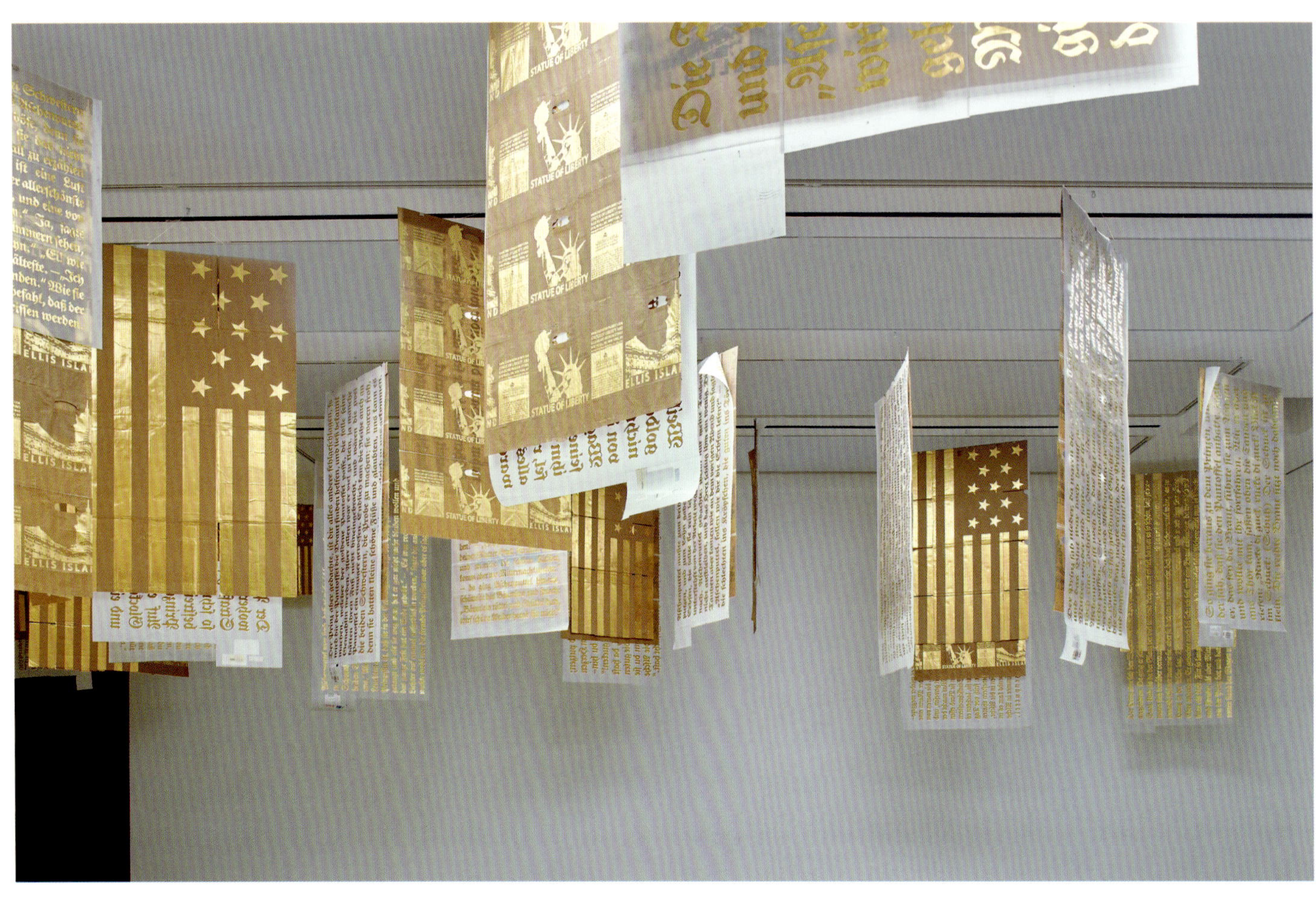

3.74 Danh Võ (Danish, born Vietnam 1975), *Massive Black Hole in the Dark Heart of Our Milky Way* (detail), 2012. Gold leaf and ink on paper. Installation view at MO.CO. Montpellier Contemporain, Montpellier, France, 2019. Ishikawa Foundation, Okayama, Japan. Courtesy of Marian Goodman, New York.

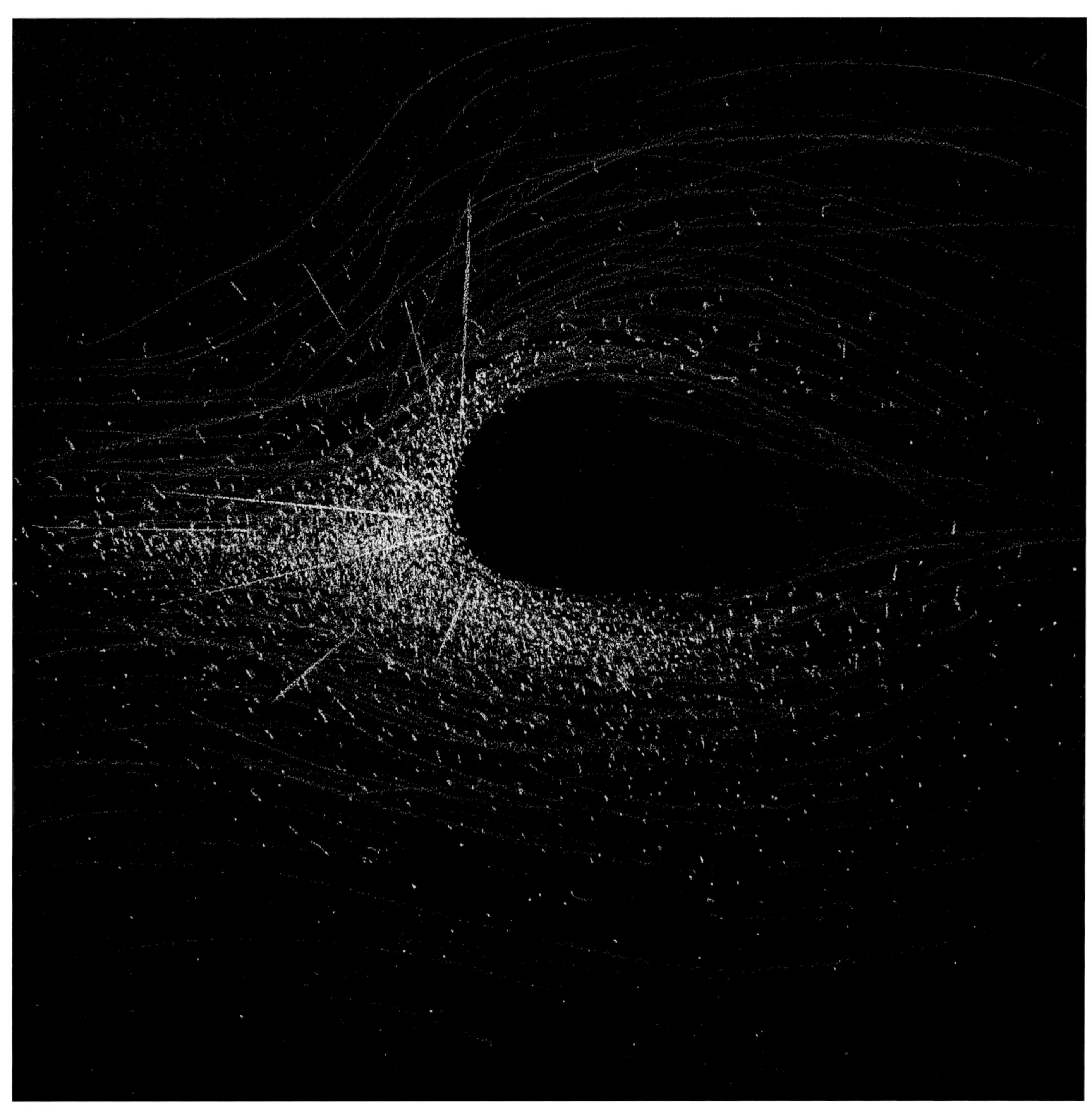

3.75 Lucas J. Rougeux (American, born 1995), *Light Particles against a Black Hole*, 2021. Charcoal and acrylic on paper, 8 × 8 in. (20.3 × 20.3 cm). Courtesy of the artist.

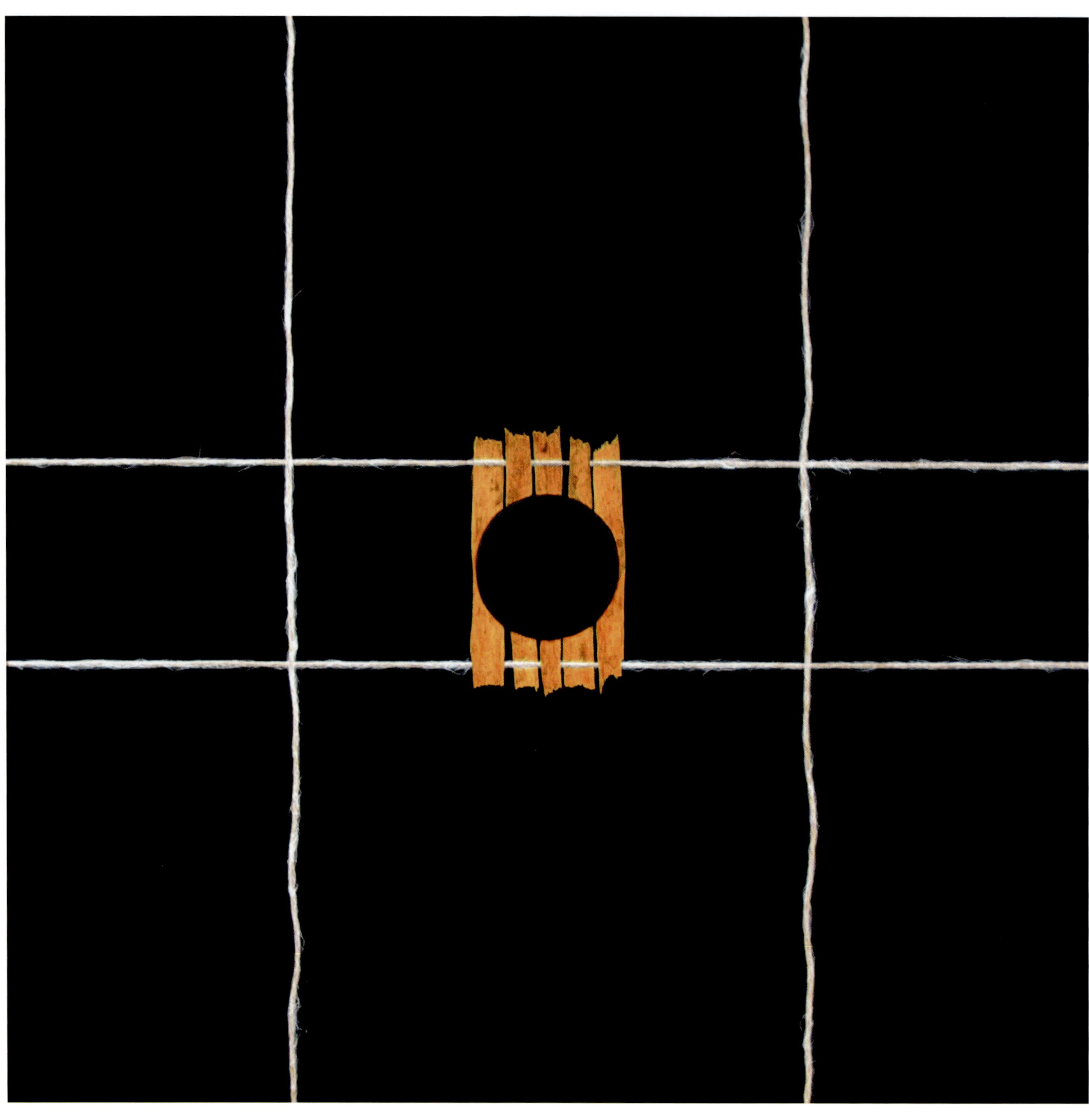

3.76 Shea Hembrey (American, born 1974), *Hatch*, 2012. Acrylic on board, 46 × 46 in. (116.8 × 116.8 cm). Private collection. Courtesy of the artist.

Hembrey was born and raised in rural Arkansas, and he makes artwork from materials one finds on a farm. In *Hatch*, Hembrey made a circular hole by burning pieces of wood so the void has ash edges. *Hatch* represents a stellar-mass black hole, with the singed edges symbolizing its event horizon.

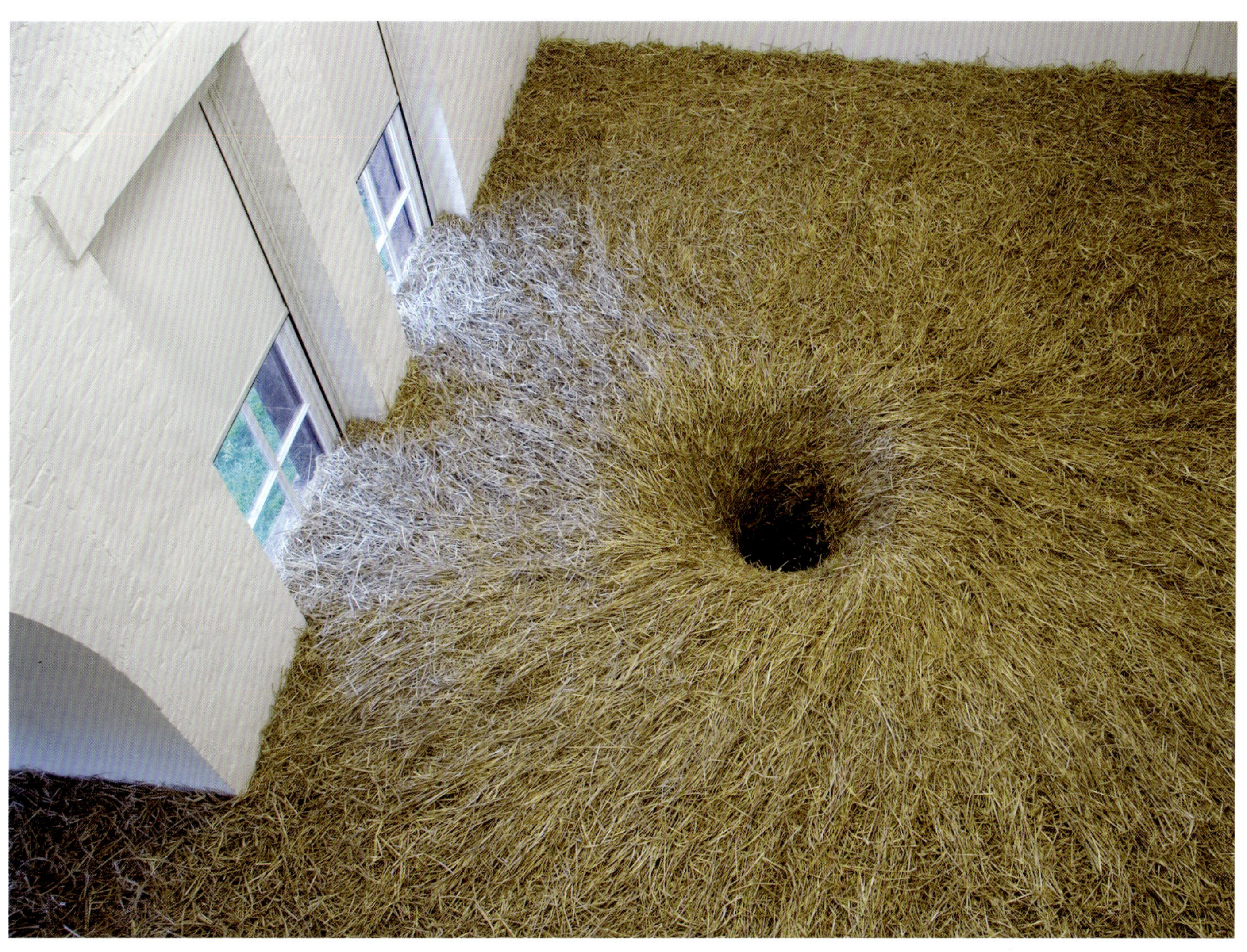

3.77 Shea Hembrey (American, born 1974), *Whirl*, 2007, from the series *Nestwork: Meditations on Nesting*. Dried grass. Installation view at Cornell University, Ithaca, New York. Courtesy of the artist.

Before entering the artworld, Hembrey studied birds and considered becoming an ornithologist. One of his earliest series of artworks was *Nestwork: Meditations on Nesting*. His installation *Whirl* consisted of a room filled with dried grass, in the center of which the artist made a deep depression. According to the artist, the void takes the form of a bird's nest as well as a black hole surrounded by an accretion disk.

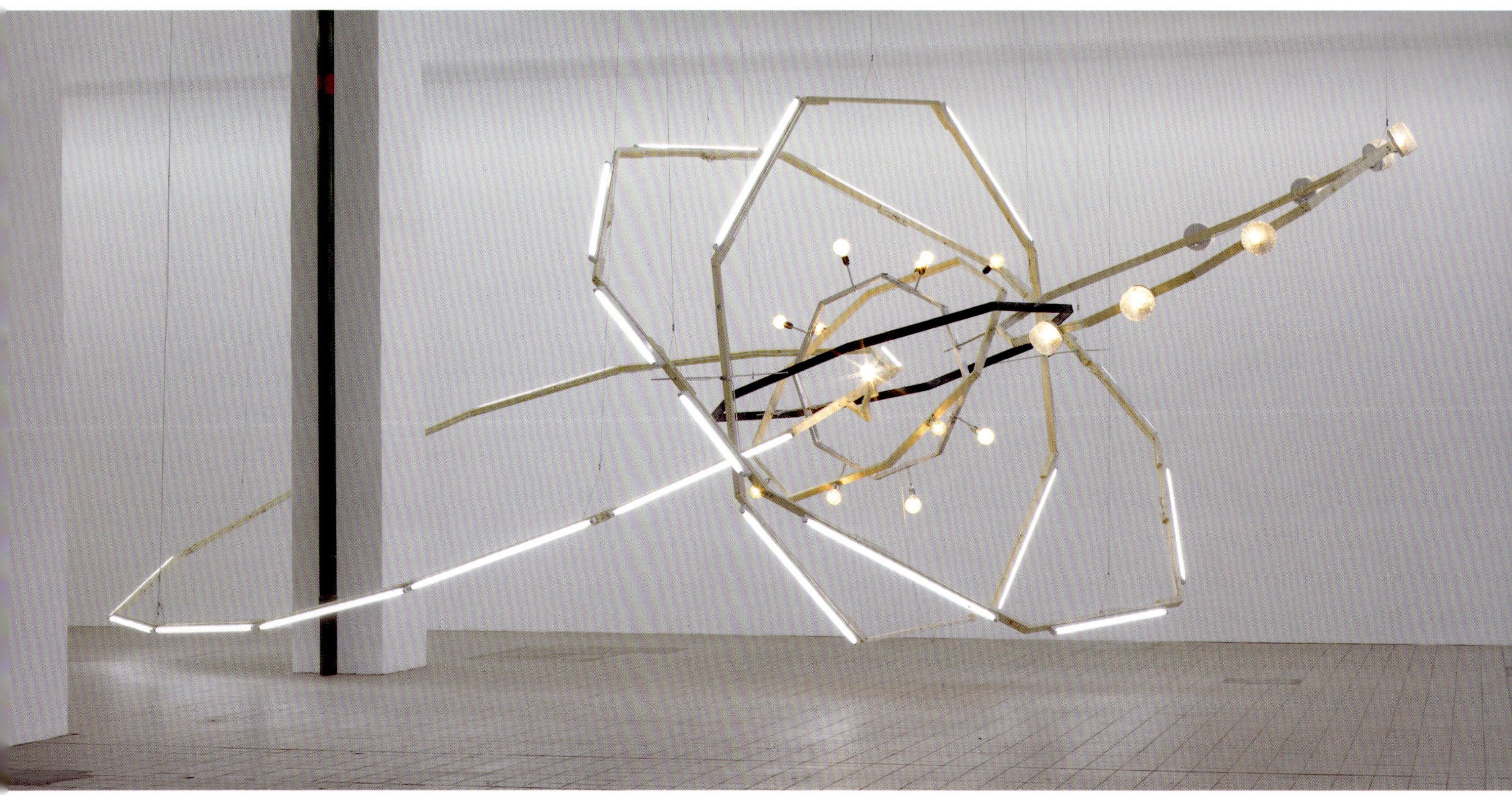

3.78 Björn Dahlem (German, born 1974), *Black Hole: M-Spheres,* 2007. Wood, lamps, lightbulbs, and neon lamps, 17 ft. 9 in. × 23 ft. 11 in. × 11 ft. 10 in. (5.4 × 7.3 × 3.6 m). Installation view at Galerie Max Hetzler, Berlin, 2007. Courtesy of Galerie Max Hetzler, Berlin, Paris, and London.

The black polygon in the center of this sculpture represents a black hole surrounded by lights, which denote stars orbiting in the galaxy.

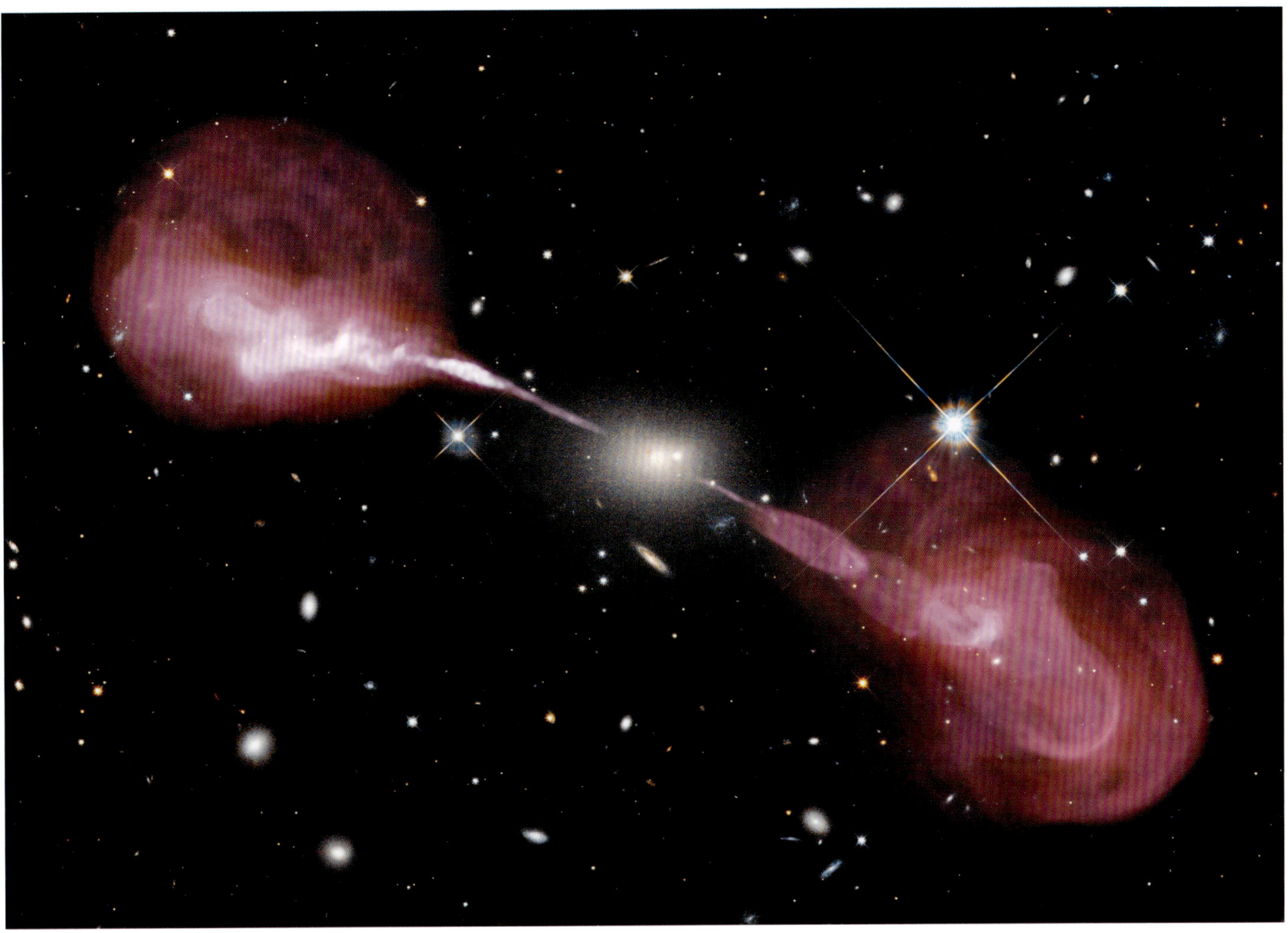

3.79 Relativistic jets from the black hole in the core of galaxy Hercules A, located two billion light-years away in the constellation Hercules. NASA, ESA, S. Baum and C. O'Dea (RIT), R. Perley and W. Cotton (NRAO/AUI/NSF), and the Hubble Heritage Team (STScI/AURA).

masses. By 2000 astronomers agreed that there is a *supermassive black hole*—with millions or billions of solar masses—at the center of all large galaxies (figure 3.78). Occasionally a galactic supermassive black hole spews forth jets of particles from its axis of rotation. Since these jets travel at close to the speed of light, they show effects of the theory of relativity and are thus called *relativistic jets* (figure 3.79), which can extend millions of light-years.

The other place supermassive black holes are found is in the early universe. By the 1970s most astronomers agreed that X-rays and other radiation from quasars are caused by matter falling into supermassive black holes. Their extreme redshift indicates that some quasars are more than thirteen billion light-years away, so they formed soon after the universe came into existence 13.8 billion years ago. Normal stars aren't any larger than about 150 solar masses, so astronomers wonder how such massive, very old black holes could have formed. Today astronomers hypothesize that these quasars were created when a primordial cloud of hydrogen and helium collapsed under its own gravity, forming a black hole directly without an intervening star—a *direct-collapse black hole*. In 2023 astronomers discovered a supermassive black hole that has all the telltale signs of a direct-collapse black hole, and it was in existence relatively soon (just 470 million years) after the Big Bang (figure 3.80).[60] A direct-collapse black hole probably formed as a giant whirlpool, recalling the one in Poe's "A Descent into the Maelström" and other artworks (figures 3.81 and 3.82).

Representations of supermassive black holes in glass are part of Josiah McElheny's sculpture *An End to Modernity* (figure 3.83). The artist

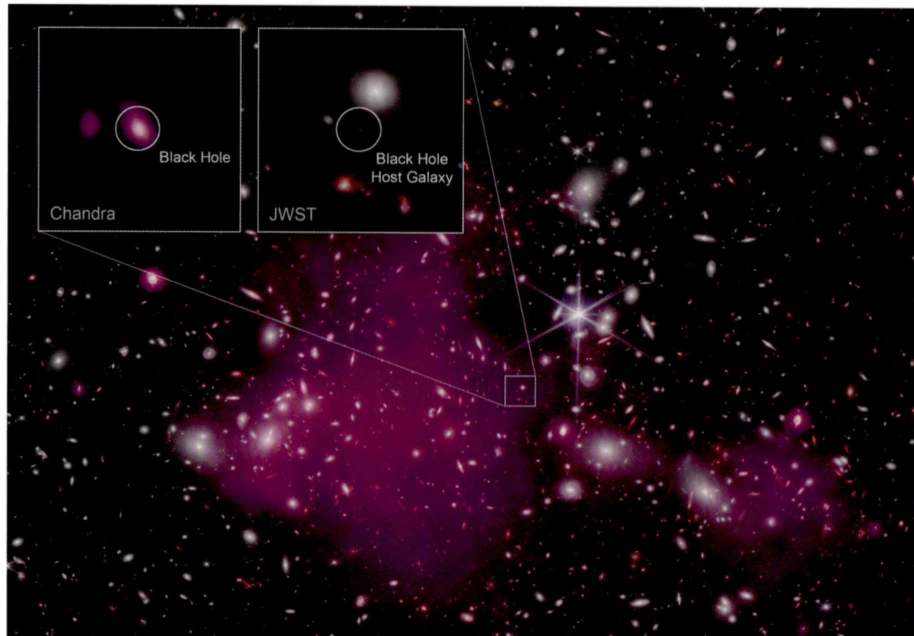

3.80 Black hole in galaxy UHZ1. X-ray: NASA/CXC/SAO/Ákos Bogdán; infrared: NASA/ESA/CSA/STScl; image processing: NASA/CXC/SAO/L. Frattare and K. Arcand.

This composite image was made with the Chandra X-ray Observatory and the James Webb Space Telescope, which records infrared radiation. The Webb was designed to observe the early universe when there were thick clouds of gas and dust, which infrared penetrates. Galaxy UHZ1 is in the direction of the cluster of galaxies named Abell 2744 (shown in the image as several hundred tiny celestial objects [galaxies in Abell 2744] and a cloud of gas recorded in X-rays [in the center of the image, slightly in the direction of the lower left, colored purple]).

Previously, the Hubble Space Telescope had shown that Abell 2744, located 3.5 billion light-years from Earth, acts as a powerful gravitational lens, and NASA decided to have the Webb look at what lies beyond it. Looking through the powerful natural magnifying lens of Abell 2744, the Webb discovered galaxy UHZ1 (the speck of light in the small circle labeled "Black Hole Host Galaxy" in box at the upper left). UHZ1 is located 13.2 billion light-years from Earth, thus much farther than Abell 2744. From the core of UHZ1, Chandra then observed X-rays: a sign of a growing supermassive black hole (labeled "Black Hole" in the box on the far left). These observations agree with theoretical predictions of a direct-collapse black hole: the large mass soon after the Big Bang, the amount of X-ray emission, and the brightness of UHZ1. Based on the energy of the X-rays, the mass of the black hole is estimated to be about equal to the mass of all the stars in its host galaxy, UHZ1. By comparison, supermassive black holes at the center of galaxies in today's nearby universe are only about one tenth of a percent of the mass of the stars in their host galaxies. The discovery of the black hole at the core of galaxy UHZ1 is important for understanding how black holes became so enormous soon after the Big Bang and how they evolve over time.

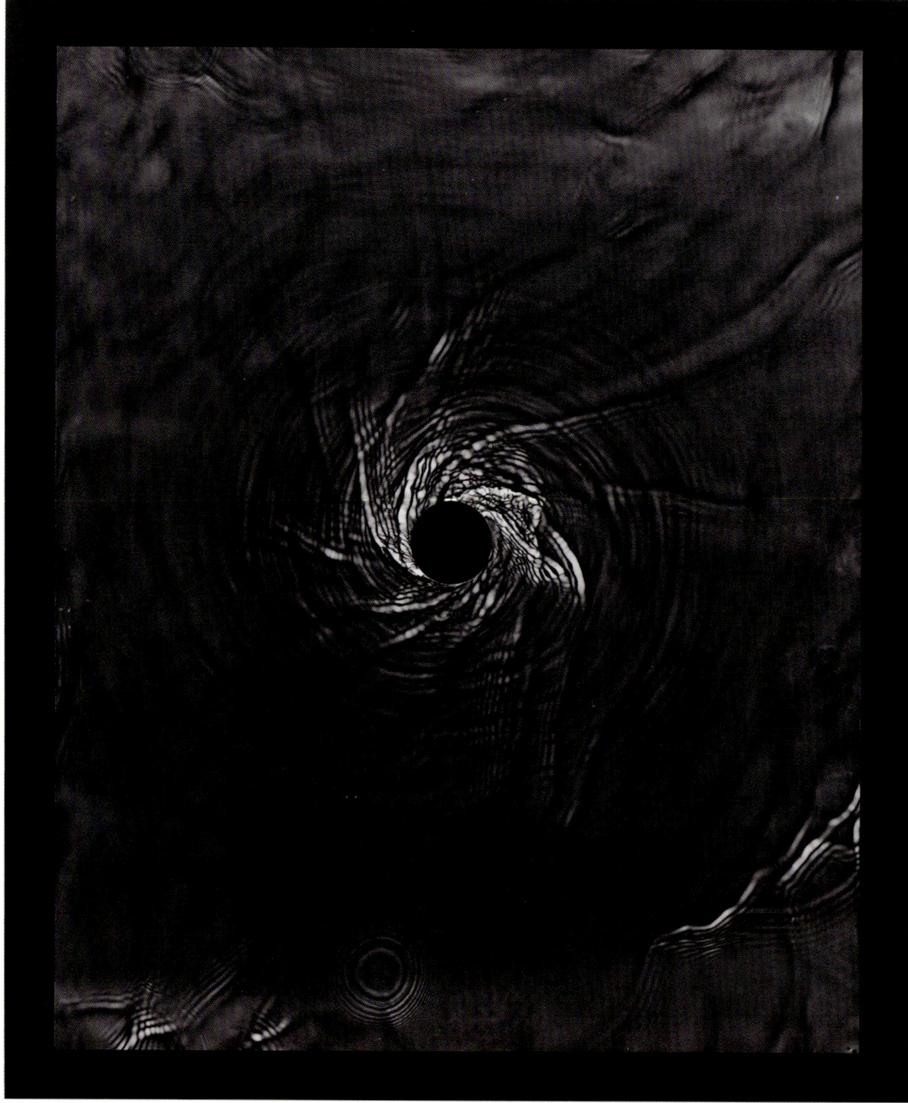

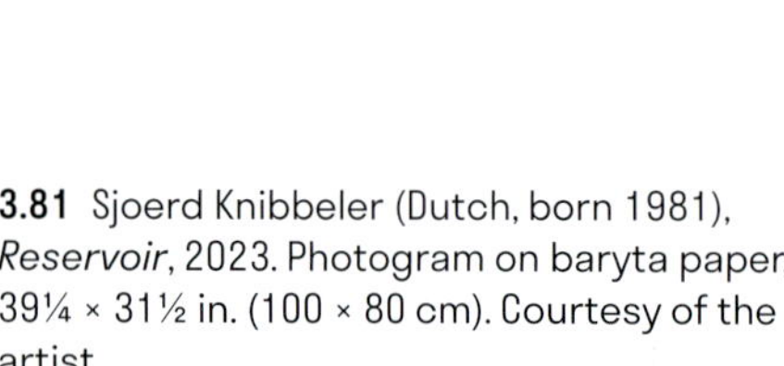

3.81 Sjoerd Knibbeler (Dutch, born 1981), *Reservoir*, 2023. Photogram on baryta paper, 39¼ × 31½ in. (100 × 80 cm). Courtesy of the artist.

Knibbeler made this photogram by capturing the pattern of light in a whirlpool of water.

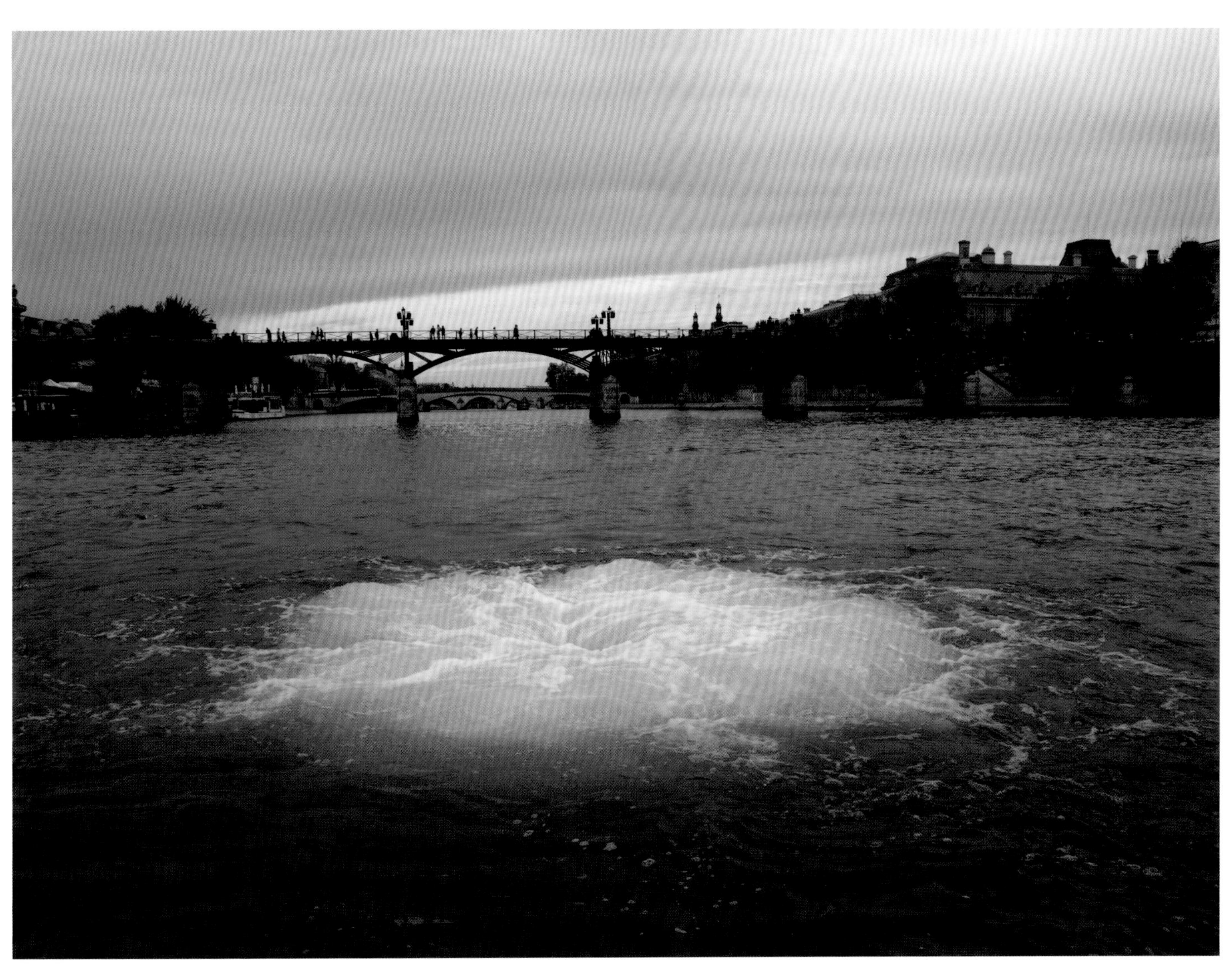

3.82 Anish Kapoor (Indian and British, born 1954), *Descension*, 2015. Installation in the Seine River, Paris, during the Nuit Blanche Art Festival, 2016.

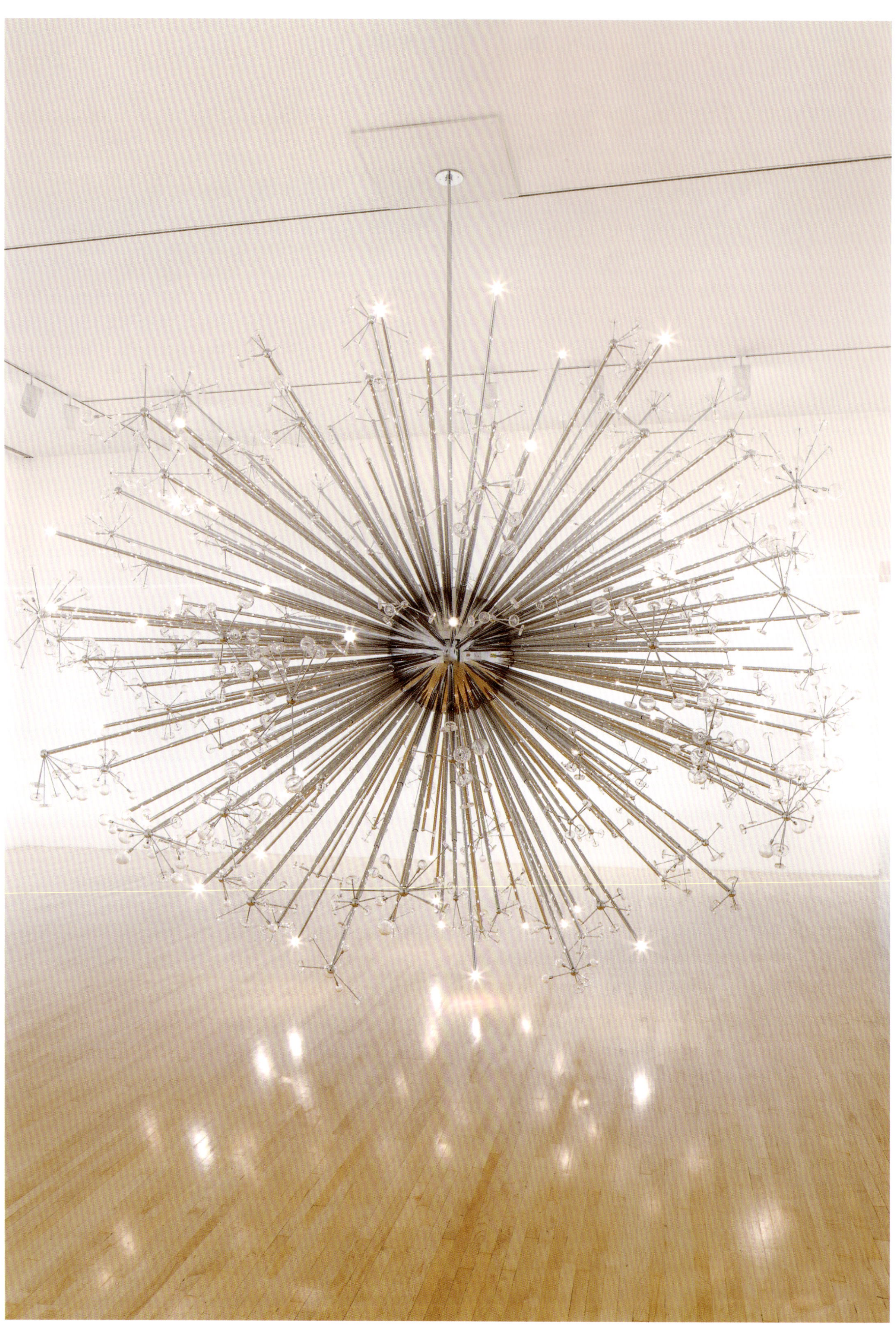

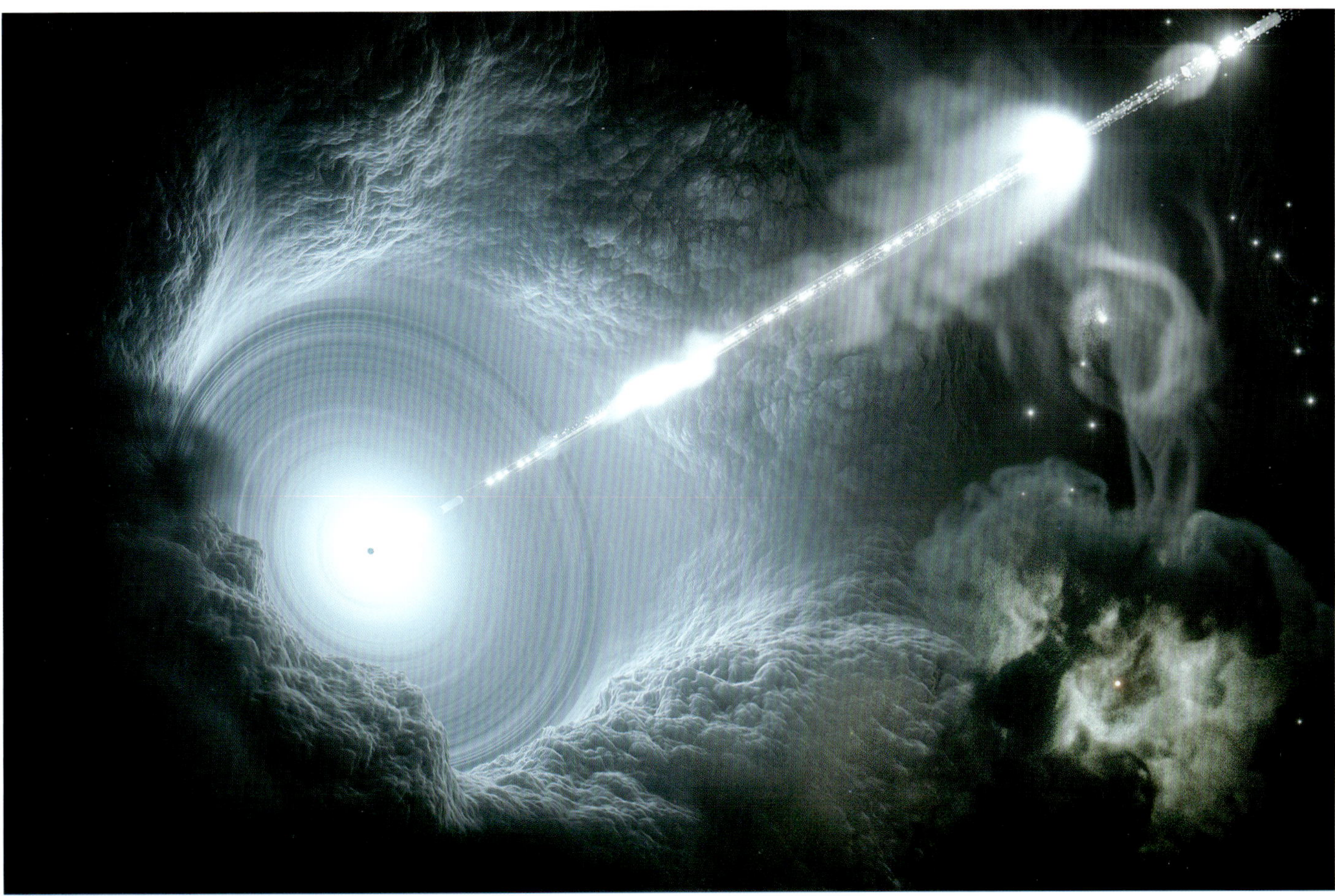

collaborated with the astronomer David Weinberg to create this representation of the universe, from the Big Bang to the present. Four groups of metal rods emerge from the center of the sculpture. The shortest represent the early universe; at the end of each rod is a glass sculpture representing the first galaxies. The third (next to the longest) group represents the era that saw the formation of supermassive black holes.

In 1983 astronomers discovered that a supermassive black hole at the center of galaxy TXS 0506+056 has a relativistic jet pointing straight at Earth. When a jet points at Earth, the black hole system is called a *blazar* (figure 3.84). Astronomers have determined that the jet of blazar TXS 0506+056 is millions of times more powerful than CERN's Large Hadron Collider, which is capable of propelling protons to 99.9999991 percent of the speed of light. Scientists determined that this cosmic particle accelerator was the source of several neutrinos recorded by the IceCube Neutrino detector in Antarctica in 2018.[61] Those accelerated neutrinos traveled 5.7 billion light-years from blazar TXS 0506+056 to Earth.

When Earthlings look at a blazar, they are looking directly into the relativistic jet, which is like looking straight into a searchlight. In 2017 astronomers did exactly that and discovered that blazar J1924-2914 twists in a spiral (figure 3.85), which is evidence of a strong magnetic field that is

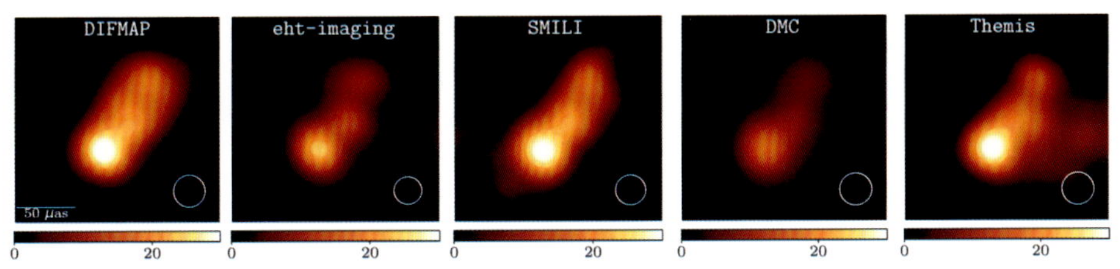

OPPOSITE

3.83 Josiah McElheny (American, born 1966), *An End to Modernity*, 2005. Nickel-plated aluminum, electric lighting, hand-blown glass, steel cable, rigging, 12 ft. 6 in. × 15 ft. ¼ in. × 15 ft. ¼ in. (3.81 × 4.58 × 4.58 m). Courtesy of the Museum of Contemporary Art, Los Angeles.

Here, "modernity" refers to the era of the Industrial Revolution, when uniform parts were mass produced. The artist has hand-blown the glass parts, so each piece is unique, and the sculpture thus represents "an end to modernity."

ABOVE

3.84 Artist's impression of the blazar TXS 0506+056. DESY, Science Communication Laboratory.

LEFT

3.85 Recordings of the blazar J1924-2914 in Sara Issaoun et al., "Resolving the Inner Parsec of the Blazar J1924-2914 with the Event Horizon Telescope," *Astrophysical Journal* 934, no. 2 (2022): fig. 4.

The same raw data from recordings of the blazar J1924-2914 was interpreted by five different algorithms, the names of which are written above the images.

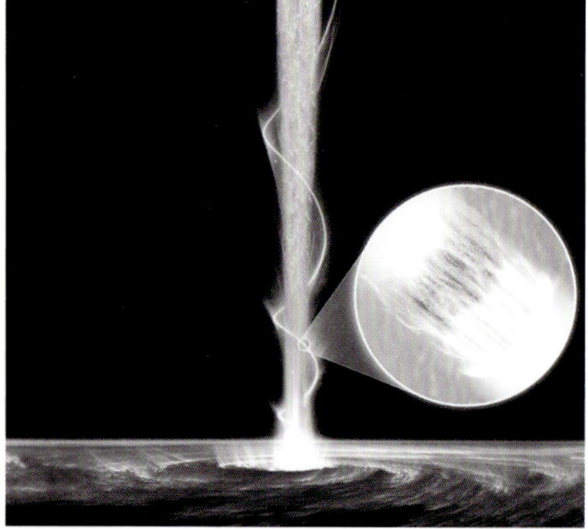

ABOVE
3.86 Artist's impression of a black hole's relativistic jet. NASA, MSFC, and Pablo Garcia.
The accretion disk of a black hole is shown at the bottom, below the relativistic jet. Researchers have determined that X-rays are generated in a sudden acceleration, or shock, of the jet that is located within the spiral magnetic field. The inset illustrates the shock front, from which the X-rays originate.

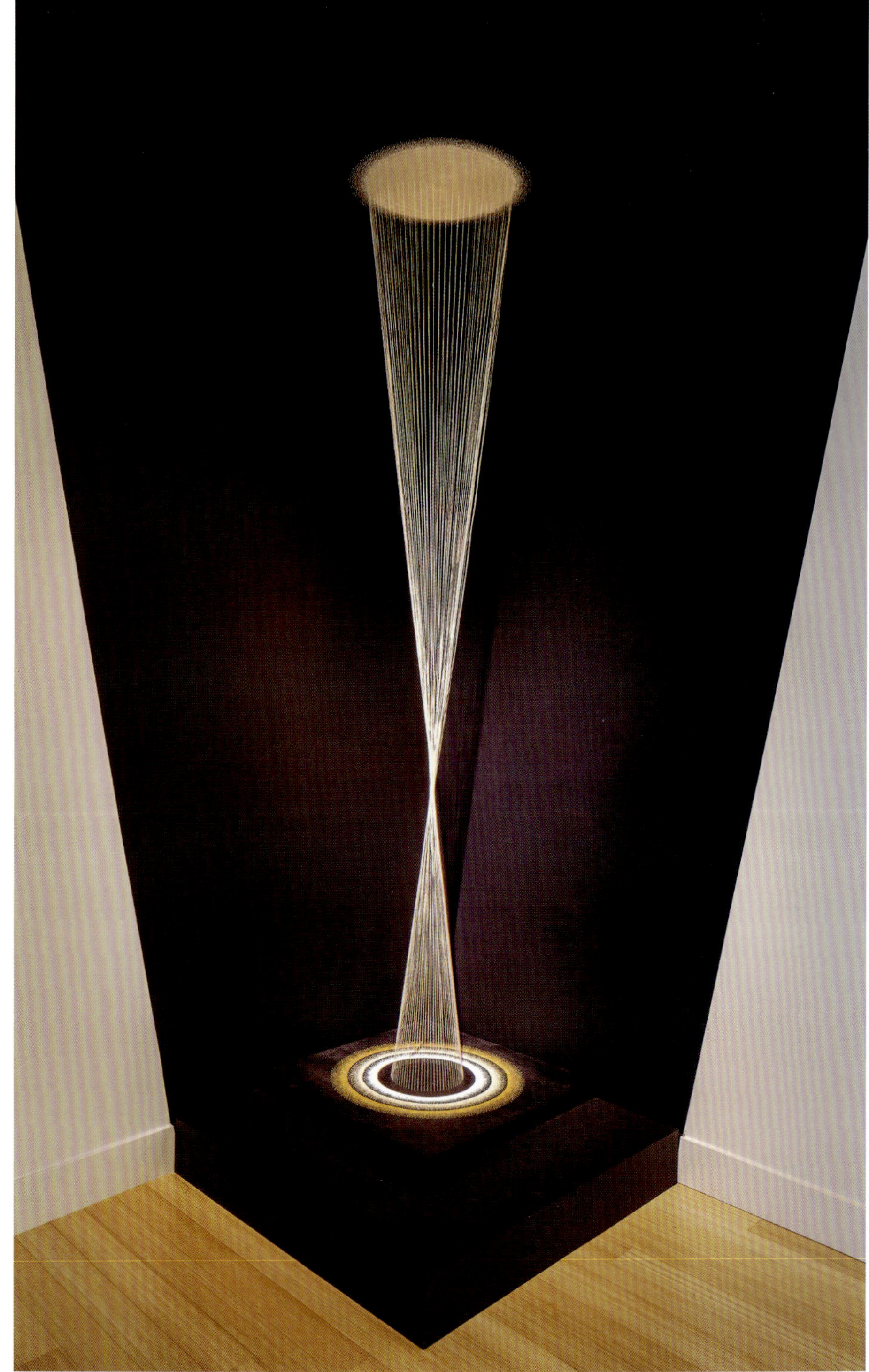

RIGHT
3.87 Melissa Walter (American, born 1976), *Black Hole*, 2017. Paint, twine, and wood, height approx. 12 ft. (3.6 m). Installation view at San Diego International Airport. Courtesy of the artist.

shaped like a torus (a smooth, doughnut-like form) near the event horizon of the black hole. To study the geometric structure of a blazar in detail, astronomers use the Imaging X-ray Polarimetry Explorer, a space observatory that was launched in 2021 and measures the polarization of X-ray light. The researchers focused on the blazar Markarian 421 and found that, in the part of the jet where particles are accelerated, the magnetic field is in the form of a helix (figure 3.86).[62]

The artist Melissa Walter twisted twine to represent the form of a helix in her sculpture *Black Hole* (figures 3.87 and 3.88). She said: "The twine represents material, such as gas, that is lucky enough to escape the gravitational pull of the black hole, but in that escape, it's jettisoned away from the black hole very quickly, and creates a jet."[63]

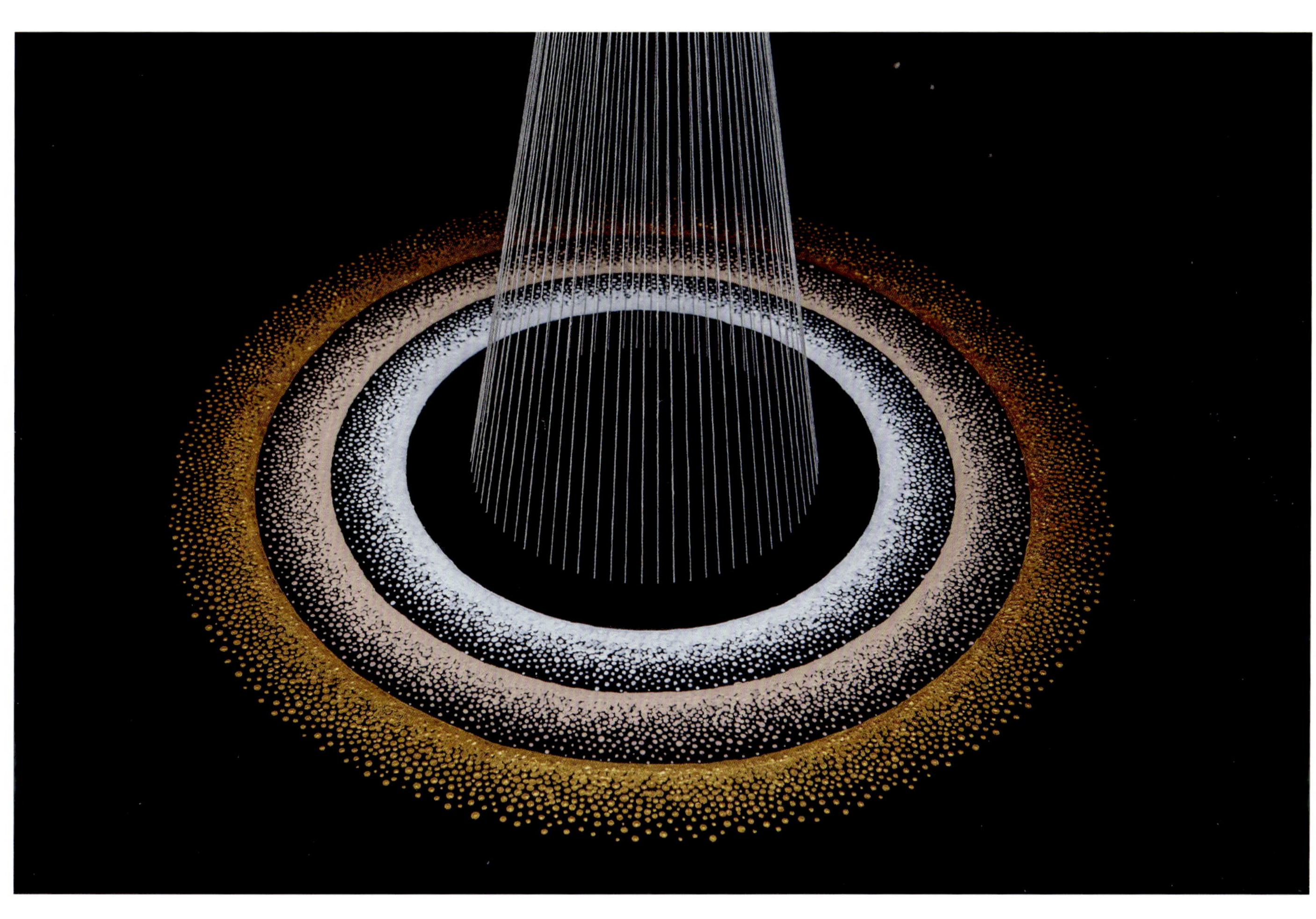

3.88 Melissa Walter (American, born 1976), *Black Hole* (detail), 2017. Courtesy of the artist.

BLACK HOLES AS A METAPHOR FOR NOTHINGNESS, SILENCE, AND BLACKNESS

For Anish Kapoor, a black hole is a metaphor for a void. Raised in India, he was born to an Iraqi Jewish mother and an Indian Hindu father. When he was sixteen, he moved from India to Israel to join a kibbutz. As a young person, he always felt that he was in an uncertain, intermediate situation: in India, he was a Jew; in Israel, he was a dark-skinned Indian. At eighteen, Kapoor left youthful feelings behind and moved to London to become an artist. There he created *Descent into Limbo*. The word "limbo" was originally a Roman Catholic term meaning a region of Hell reserved for the unbaptized. Kapoor's work is titled after an Italian Renaissance painting by Andrea Mantegna, which shows Christ leaning on his staff and about to descend into limbo, shown as a cylindrical hole (figure 3.89). In Mantegna's painting, Christ summons virtuous deceased persons, such as Noah and Moses, who were not baptized because they lived long before Christ. Figuratively, being "in limbo" means being in an intermediate situation. To see Kapoor's work, viewers entered a building in which there appeared to be a flat black circle on the floor. In fact, it was a cylindrical hole 8 feet (2.4 m) deep that was painted black inside—literally a black hole (figure 3.90). Kapoor has stated: "I am interested in the void, the moment when it's not a hole. It is a space full of what isn't there."[64]

BELOW

3.89 Andrea Mantegna (Italian, 1431–1506), *Descent into Limbo*, 1492. Tempera on panel, 15¼ × 16½ in. (38.8 × 42.3 cm). Private collection.

OPPOSITE

3.90 Anish Kapoor (Indian and British, born 1954), *Descent into Limbo*, 1992. Concrete and pigment; dimensions of the building: 19 ft. 8 in. × 19 ft. 8 in. × 19 ft. 8 in. (6 × 6 × 6 m). Installation view at Serralves, Porto, Portugal, 2018.

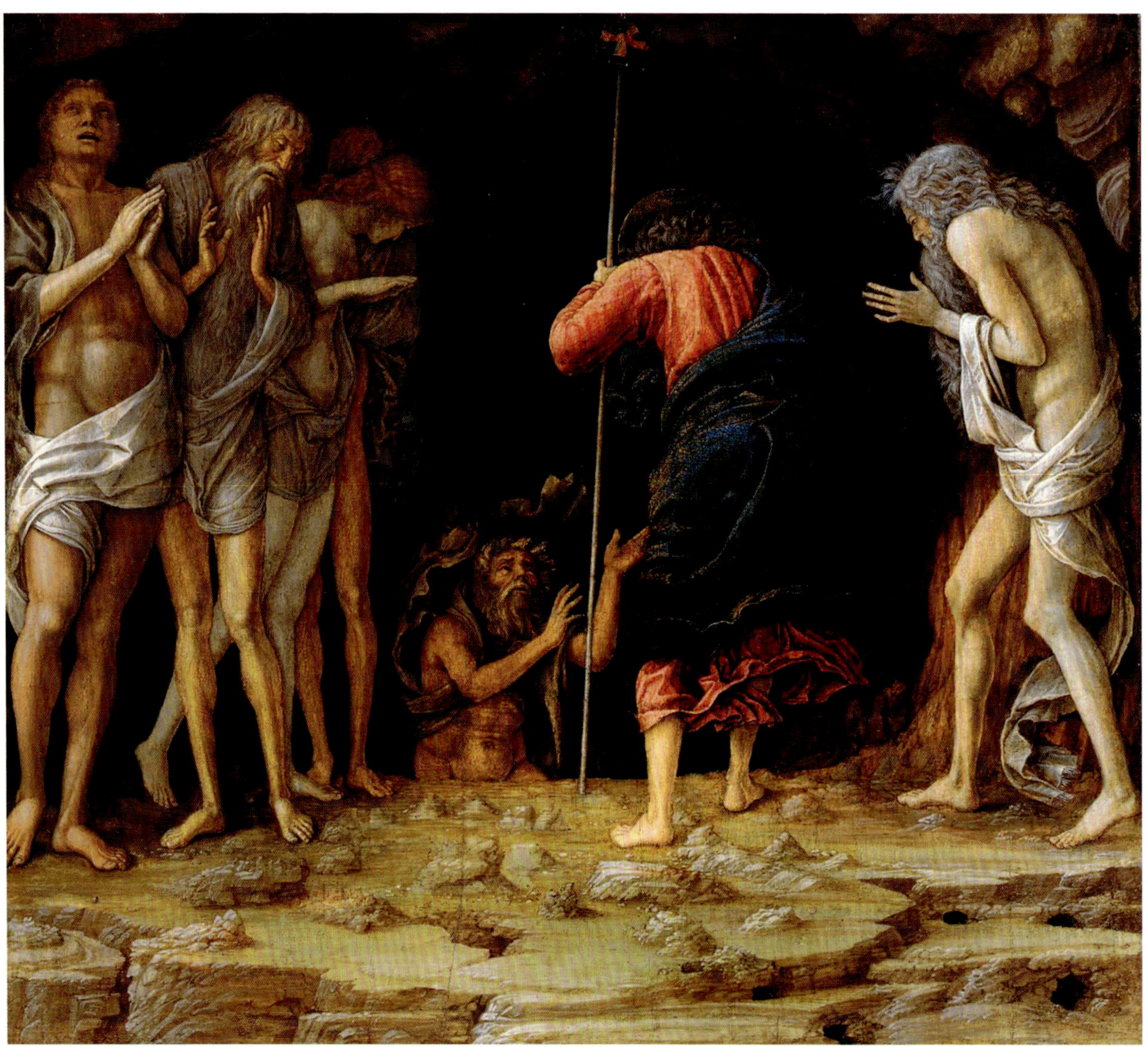

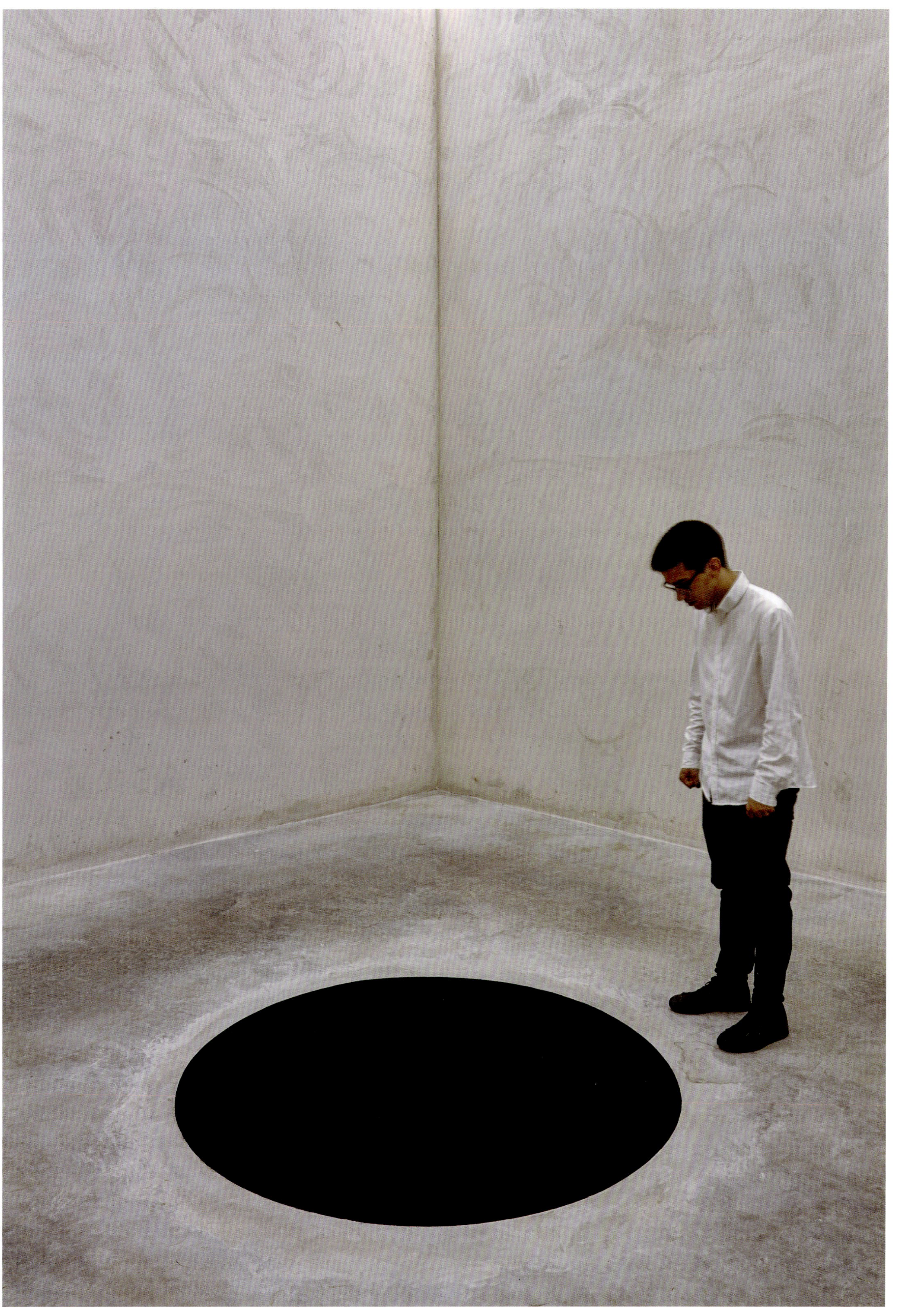

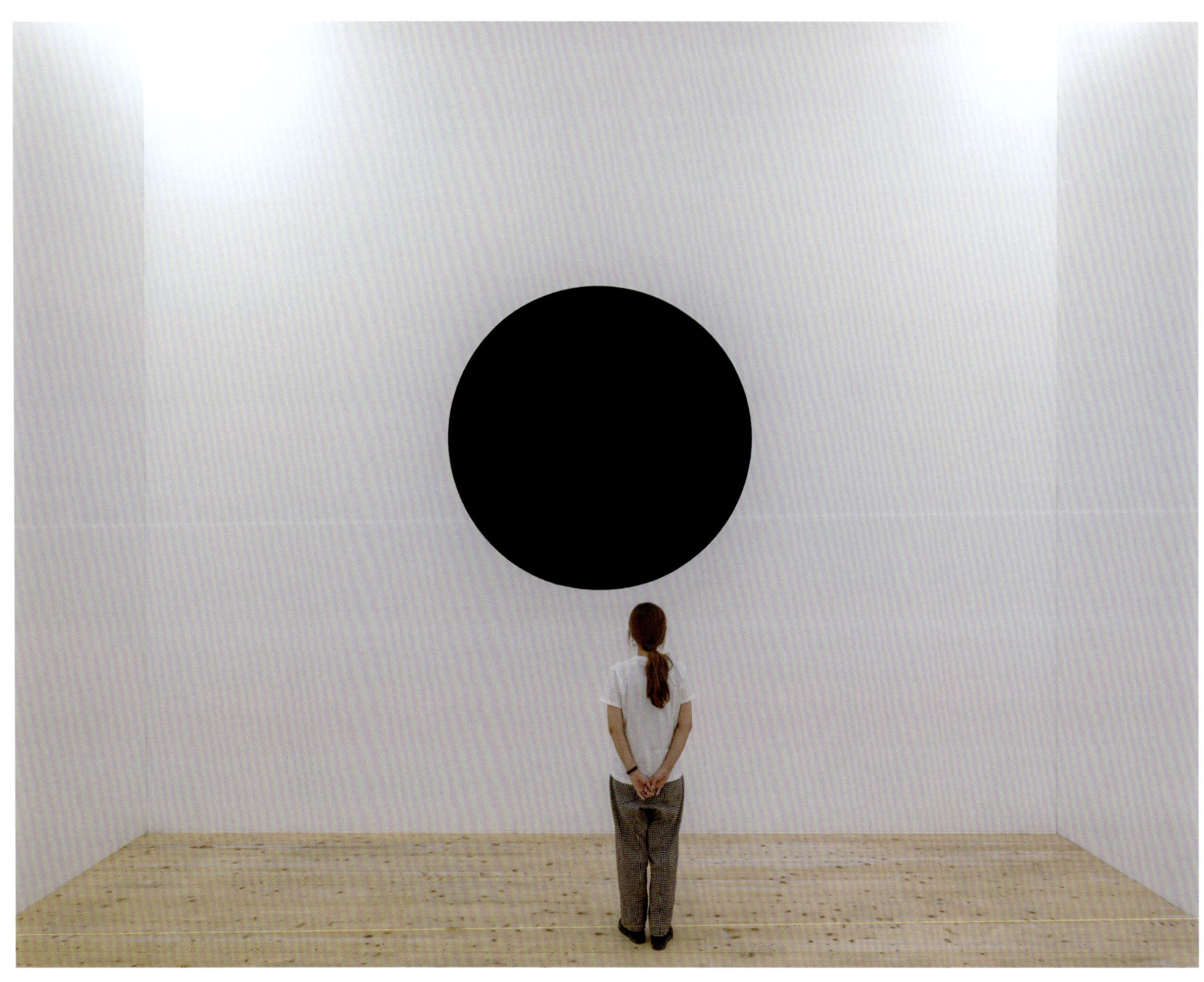

3.91 Anish Kapoor (Indian and British, born 1954), *Void Pavilion VI*, 2018. Wood, concrete, and pigment; dimensions of the pavilion: 19 ft. 8 in. × 19 ft. 8 in. × 30 ft. 4 in. (6 × 6 × 12 m). Installation view at Beppu Park, Noguchibaru, Japan, 2018.

In 2018 Kapoor created *Void Pavilion VI*—a pavilion of nothingness—that had a circular black painting in it (figure 3.91). The painting represents a void.

Vantablack is a coating that absorbs 99.96 percent of visible light. It was developed to coat surfaces inside optical telescopes so reflected light wouldn't interfere with the celestial light being recorded. Vantablack is made of vertical carbon nanotubes (the brand name is a contraction of *vanta*—vertically aligned nanotube arrays—and *black*). When light strikes a surface coated with Vantablack, it becomes trapped because light is continually deflected in the nanotubes and eventually dissipates as heat. After Vantablack was released in 2014, Kapoor began experimenting with it. When a viewer stands before an object coated with Vantablack, such as *Non-Object Black* (figure 3.92), the surface is as good as invisible. Vantablack can only be applied in a laboratory setting by specialists. Artworks coated with Vantablack are small and have to be protected in an acrylic vitrine because the surface is fragile and toxic.

Kapoor is an artist who—as far as I know—didn't make a conscious decision to create artwork about black holes, but he intuits (whether consciously or not) qualities of black holes from his cultural matrix. Indeed, Kapoor said that Vantablack is "blacker than a black hole."[65] Albert Einstein once made a remark about how artists think that applies to Kapoor and his work: "If we reproduce in the language of logic what we see and experience, then we are practicing science, but if we communicate through forms whose connections are inaccessible to the conscious mind, but we intuitively recognize them as meaningful, then we're making art."[66]

Unlike Kapoor, Govinda Sah *did* make a conscious decision to make artwork about black holes.[67] Sah, who is from Nepal, painted an accretion disk and cut a hole in the center to represent a black hole, which is invisible because there's literally nothing there (figure 3.93). He titled the painting

3.92 Anish Kapoor (Indian and British, born 1954), *Non-Object Black*, 2019. Mixed media, 15¼ × 15¼ × 4¾ in. (39 × 39 × 12 cm).

3.93 Govinda Sah (Nepalese, born 1974), *Trap Within*, 2009. Oil, acrylic, and collage on canvas, 47¼ × 53 in. (120 × 135 cm). Courtesy of October Gallery, London.

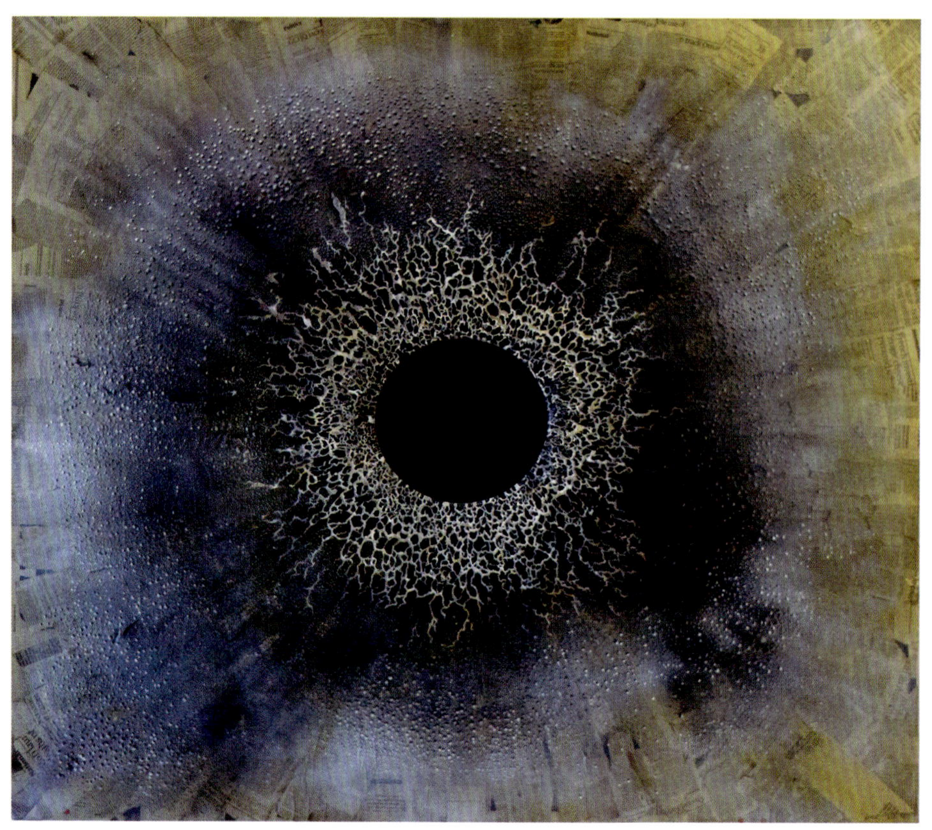

3.94 Govinda Sah (Nepalese, born 1974), *Nothing, Everything*, 2016. Oil and acrylic on canvas, 57¾ × 56¼ in. (147 × 143 cm). Courtesy of October Gallery, London.

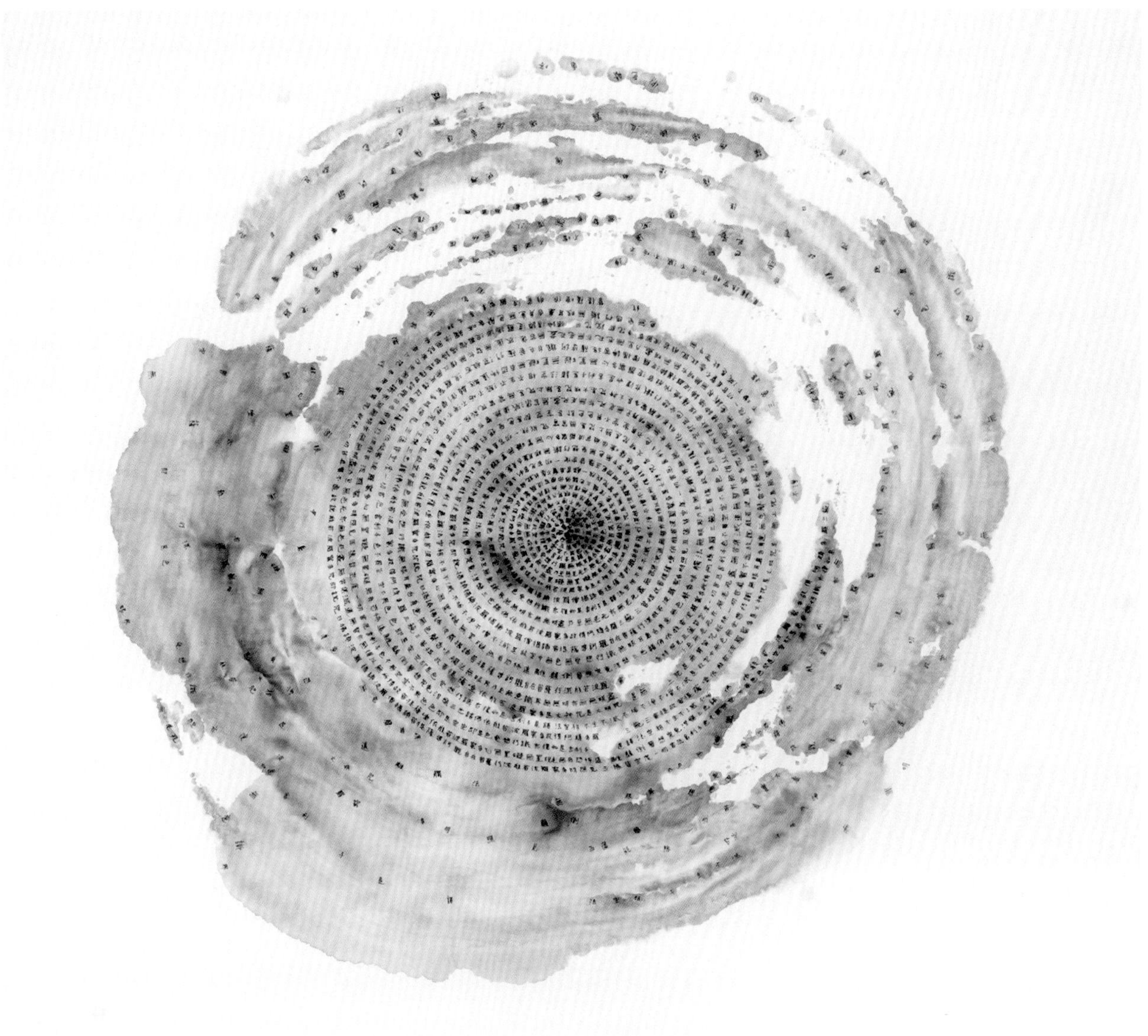

Trap Within because light is stuck inside a black hole. Sah also makes paintings about themes in Eastern philosophy, such as *Nothing, Everything* (figure 3.94), which refers to the notion that nothingness is the source of everything.

The relation of nothingness and everything is also the theme of *We Came Whirling Out of Nothingness I-I* (figure 3.95), addressing the Buddhist belief that the natural world emerged from nothingness. The work was made by Charwei Tsai, who created her ink painting as a flat abstract shape in the modernist tradition and then fused that shape with Buddhist thought by making the central whirlpool composed of the oft-chanted words from the Heart Sutra: "Form is emptiness, emptiness is form."

Toshikatsu Endō is associated with the Japanese style called Mono-ha ("school of things"), which combines Western geometric abstraction with meditations on nothingness. Endō burned a large wooden sphere, as shown in a photograph documenting how the sculpture was made (figure 3.96). Like a black hole, the burst of energy from the fire ended in a void, as the title of the finished sculpture suggests (figure 3.97).[68] Another Japanese artist, Chiharu Shiota, burned a piano and dozens of wooden chairs and then strung them together in an installation. This artwork involved another fiery burst of energy that ends *In Silence*, as she titled her artwork (figure 3.98). The heart of a black hole contains unimaginable energy, but

3.95 Charwei Tsai (born 1980), *We Came Whirling Out of Nothingness I-I*, 2014. Watercolor and ink on rice paper, 26½ × 26¾ in. (67.5 × 68 cm). Courtesy of TKG+, Taipei.

FOLLOWING PAGES, LEFT
3.96 Toshikatsu Endō (Japanese, born 1950), *Void-Blackening: The Documentation of Firing 005*, 2015. Digital print, 13¾ × 10¼ in. (35 × 26.5 cm). Courtesy of SCAI The Bathhouse, Tokyo.

FOLLOWING PAGES, RIGHT
3.97 Toshikatsu Endō (Japanese, born 1950), *Void-Blackening*, 2015. Wood, iron, cement, tar, and fire, 45¾ in. (116 cm) diameter. Courtesy of SCAI The Bathhouse, Tokyo.

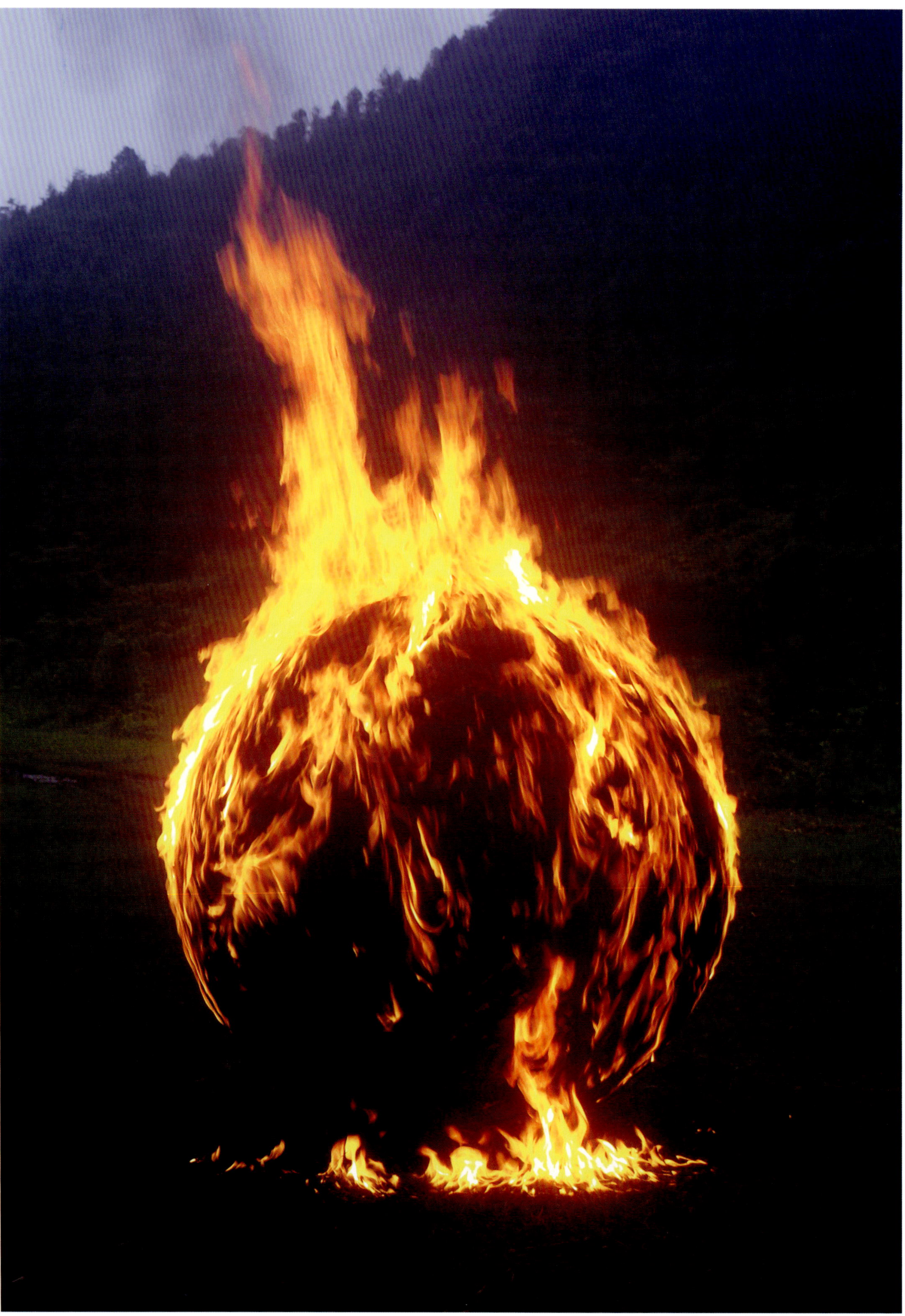

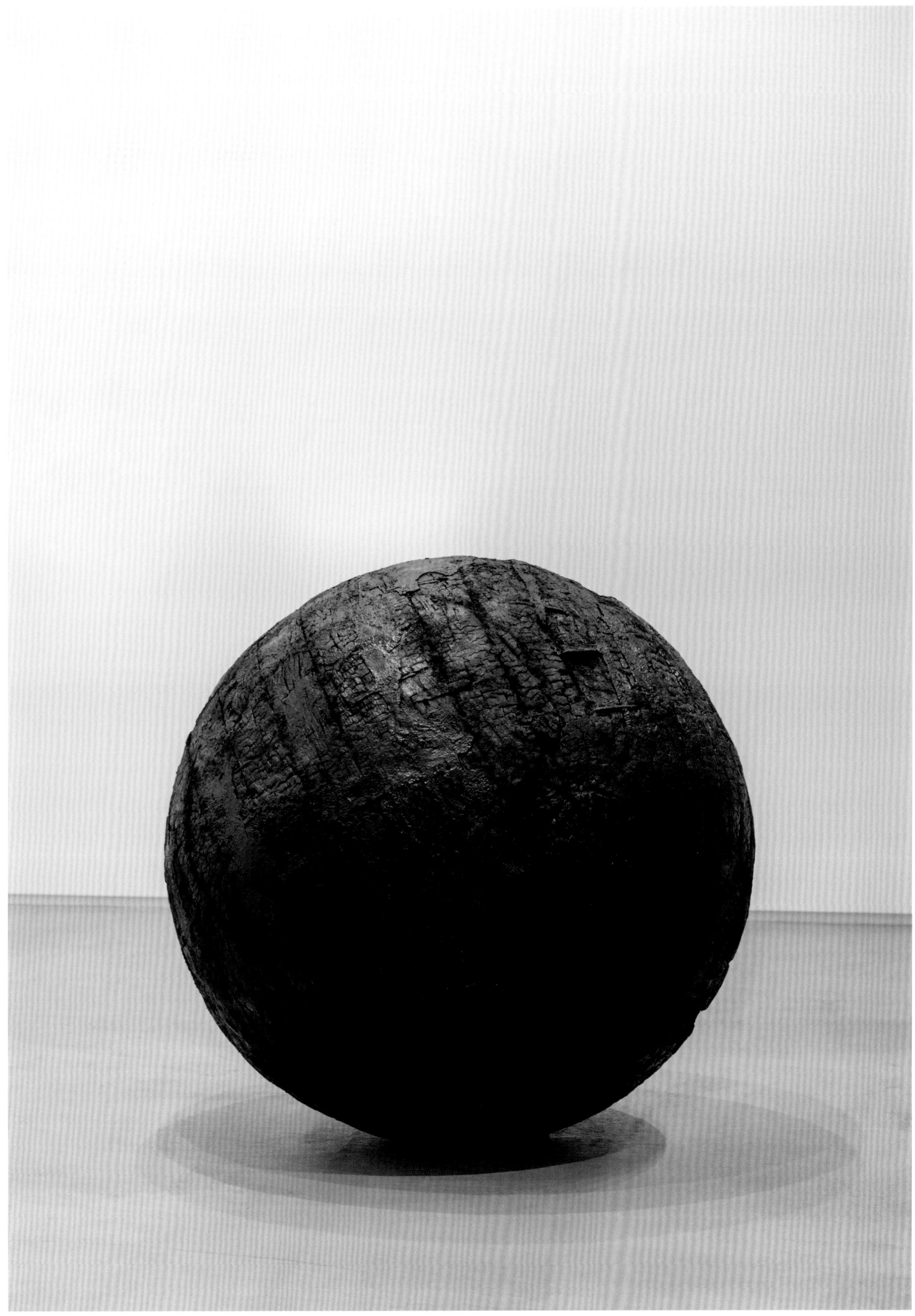

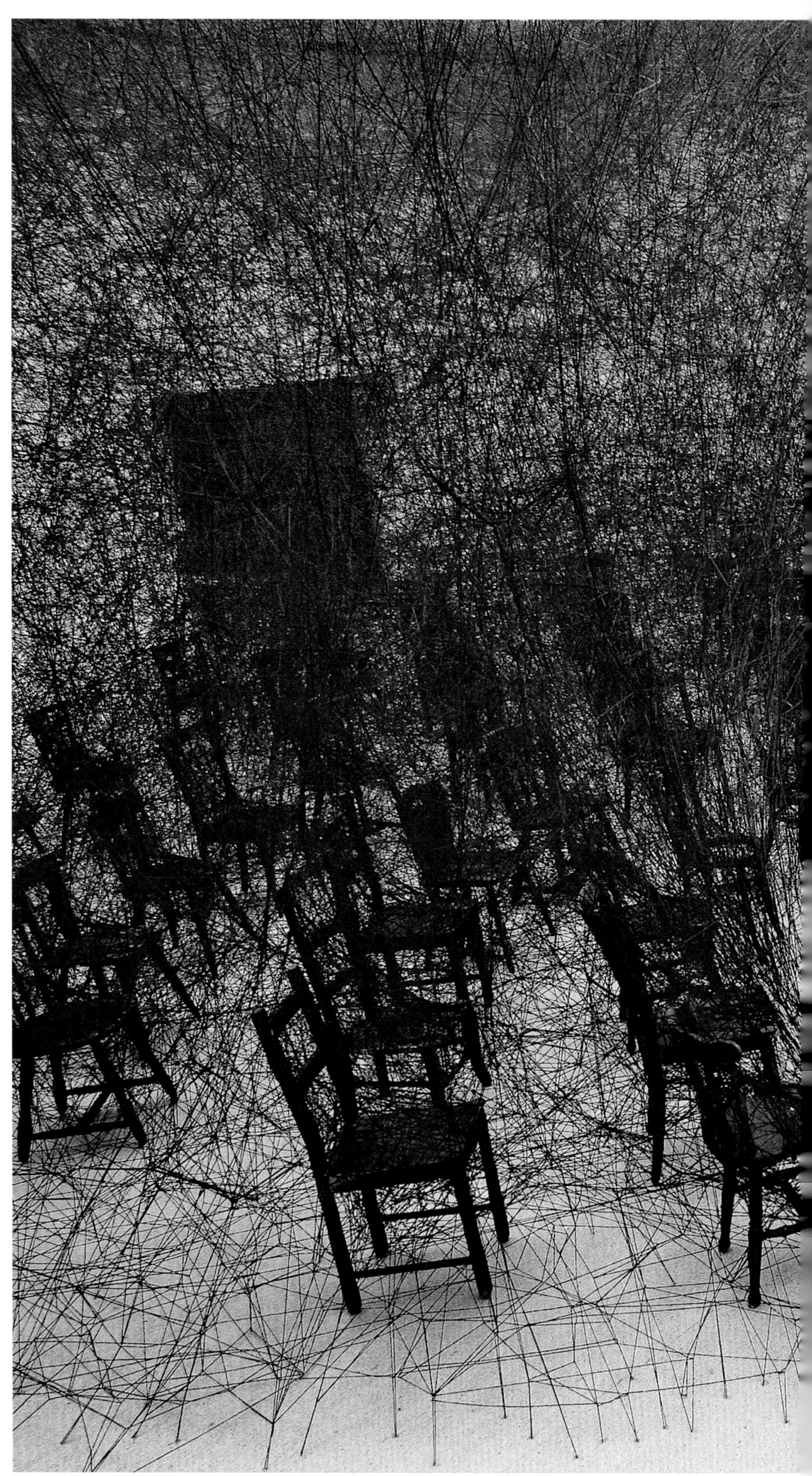

3.98 Chiharu Shiota (Japanese, born 1972), *In Silence*, 2008. Black wool, burned grand piano, and burned chairs. Installation view at Pasquart Kunsthaus Centre d'Art, Biel, Switzerland. Courtesy of Chiharu Shiota Studio and Blain|Southern, Berlin.

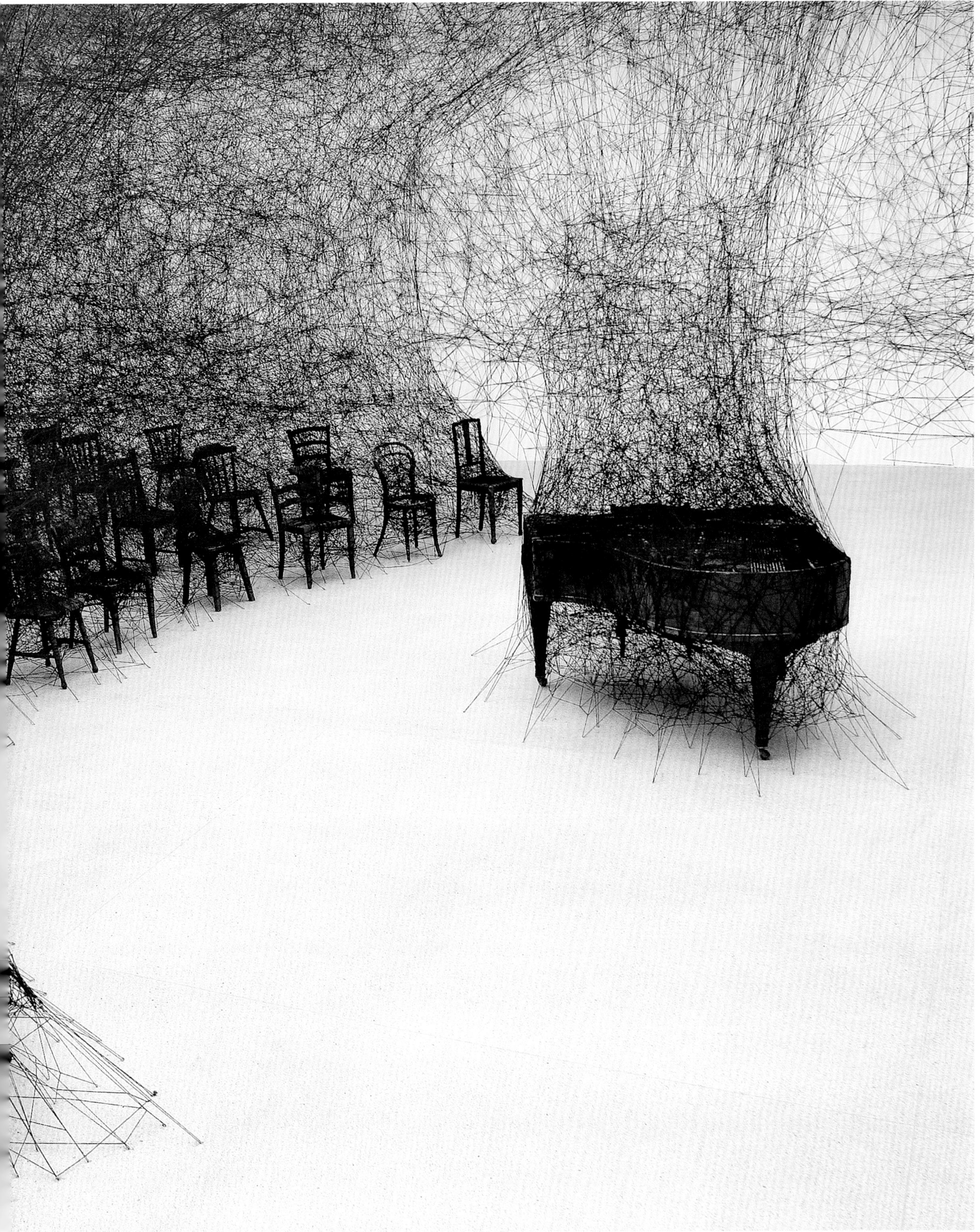

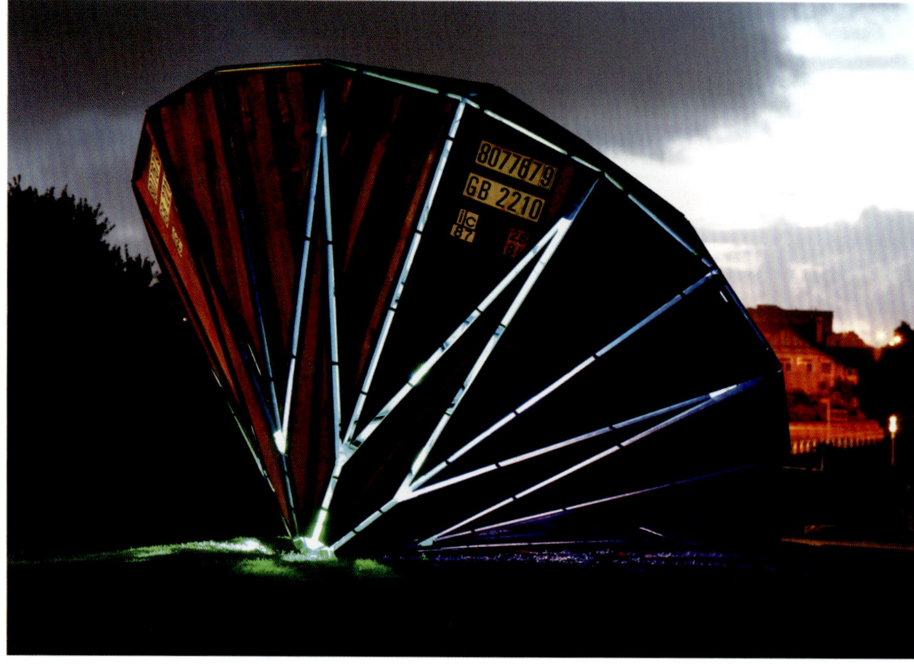

3.99 Yin Xiuzhen (Chinese, born 1963), *Black Hole*, 2010. Used shipping container, LED lights, 13 ft. 5 in. × 10 ft. 10 in. × 8 ft. 10 in. (4.10 × 3.30 × 2.70 m). © Yin Xiuzhen. Courtesy Pace Gallery, New York.

nothing on the inside can communicate with the outside, where a black hole appears serene and silent.

The Chinese artist Yin Xiuzhen sees black holes as a metaphor for the emptiness of objects that don't contribute to the meaning of life: "The black hole can serve as a stand-in for an inescapable predicament. Human desire is like a black hole, devouring everything."[69] Her sculpture, made from the wooden walls of a shipping container, takes the form of a cut diamond (figure 3.99). She hopes that by titling the work *Black Hole* she will encourage people to think about the human desire for luxury objects (diamonds) in a consumer society (represented by the shipping container).

Black holes can also suggest a hidden place filled with mystery and potential—the creative mind of Japanese artist Tadanori Yokoo. He created a poster (figure 3.100) for an exhibition of his graphic design work from the 1960s to the 1980s. Rather than exhibit finished work, the gallery showed rough sketches, dashed notes with creative ideas, and layouts for books, advertisements, and posters. The artist and gallerist titled the exhibit *My Black Holes* because they wanted the viewer to enter "into the realm of nothingness, like a black hole," where they "will be drawn, absorbed, and disappear into his never-before-seen creative space."[70]

Blackness of the black hole is emphasized in one work from a trilogy by choreographer Shamel Pitts: *Black Box*, *Black Velvet*, and *Black Hole*. In the performance of *Black Hole* (figure 3.101), a trio of dancers, their bodies bronzed, emerge from an abstract form on the floor and move as one. According to Pitts, the performance "proposes to use the idea of the transformational environment of a black hole to create an atmosphere of mystery."[71] In the style of Afrofuturism, the dancers explore the convergence of the African diaspora with the science of black holes.

The adjective "black" in black hole resonates with another artist of African descent, Kara Walker. In the 1990s she revived the medium of the paper silhouette, which was popular in nineteenth-century America, where slavery was practiced until 1865. Walker's grim silhouettes of debased Black women, such as the seminude figure tumbling in midair in *African/American* (figure 3.102), are a jarring contrast to decorous silhouettes of corseted white women with charming parasols from the antebellum South. In 2021

OPPOSITE
3.100 Tadanori Yokoo (Japanese, born 1936), *My Black Holes*, 2023. Poster for an exhibition at the Ginza Graphic Gallery, Tokyo. Printed ink on paper. Courtesy of the artist.

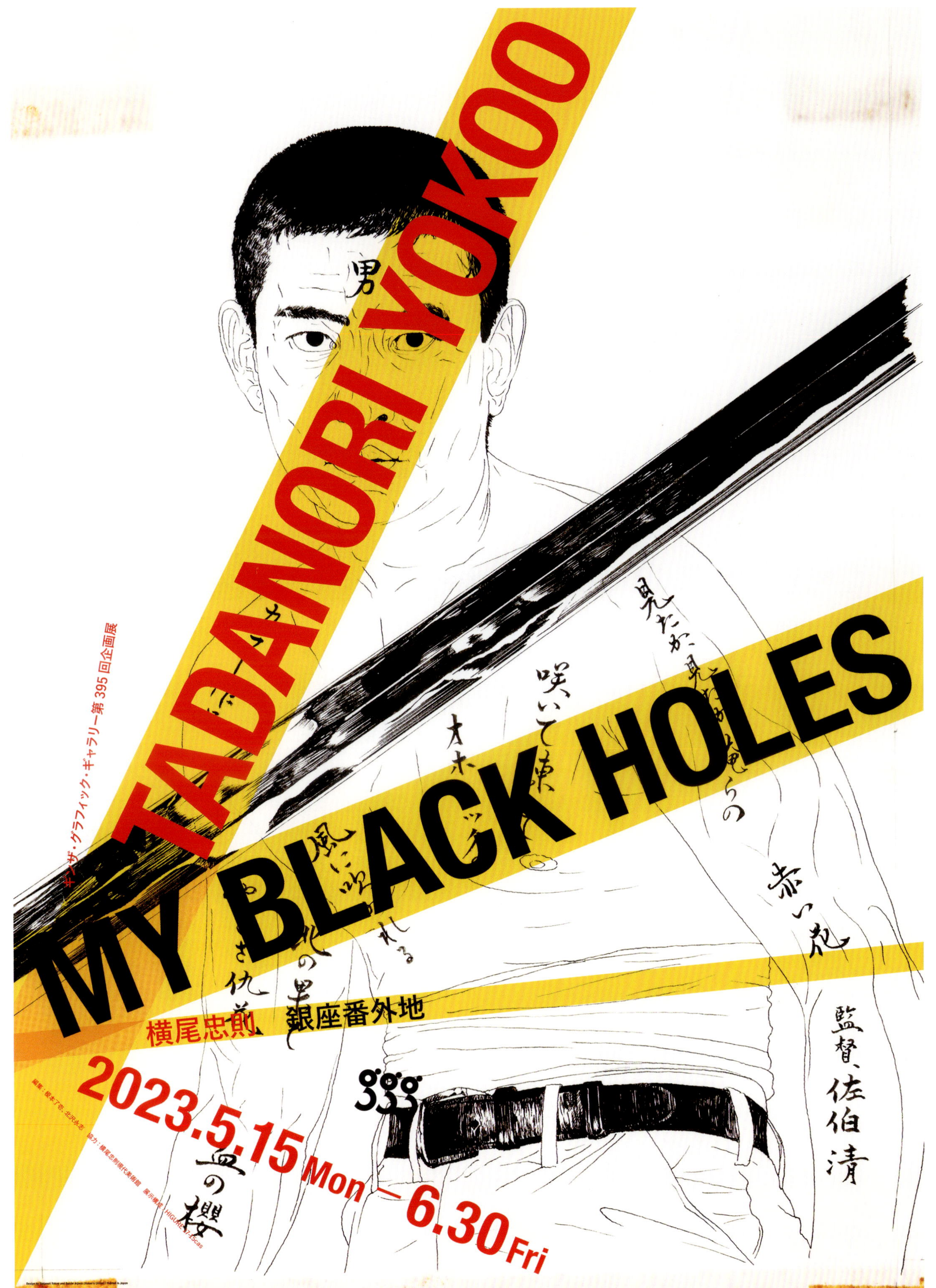
TADANORI YOKOO
MY BLACK HOLES
ギンザ・グラフィック・ギャラリー第395回企画展
横尾忠則　銀座番外地
2023.5.15 Mon — 6.30 Fri
ggg
赤い花
監督、佐伯清
血の櫻
編集：榎本了壱、北沢永志　協力：横尾忠則現代美術館　展示構成：HIGURE 17-15cas

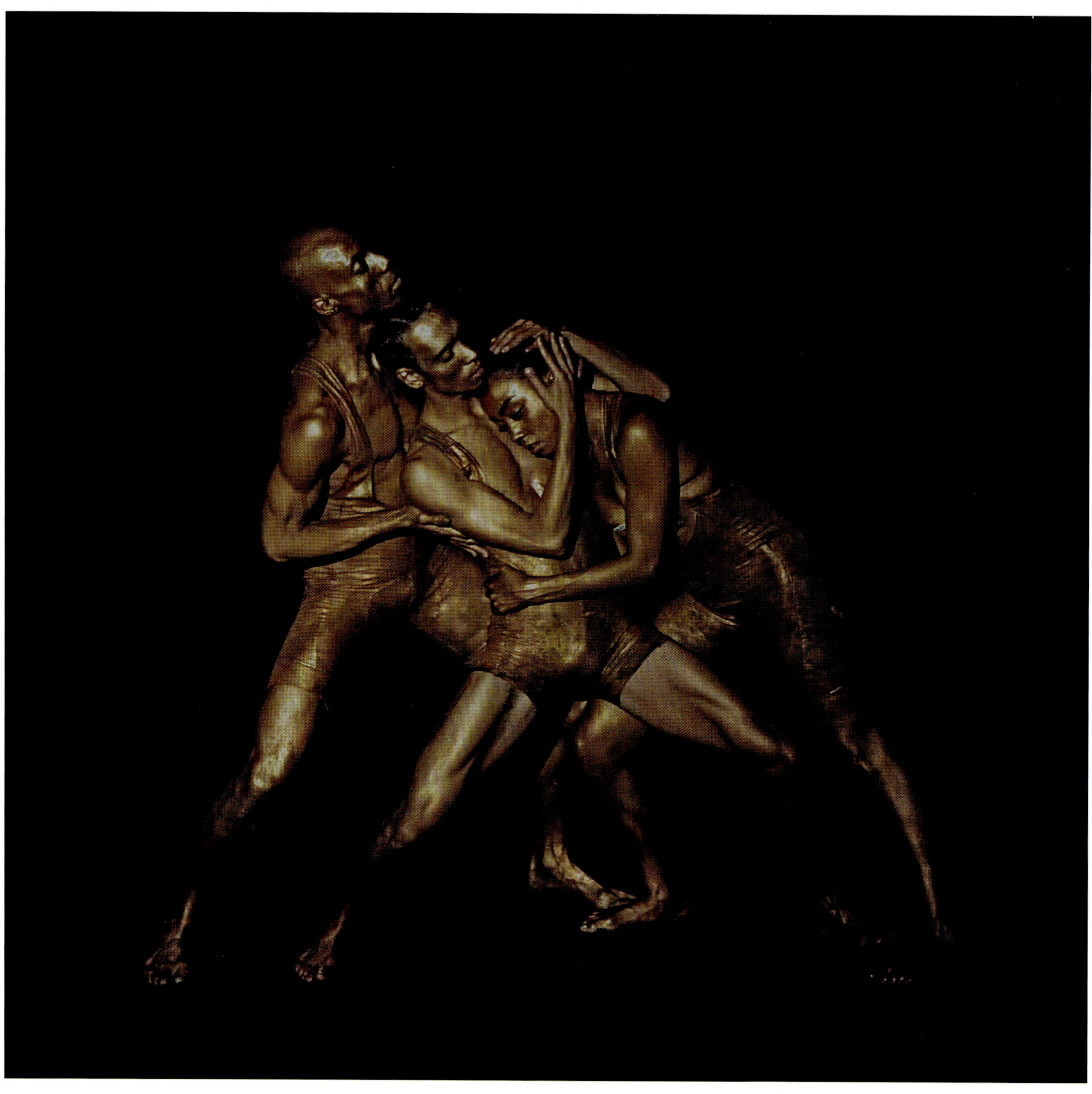

3.101 Shamel Pitts (American, born 1985), *Black Hole*, 2018, from the series *Trilogy and Triathlon*. Still from a performance; from left: Shamel Pitts (choreographer), Tushrik Fredericks, and Marcella Lewis.

the Kunstmuseum Basel showed a group of 600 drawings by Walker that the artist titled *A Black Hole Is Everything a Star Longs to Be*. What does Walker think about black holes? She has said: "Astronomically a black hole tears apart the known universe; it shakes the foundations of what science can know (and is thus ironically relegated to being 'black') and it is the potential fate of every star in the known heavens." She went on to talk about the exhibition: "The haunts from my archive should come out and comingle with new drawings, and perhaps *supernova* into a Pandora's boxlike profusion that might tear at my own known universe. That universe includes Art and Identity Politics, Narrative Impulse, Figuration, Abstraction, Vernacular vs. Fine Art, History Painting, political art movements like the Black Arts Movement or Third-Wave Feminism, ideas about the Personal vs. the Collective, debates about Drawing vs. Painting (vs. choosing to do neither) and many more cosmologies. For me, every scrap of paper is an event horizon—the boundary between the ordered world and chaos."[72]

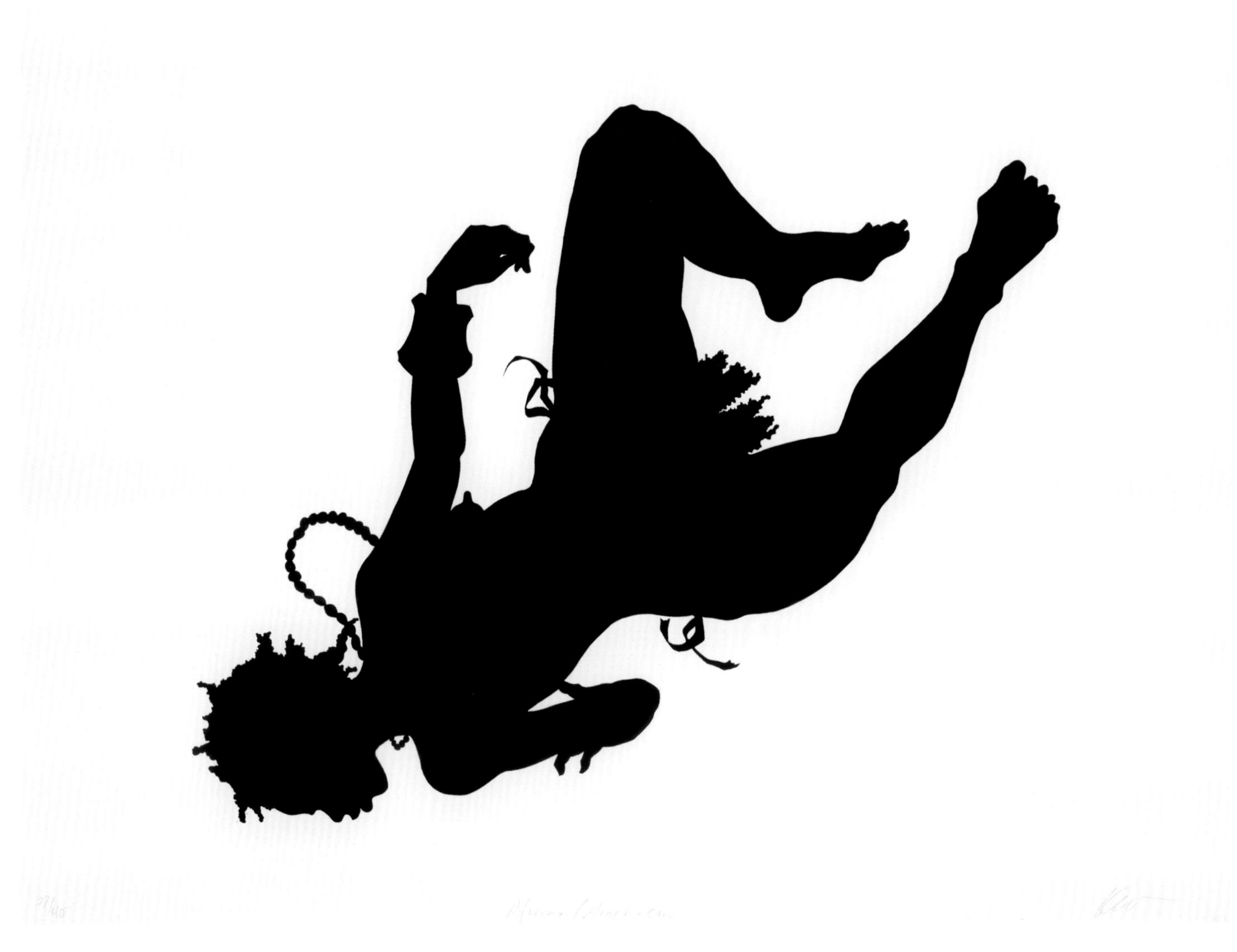

3.102 Kara Walker (American, born 1969), *African/American*, 1998. Linoleum cut, 46 × 60½ in. (116.8 × 153.7 cm). Museum of Modern Art, New York, Ralph E. Shikes Fund. Courtesy of Sikkema Jenkins & Co. and Sprüth Magers, New York.

GRAVITATIONAL WAVES CAUSED BY MERGING BLACK HOLES

As a consequence of the general theory of relativity, Einstein predicted that a cataclysmic event would cause a detectable ripple in spacetime—a *gravitational wave*. It's a measure of scientists' high esteem for Einstein that in the 1960s they began planning to build an instrument—a large interferometer—capable of testing Einstein's prediction. The Laser Interferometer Gravitational-Wave Observatory (LIGO) splits a *laser beam* (a narrow beam of light waves *in phase*—with their peaks all lined up). LIGO sends the two beams down perpendicular arms that are 2.5 miles (4 km) long (figure 3.103). The laser beams are reflected back from mirrors positioned at precisely the same distance from the beam splitter and rejoined so that any displacement can be observed. Scientists reason that if a gravitational wave rolls over the perpendicular arms, the wave will compress one arm and stretch the other. Like sound waves, gravitational waves fade with distance. Depending on its origin, a gravitational wave could be very small by the time it reaches Earth. Therefore, scientists designed LIGO so it can detect a change in the length of an arm by as little as 1/10,000 the diameter of a proton. The mirrors are suspended in a vacuum to keep them isolated from disturbances from the environment. LIGO is located in the United States and operated by a collaboration of more than a thousand scientists in twenty countries. There are actually two LIGOs—one in Livingston, Louisiana, and a twin facility about 1850 miles (3000 km) to the northwest, the Hanford site in the state of Washington—so that the instruments in one can check the other. If one LIGO senses a vibration but the other does not, then the movement may have been caused, for example, by a passing truck. Gravity waves travel at the speed of light, so if one LIGO gets a reading but the other does not within seven milliseconds, then it is a false alarm.

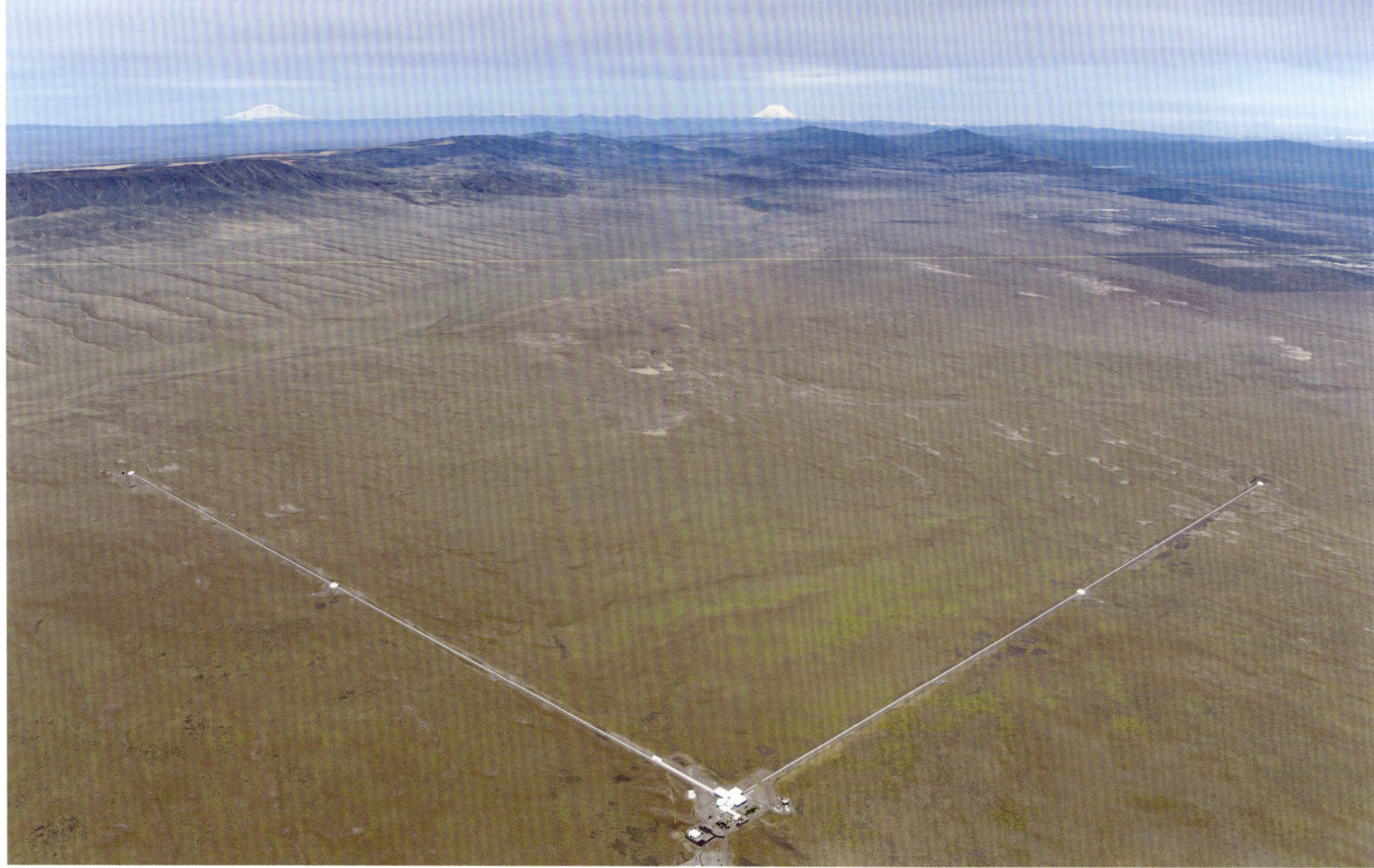

3.103 Aerial view of the Laser Interferometer Gravitational-Wave Observatory (LIGO) Hanford, Richland, Washington. David Wyatt, Caltech, MIT, and LIGO Laboratory.

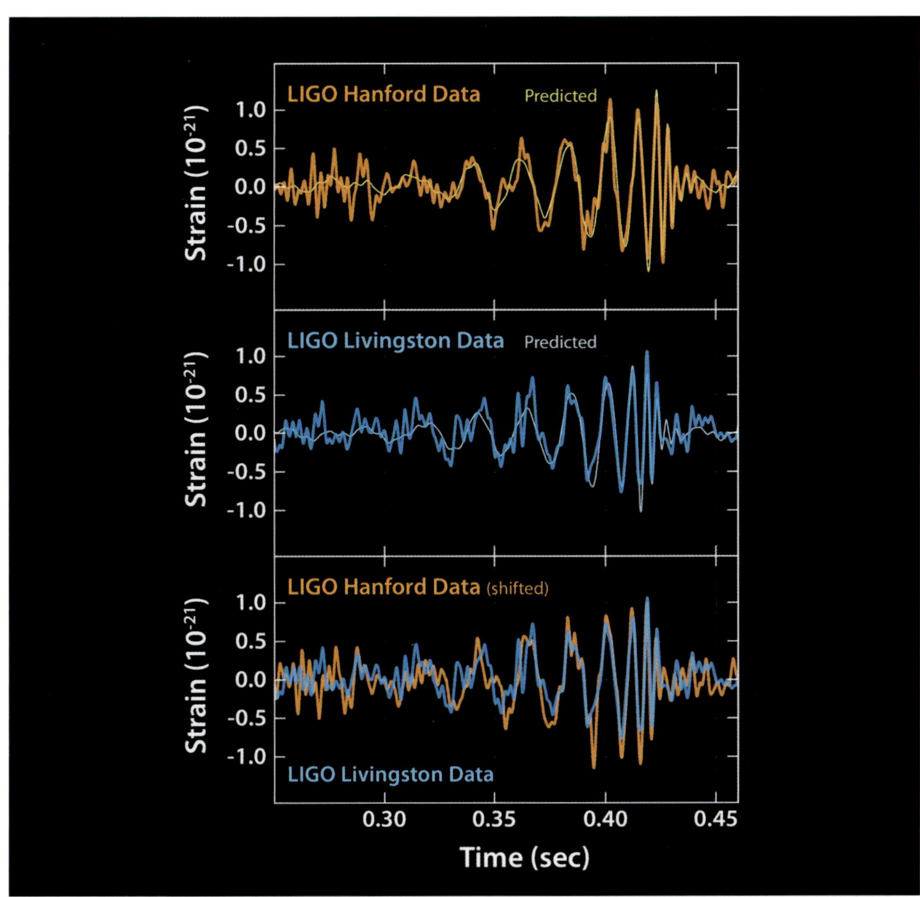

3.104 Gravitational-wave signal recorded by LIGO detectors. Caltech, MIT, and LIGO Laboratory.

These three graphs show the signals of gravitational waves caused by two merging black holes recorded by LIGO detectors at the Hanford site in Washington and Livingston, Louisiana, on September 14, 2015. The top two plots show the signals recorded at Hanford and Livingston, respectively, along with the predicted waveforms of what two merging black holes should look like according to the equations in Einstein's general theory of relativity. Time is plotted on the X-axis, and strain (the fractional amount by which distances are distorted) on the Y-axis.

The graph on the bottom compares data from the twin observatories, confirming that they both witnessed the same event. The Hanford data were shifted to correct for the time it took for the signal to travel between the two detectors. The Hanford data have also been inverted for comparison because of the different orientations of the detectors at the two sites.

On September 14, 2015, the Livingston LIGO detected a gravitational wave, and seven milliseconds later the Hanford LIGO detected it as well (figures 3.104 and 3.105). Astronomers determined that the gravitational wave was caused by the collision of two black holes more than a billion years ago (figures 3.106 and 3.107). From the signals observed by LIGO, scientists calculated that the black holes were twenty-nine and thirty-six solar masses, respectively, but neither event horizon was larger than 125 miles (200 kilometers) in diameter. (For comparison, the diameter of the sun is about 865,000 miles [1.4 million km]). In a few tenths of a second, the black holes merged to form a single one of around sixty-two solar masses. The collision radiated energy in the form of gravitational waves equivalent to three solar masses—more energy than all stars in the visible universe emitted as light in that same fraction of a second. Several artists created works in response to LIGO's recording of the colliding black holes (figures 3.108–113).

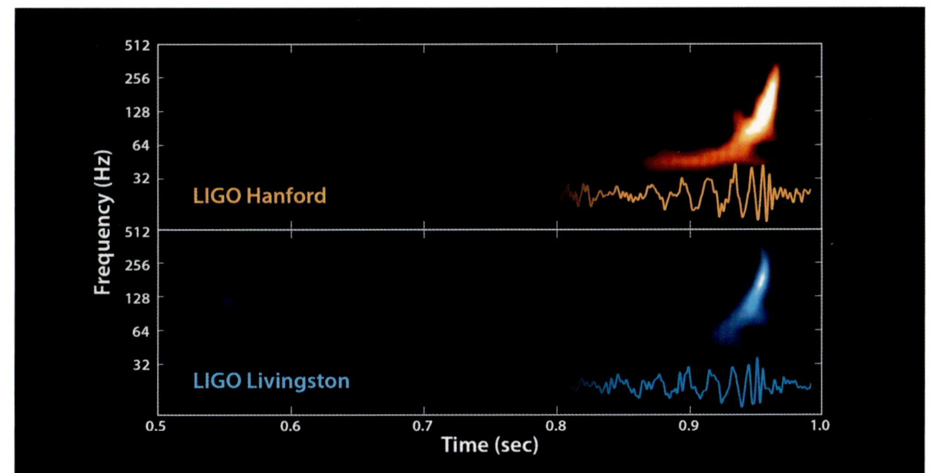

3.105 The sound of two black holes colliding. Caltech, MIT, and LIGO Laboratory.

Scientists have converted the gravitational waves caused by two black holes colliding into sound waves that humans can hear. As the black holes spiral in closer and closer, the frequency of the gravitational waves increases rapidly, resulting in sound waves that rise quickly, like the chirp of a bird.

3.106 David Sipress, "Daily Cartoon: Friday, February 12," *The New Yorker* website, February 12, 2016, https://www.newyorker.com/cartoons/daily-cartoon/friday-february-12th-gravitational-waves.

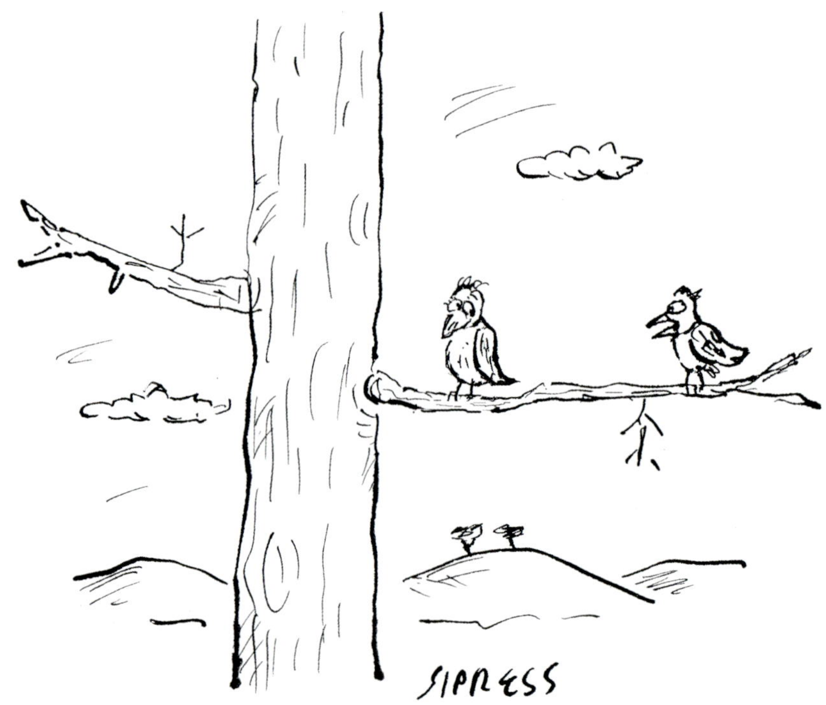

"Was that you I heard just now, or was it two black holes colliding?"

3.107 Artist's impression of two black holes merging into one. Simulating eXtreme Spacetimes.

In this rendering of the event recorded by LIGO in 2015, one black hole is slightly larger than the other. They spiral toward each other, and soon there will be a single black hole.

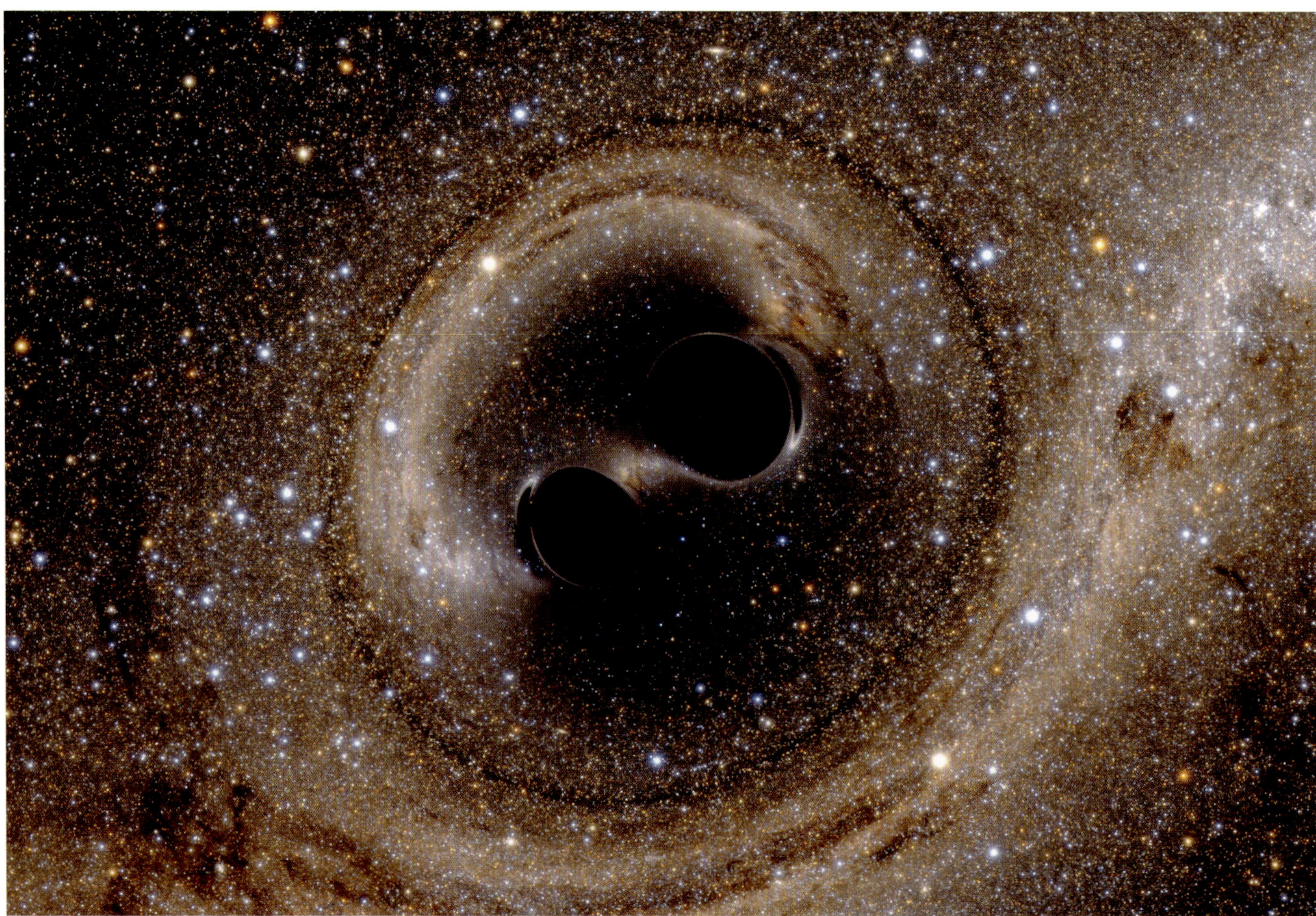

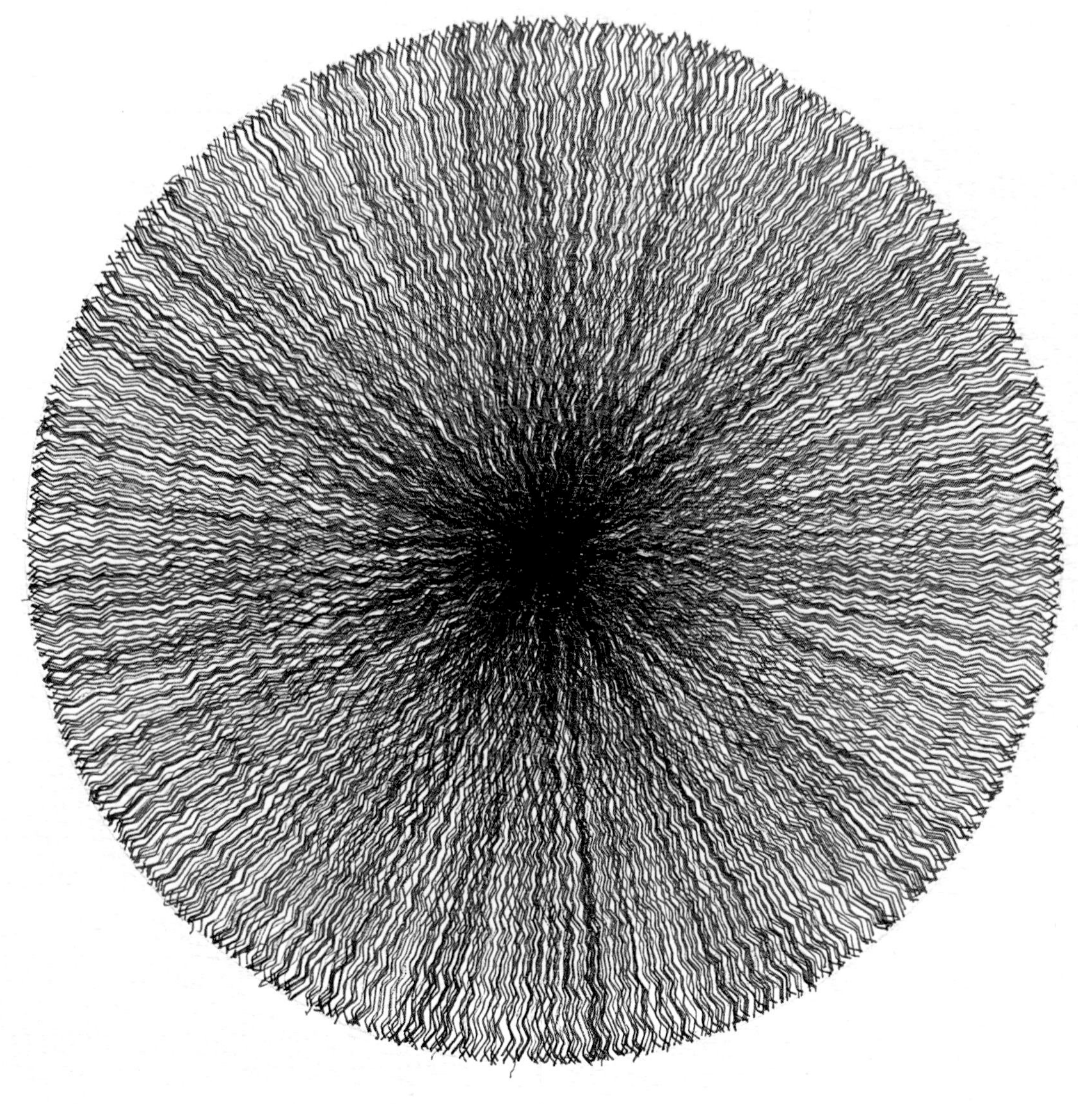

3.108 Shanthi Chandrasekar (American, born India 1967), *Black Hole—Gravitational Waves*, 2022. Pen on paper, 9 × 9 in. (22.8 × 22.8 cm). Courtesy of the artist.

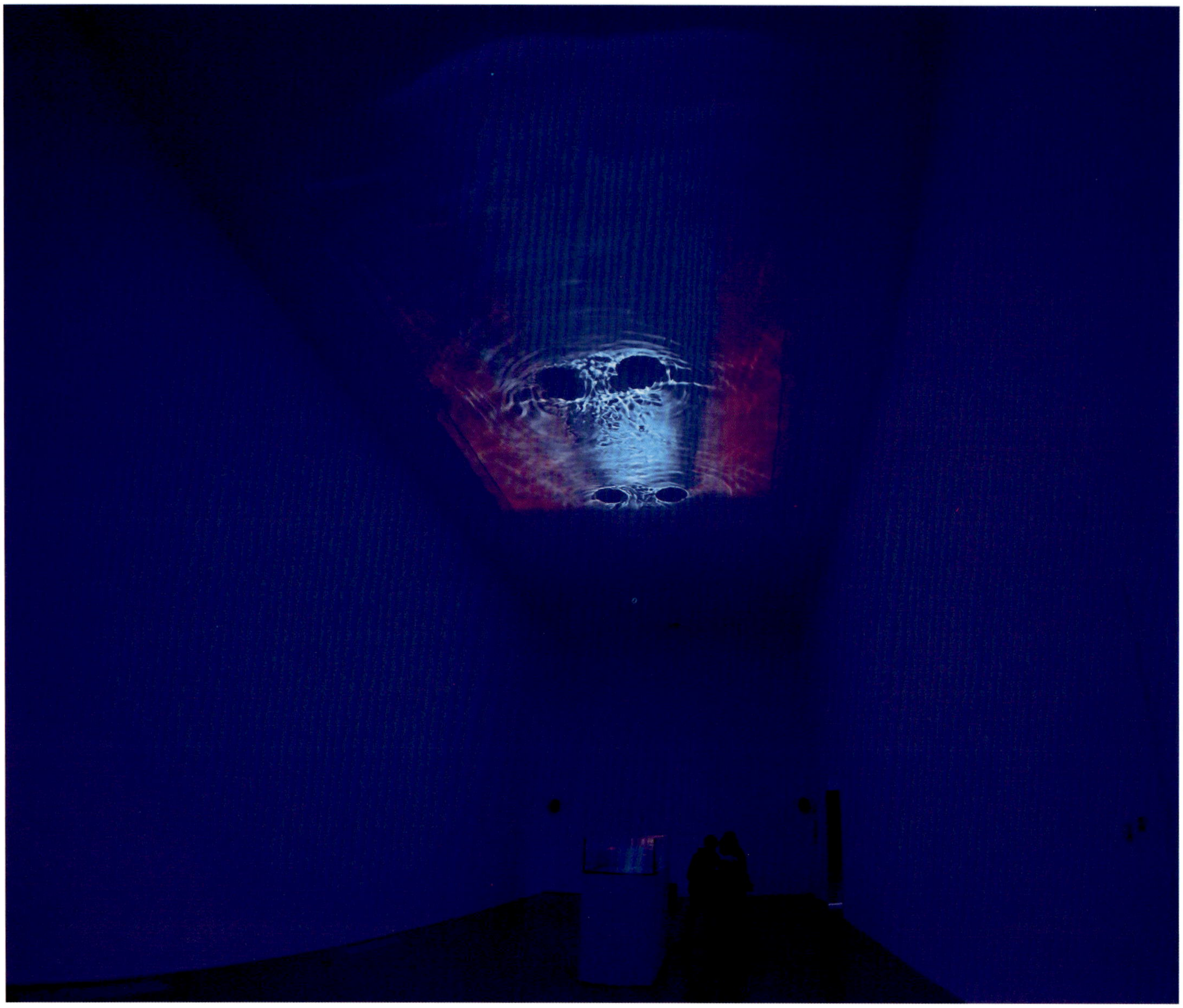

3.109 Evelina Domnitch (Belarusian, born 1972) and Dmitry Gelfand (Dutch, born 1974), *ER=EPR*, 2017. Water, laser beam, speakers. Installation view at GAMeC, Bergamo, Italy, in collaboration with composer William Basinski, artist/physicist Jean-Marc Chomaz, and LIGO. Courtesy of the artists.

In this artwork, Domnitch and Gelfand created representations of black holes as rotating vortices in a body of water, where some of the vortices merge. A laser beam illuminates the water from below, projecting the black holes/vortices onto the ceiling. Visitors saw black holes merging while they heard music composed by Basinski that included LIGO's recording of gravitational waves converted to sound waves.

The title, *ER=EPR*, refers to two ideas discussed by Einstein in publications of 1935: wormholes and entanglement. One essay proposed that black holes that are far apart were connected by "Einstein-Rosen bridges" (today known as wormholes), which cause distant objects to be close together. Einstein coauthored that publication, "The Particle Problem in the General Theory of Relativity," with Nathan Rosen, and the idea became known as "ER" after the initials of the authors' last names. The second paper, "Can Quantum-mechanical Description of Physical Reality be Considered Complete?," was written by Einstein, B. Podolsky, and Rosen. It addressed entangled particles with paired properties: if one is measured spinning up, the other is instantaneously measured spinning down, even if the particles are far apart. This seems to violate the rule that nothing—not even information—can travel faster than the speed of light. Therefore, the authors wrote, entanglement creates a paradox for quantum mechanics, which is known as the "EPR paradox" after the initials of the authors. For more than a half century, wormholes and entanglement were considered separate topics. Then, in 2013, the physicists Juan Maldacena and Leonard Susskind published a paper, "Cool Horizons for Entangled Black Holes," in which they put forth a hypothesis, called "ER=EPR," proposing that wormholes connect entangled black holes.

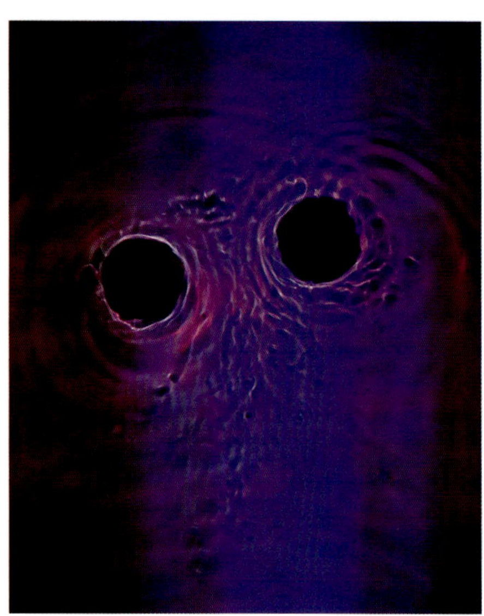

3.110 Evelina Domnitch (Belarusian, born 1972) and Dmitry Gelfand (Dutch, born 1974), *ER=EPR* (detail), 2017. Installation view at Gropius Bau, Berlin. Courtesy of the artists.

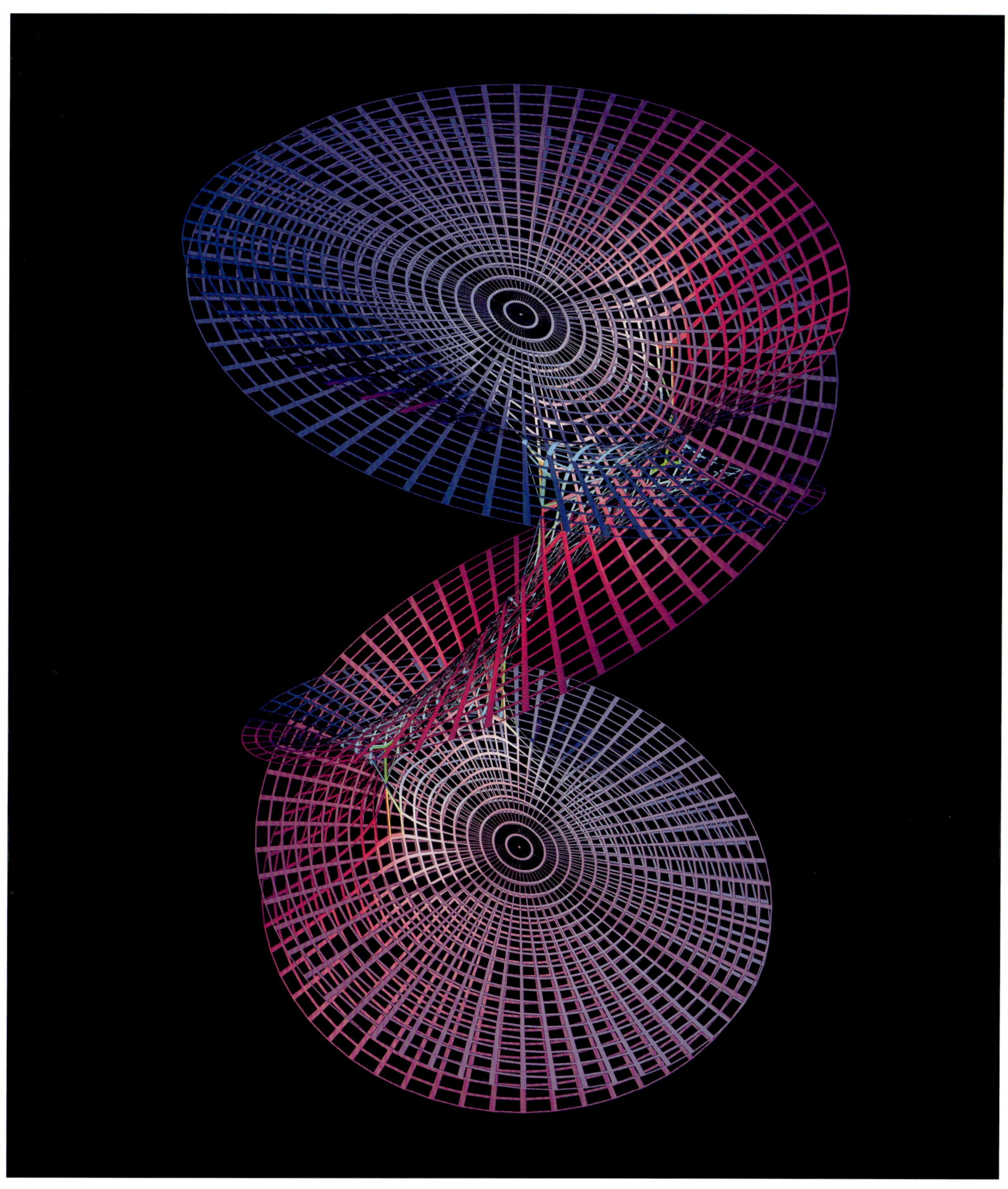

3.111 Eric Heller (America, born 1946), *Black Holes Merging*, 2020. Digital image. Courtesy of the artist.

Heller is a physicist who studies wave phenomena in quantum mechanics, acoustics, and oceanography. He's also a practicing artist who creates digital images about scientific subjects. In this example, he imagined the pattern two black holes might make when they spiral into each other.

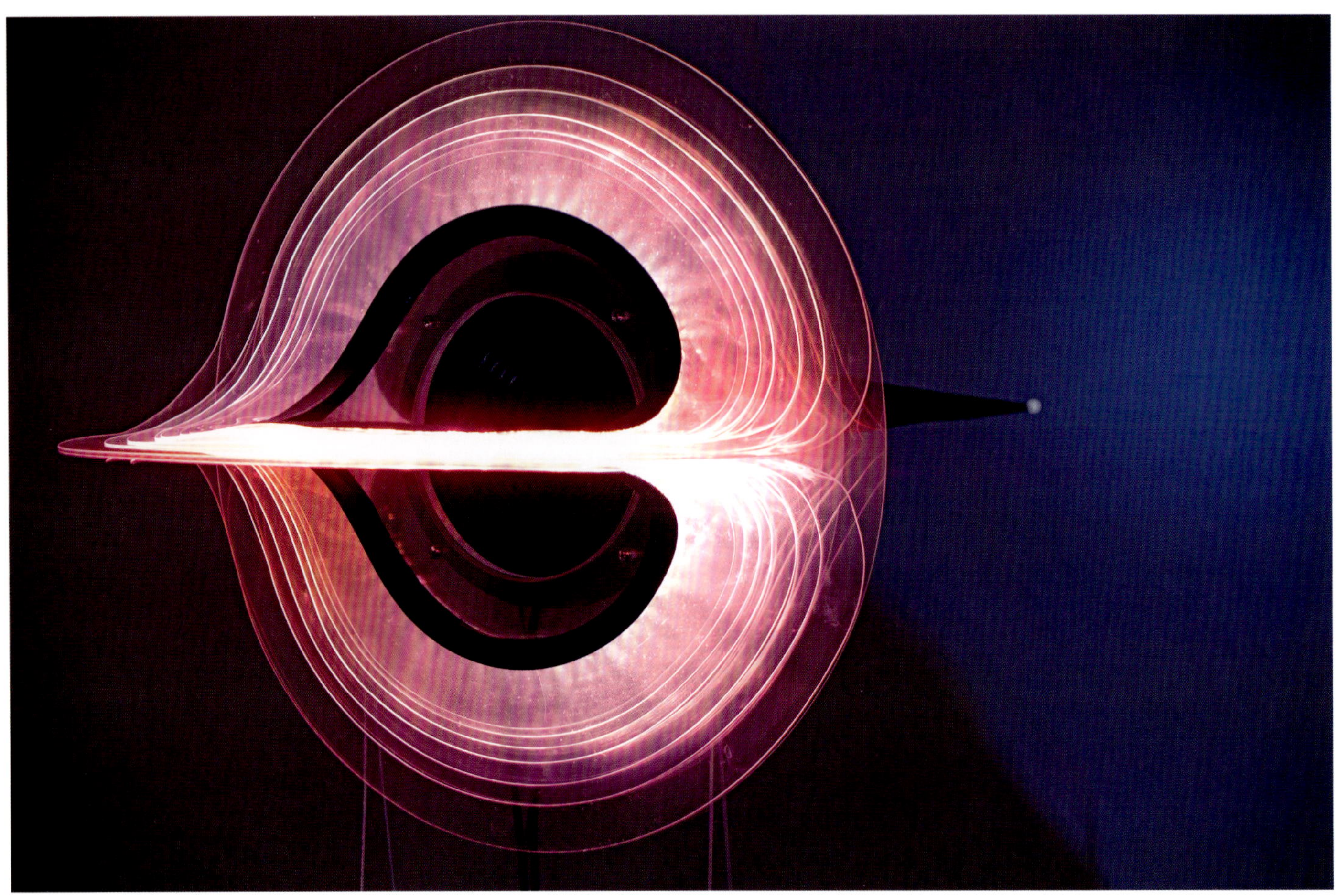

3.112 Yiannis Kranidiotis (Greek, born 1974), *Black Hole*, 2019. Plastic sheets, LEDs, speaker, gravitational-wave data, custom software, 19½ × 31½ × 43¼ in. (50 × 80 × 110 cm). Courtesy of the artist.

Kranidiotis symbolized the curvature of spacetime around a black hole by bending circular plastic sheets. He then put a speaker at the center of the sculpture, where the metaphorical black hole is located. The speaker broadcasts the gravitational waves that LIGO converted into sound. Projecting from the center of the sculpture is a black cone (visible on the right), at the end of which is a point of light, symbolizing the singularity.

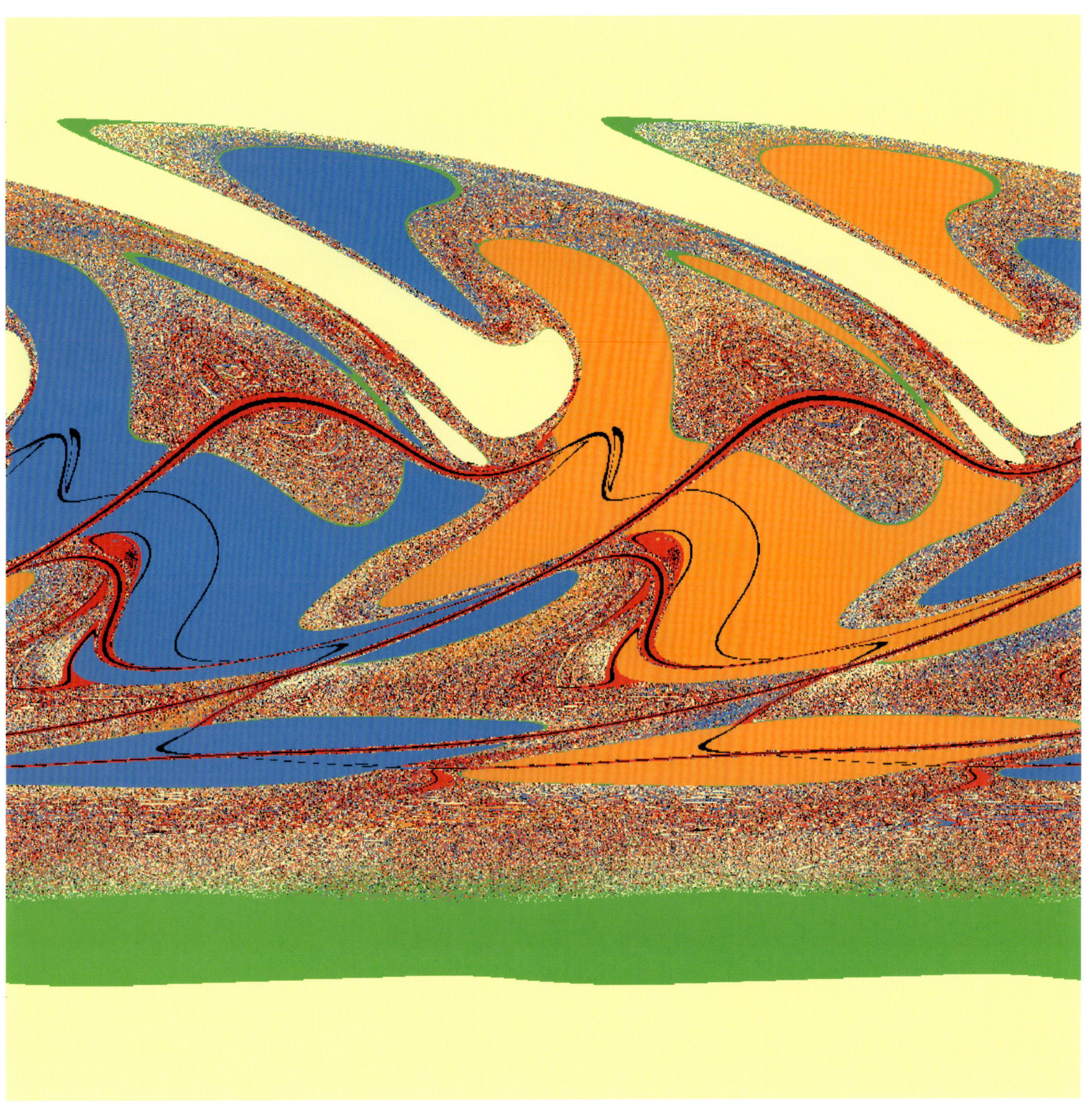

3.113 Johan Samsing (Danish, born 1984), *Colorful Black Holes*, 2023. Digital image. Courtesy of the artist.

Samsing is an astrophysicist, and this colorful image is based on data showing a simulation of three black holes. Each pixel in the image shows the result of a gravitational simulation between a binary pair and a single black hole, where the horizontal axis is the initial binary phase, and the vertical axis is the impact parameter. The three black holes often move around each other dramatically before an end state is reached. The red and black lines indicate a collision of two black holes. The remaining colors indicate different ways for each black hole to survive the interaction with the others.

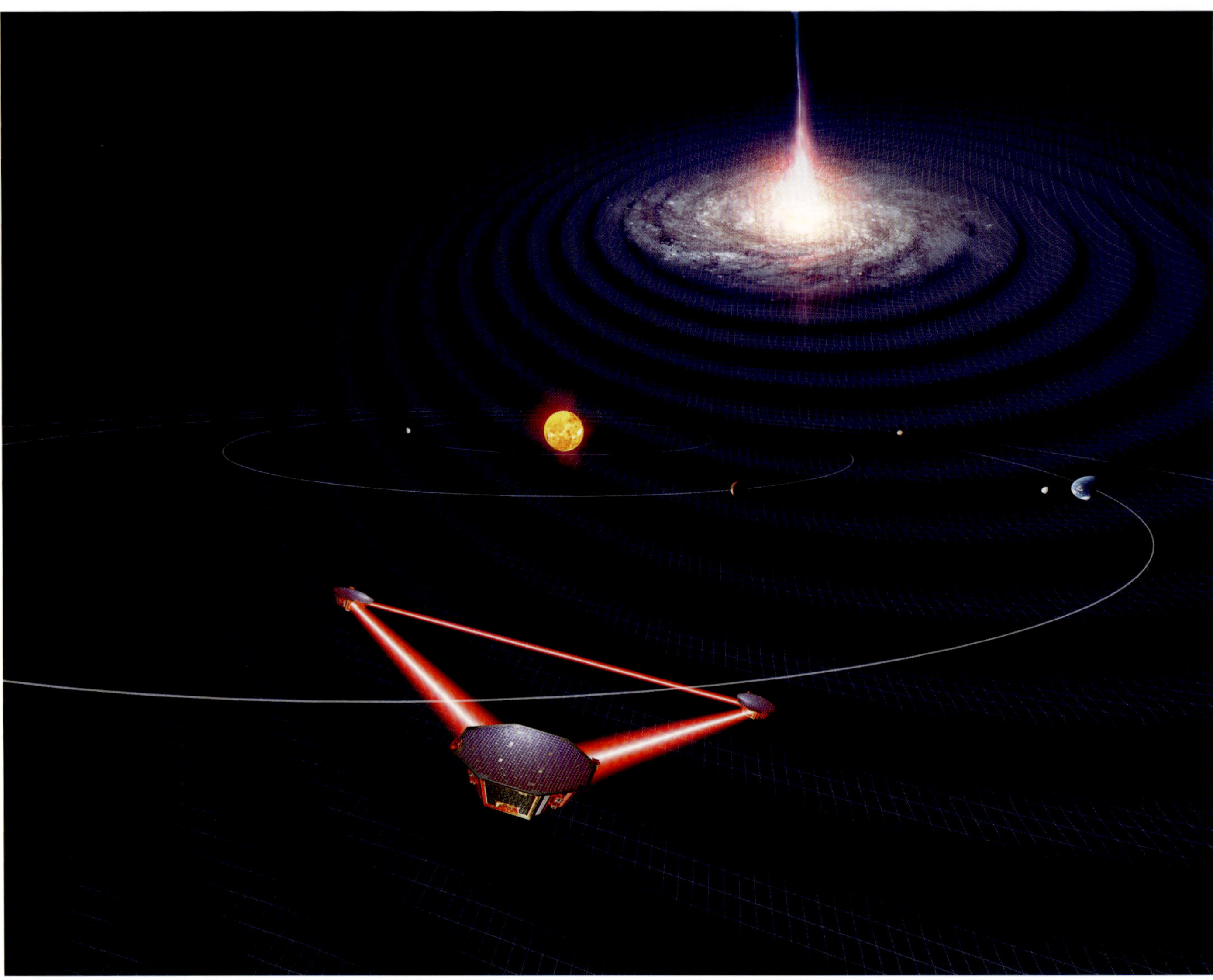

3.114 Artist's impression of the Laser Interferometer Space Antenna (LISA). University of Florida and Simon Barke.

The European Space Agency, in collaboration with NASA, plans to build a gravitational-wave observatory in outer space. The Laser Interferometer Space Antenna (LISA) will consist of three spacecraft—one for the source of light and the beam splitter and two carrying the mirrors. The three spacecraft will fly in an equilateral triangle linked by laser beams (figure 3.114). Each side of the triangle will be more than a million miles (1.6 million km) long, more than twice the distance between Earth and the moon. LISA will be an immense interferometer dedicated to the detection of gravitational waves.

In 2023 an international consortium of researchers, the North American Nanohertz Observatory for Gravitational Waves (NANOGrav), revealed compelling evidence for the existence of gravitational waves reverberating across the universe. The data was collected over the course of fifteen years by radio telescopes around the world. The researchers observed an array of rapidly rotating neutron stars known as *millisecond pulsars* because they rotate with a period between 1 and 10 milliseconds. (As of 2025, the fastest-spinning pulsar known spins 716 times per second.) The rotation rate of millisecond pulsars is extremely precise, so they function as cosmic clocks; at each rotation, they give off radio waves. The scientists observed pairs of pulsars located in the Milky Way, within a few thousand light-years from Earth. Scientists reasoned that if a gravitational wave passed through the galaxy, it would warp the local spacetime and produce a small but detectable

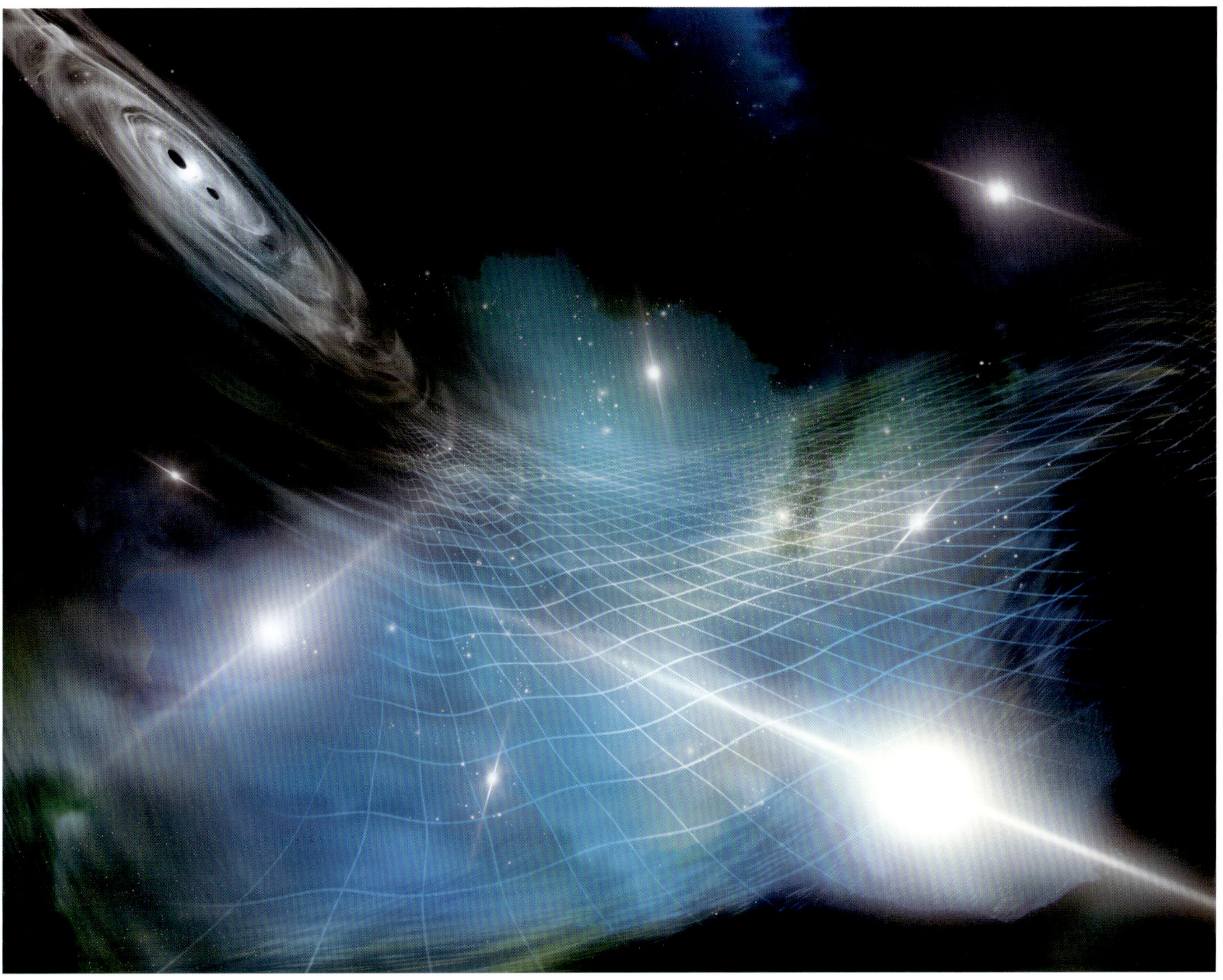

change in the arrival time of the pulses from the millisecond pulsars recorded by radio telescopes on Earth. The NANOGrav scientists found exactly this. In contrast to LIGO's 2015 recording of gravity waves caused by the collision of two black holes, the NANOGrav researchers found gravity waves emanating from everywhere all the time, a so-called *gravitational-wave background* (figure 3.115). The scientists hypothesize that these gravitational waves were caused by thousands or millions of pairs of supermassive black holes as they merged in the early universe over ten billion years ago. Alternatively, the pattern could be an echo of the Big Bang, similar to the cosmic microwave background that formed 380,000 years after the Big Bang. If the gravitational-wave background is from the Big Bang, it would have been emitted almost instantaneously and would provide a way to probe the primordial universe.

3.115 Artist's impression of an array of millisecond pulsars being affected by gravitational ripples produced by the collision of a pair of supermassive black holes in a distant galaxy. Aurore Simonnet and NANOGrav.

IMMERSIVE ART ABOUT BLACK HOLES

Artists create immersive art—artworks the viewer can walk into—to enhance the immediacy of the experience. Between 2014 and 2015 Ryoji Ikeda was artist-in-residence at CERN, where he created *Micro/Macro* about the relationship between the subatomic realm (a pattern of particles shown as a video projection on the floor) and the universe (celestial objects projected on the wall; figure 3.116). To make *Point of No Return* in 2018 (figure 3.117), Ikeda painted a black circle on the wall and projected light around it, emphasizing the event horizon—the point of no return.

In 2016 the choreographer Wen-chi Su was also an artist-in-residence at CERN, where she met the theoretical physicist Diego Blas, and they discussed the meaning of gravity in dance and astronomy. Su imagined what happens when a body falls into a black hole. Together with her production team, she directed a film in which the sets were animations (figure 3.118 *left*) and the movements of the dancer were captured by motion sensors (figure 3.118 *right*). Additionally, a surround-sound system immersed the audience in a three-dimensional sound field.

The American artist Yambe Tam, who merges Western science with Chinese philosophy, has said: "Black holes are a reoccurring theme in my practice. Beyond my interest in theoretical physics, I see connections to the Buddhist philosophical concept of the void/emptiness/nothingness, which is shared more widely with other Eastern spiritual traditions. Rather than signifying a negative space or absence of something, void/emptiness/nothingness is a space of infinite potentiality. It is during the practice of zazen [silent meditation] that I most feel an embodied sense of this—the emptying of oneself, or dissolution of form and ego into pure being."[73] The

3.116 Ryoji Ikeda (Japanese, born 1966), *Micro/Macro*, 2015. Digital projections. Installation view at Carriageworks, Sydney, Australia, 2018.

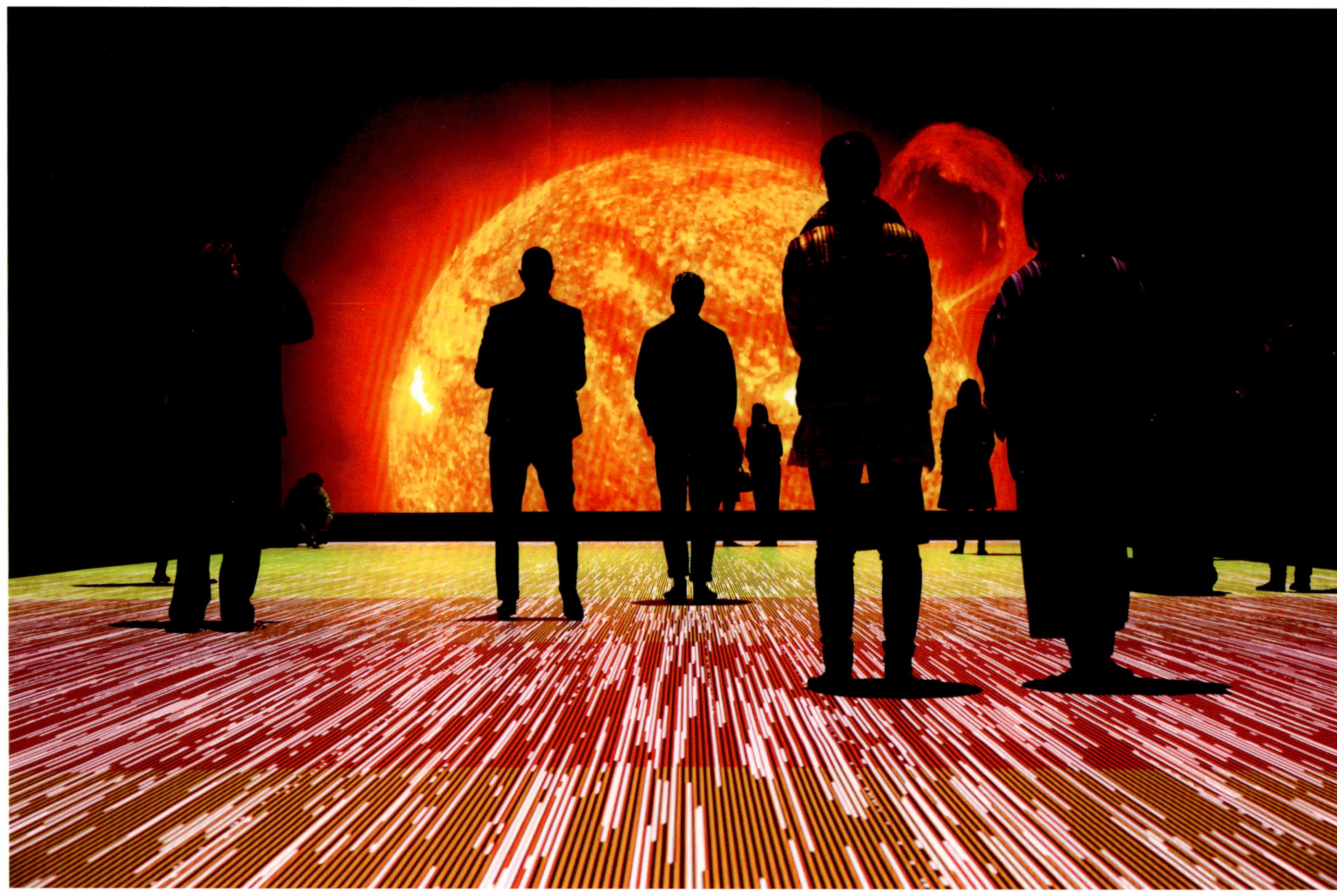

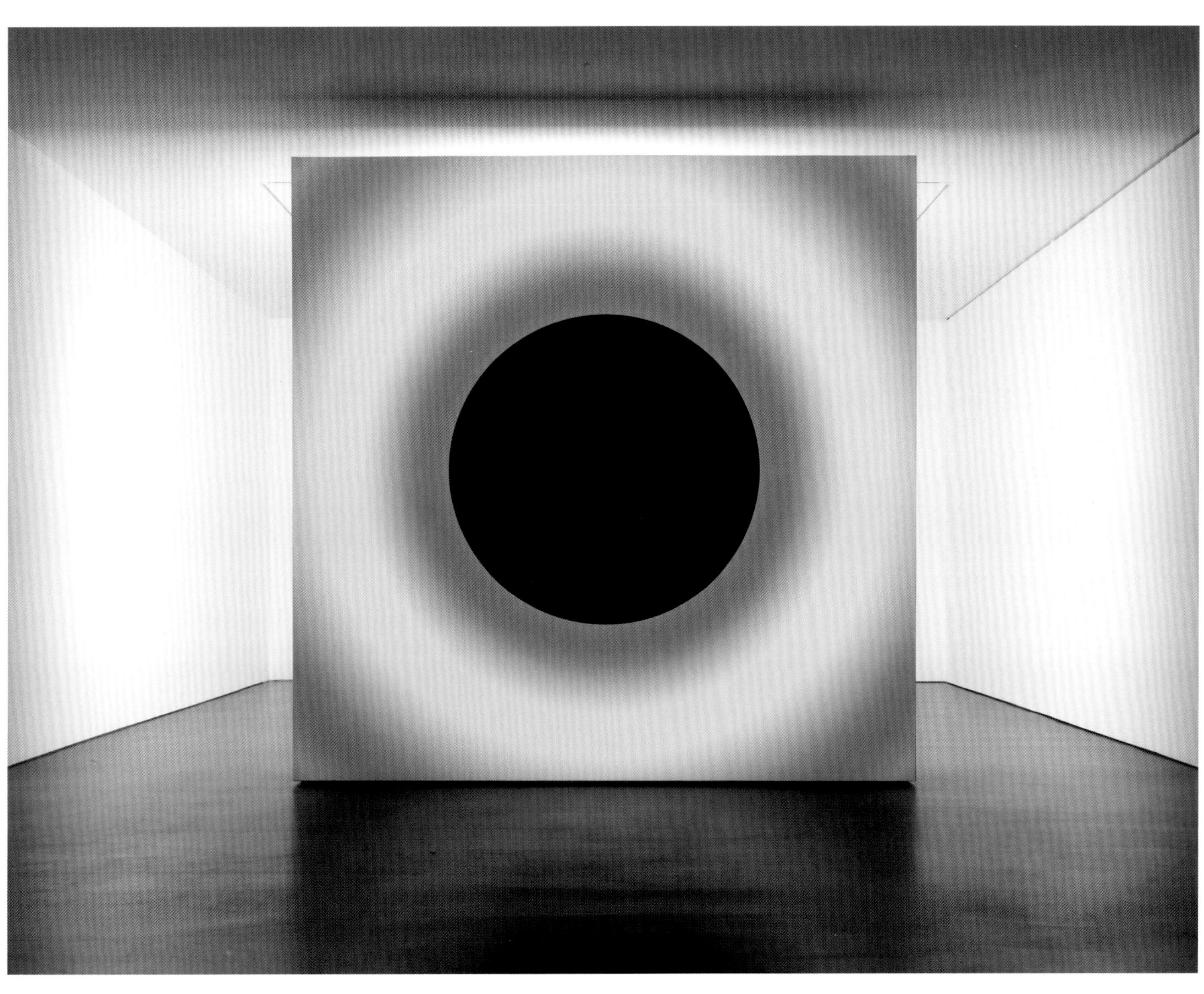

3.117 Ryoji Ikeda (Japanese, born 1966), *Point of No Return*, 2018. DLP projector, computer, speakers, HMI lamp. Installation view at Esther Shipper, Berlin, 2023. Courtesy of Almine Rech Gallery, New York, and Esther Schipper, Berlin.

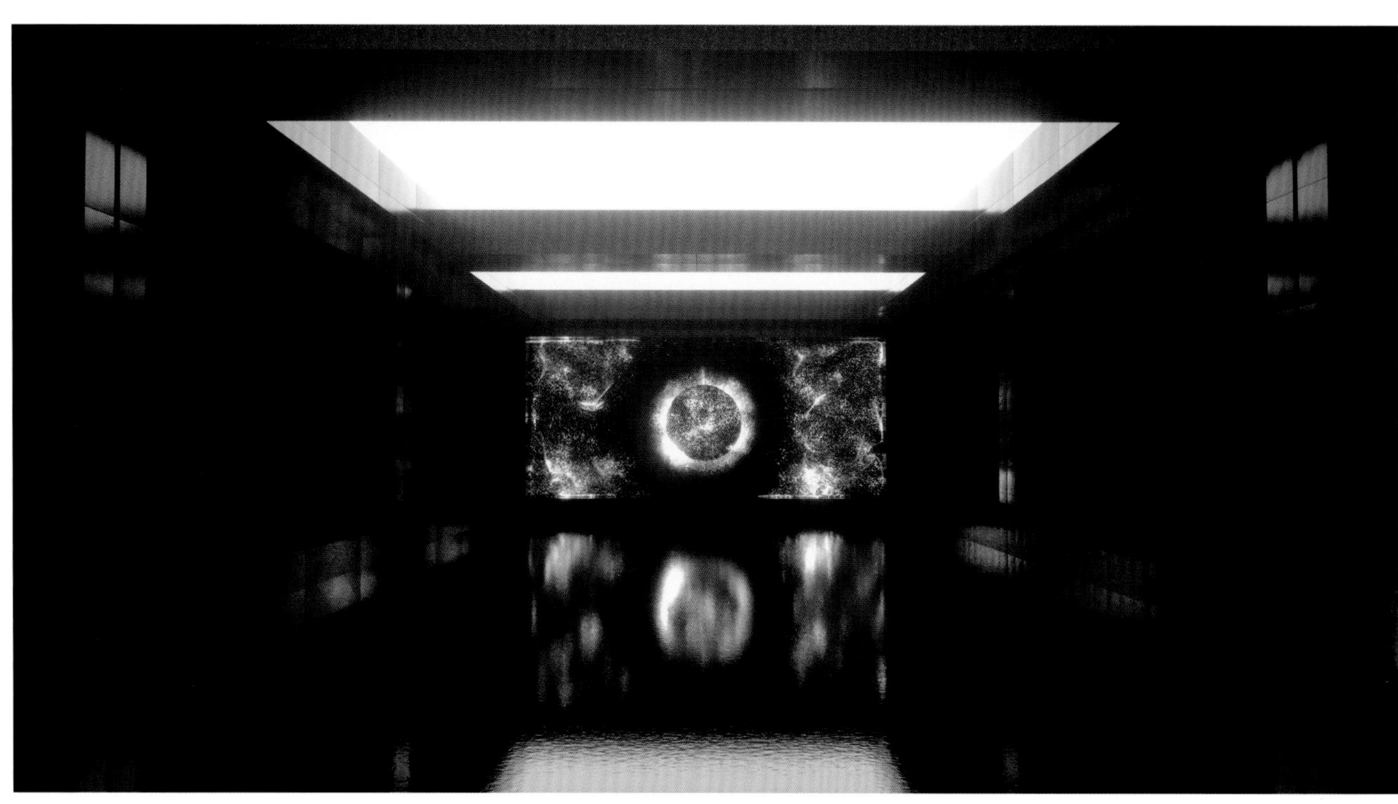

3.118 Wen-chi Su (born 1977), *Black Hole Museum + Body Browser*, 2022. Stills from VR animation. Animation of *Black Hole Museum* by Wen-yee Hsieh; animation of *Body Browser* by Yu-jie Huang; music and sound by Ping-sheng Wu; VR program integration/motion capture by Yu-jie Huang; scenography by Huei-ming Chang; choreography by Wen-chi Su and Li-wei Tu; dancer was Li-wei Tu; production by YILAB, CLAB. Courtesy of YILAB.

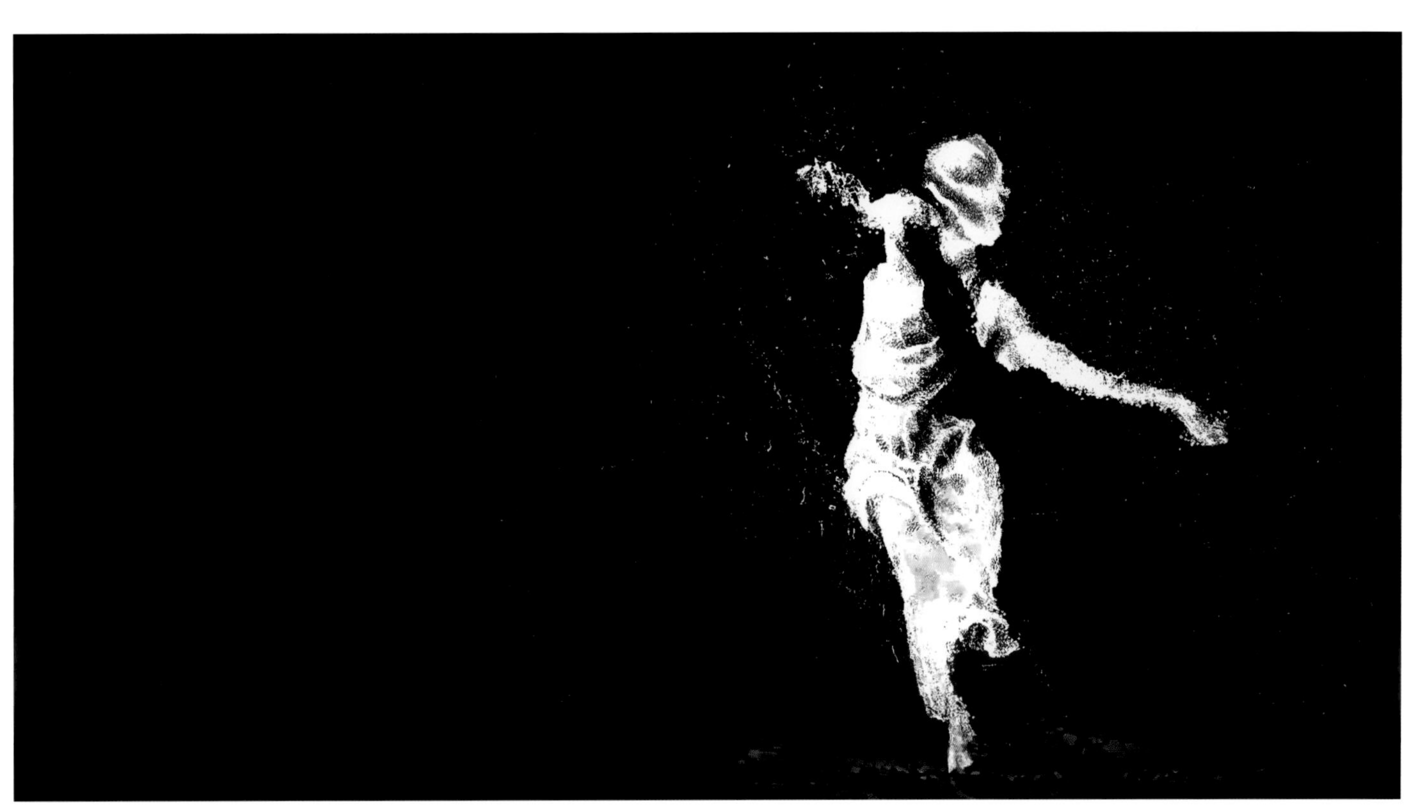

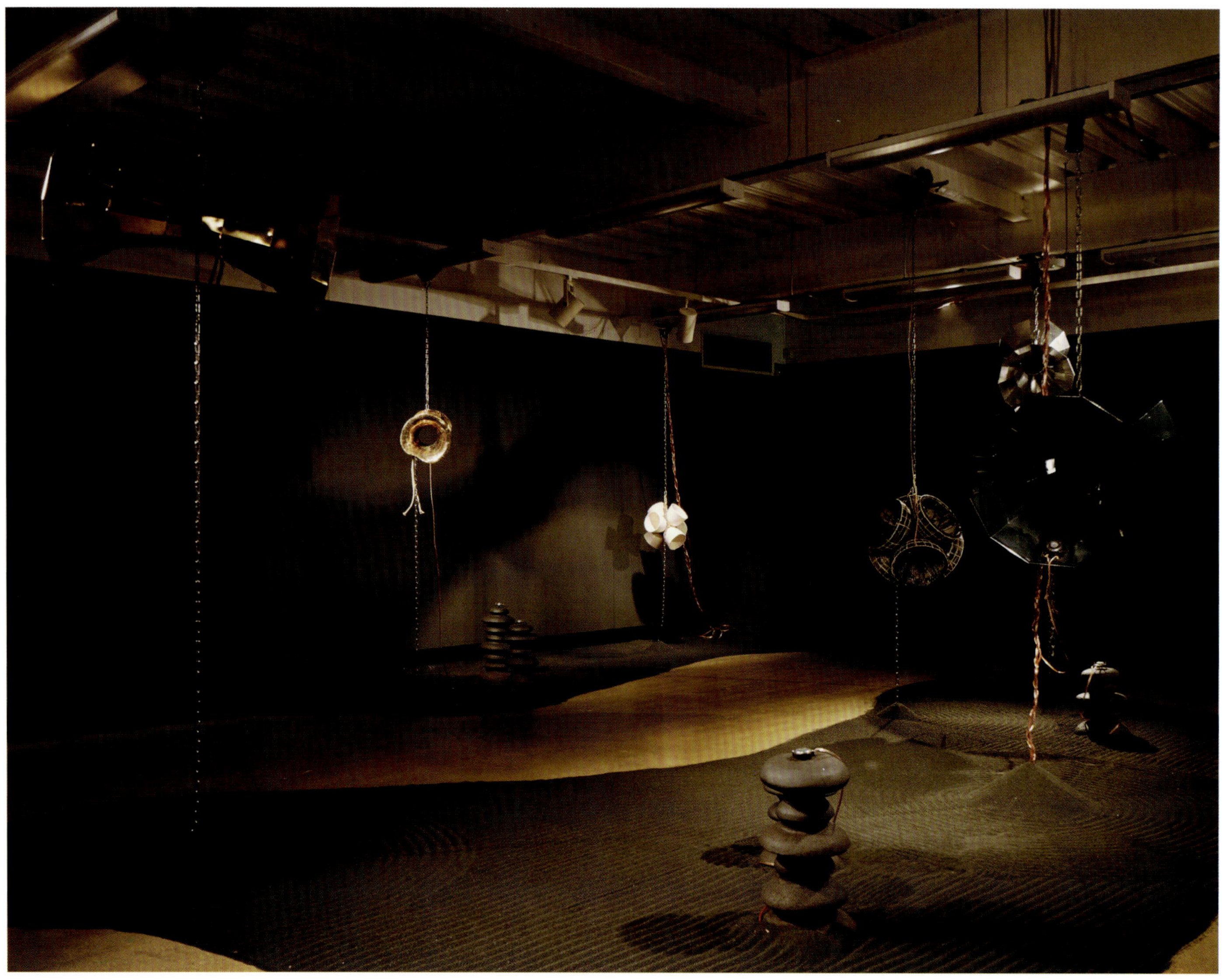

3.119 Yambe Tam (American, born 1989), *Cosmic Garden*, 2021. Cast bronze, brass, powder-coated steel, aluminum, Jesmonite, steel chains, hemp rope, audio exciters, feedback microphones, stoneware, volcanic sand. Installation view at Attenborough Arts Centre, Leicester, England. Courtesy of the artist.

eighth-century poem by Li Po, "Zazen on Ching-t'ing Mountain" recalls Tam's outlook:

> The birds have vanished down the sky.
> Now the last cloud drains away.
>
> We sit together, the mountain and me,
> until only the mountain remains.[74]

Tam's *Cosmic Garden* was created to resemble a Buddhist dry garden (figure 3.119). From the ceiling hang several of the artist's sculptures that take the form of bells. One of these sculptures, *Wormhole Bell* (figure 3.120) has feedback microphones that turn the object into a self-resonating instrument, which helps induce a deep state of meditation. In astronomy, a *wormhole* is a hypothetical tunnel that connects separate regions of spacetime. Tam says: "To me, black holes and the speculative, double-ended form of the wormhole are symbols of transformation—whether the breakdown of classical Newtonian physics to general relativity or the spiritual transcendence one feels in contemplative practices like zazen. Physically, traveling into a black hole is obliteration—a return to pure atomic matter. However, in more philosophical and spiritual terms, a wormhole is an unknowable space of no return, a portal to another side of reality."[75]

OPPOSITE
3.120 Yambe Tam (American, born 1989), *Wormhole Bell*, 2018. Cast bronze, 11¾ × 11¾ × 14 in. (30 × 30 × 36 cm). Private collection.

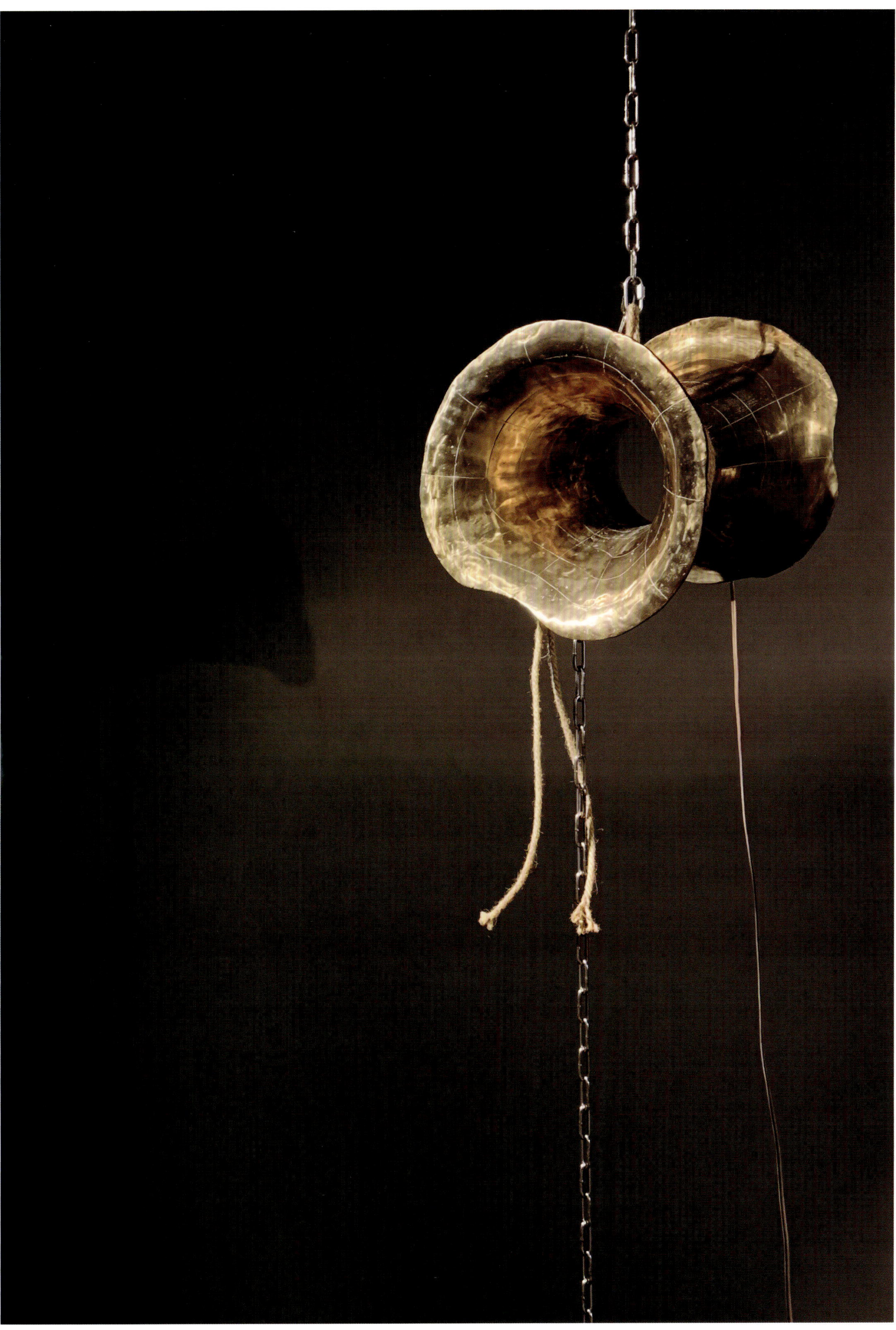

3.121 Xu Bing (Chinese, born 1955), *Table of Square Word Elements* (detail) from *An Introduction to Square Word Calligraphy*, 1994–1996. Woodblock hand-printed book, water-based ink on grass paper, approx. 15⅜ × 9 × 1 in. (39 × 23 × 2.7 cm). Courtesy of Xu Bing Studio. © Xu Bing.

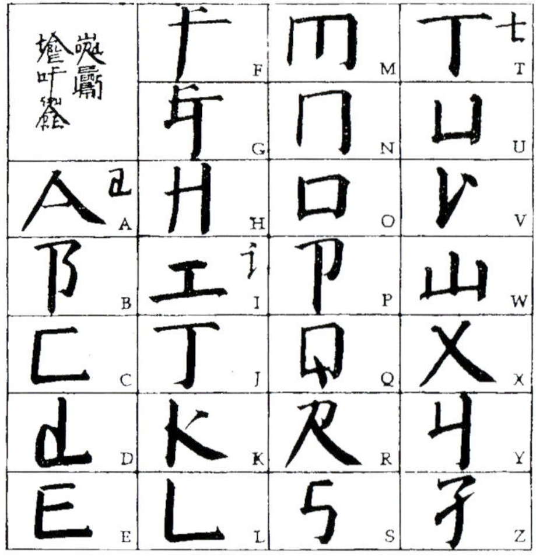

Transformation and translation are themes in the work of Xu Bing, a Chinese artist with studios in Beijing and New York who fuses the Chinese language with Western modernism. In Square Word Calligraphy, a type of writing invented by Xu, the Western alphabet is written to resemble Chinese characters (figure 3.121). Figure 3.122 features a work with four words written in Square Word Calligraphy that can be read as "ART FOR THE PEOPLE." For the 2022 sculpture *Gravitational Arena*, Xu began by selecting a quotation about language from Austrian philosopher Ludwig Wittgenstein (figure 3.123 *top*),[76] which he then rendered in Square Word Calligraphy (figure 3.123 *bottom*). Next, the artist applied gravitational forces at work in a black hole to his translation of Wittgenstein, transforming it into a verbal singularity (figure 3.124). This then became the shape of the finished sculpture (figure 3.125). In figure 3.126 we see the sculpture installed in the Museum of Art Pudong in Shanghai. It's several stories high, and the metaphorical singularity touches the floor, which is a mirror. Looking down from a few floors up, we see the view in figure 3.127. The mirror gives the singularity the appearance of having up-down symmetry as if it were a wormhole, which you can see in the artist's drawing for the sculpture (figure 3.128). Of *Gravitational Arena*, Xu says, "The shape and mirror reflection of the work resemble the 'wormhole model.'…I hope this piece evokes the viewer's thinking on the tensions, interactions, and wrestling-like relationships among different civilizations."[77] In Xu's sculpture, different civilizations are represented by Eastern and Western letter forms, with the wormhole symbolizing translation.

3.122 Xu Bing (Chinese, born 1955), *Square Word Calligraphy: Art for the People*, 2011. Water-based ink on grass paper, mounted on thick archival paper, 13¾ × 43¼ in. (34.8 × 109.8 cm). Courtesy of Xu Bing Studio. © Xu Bing.

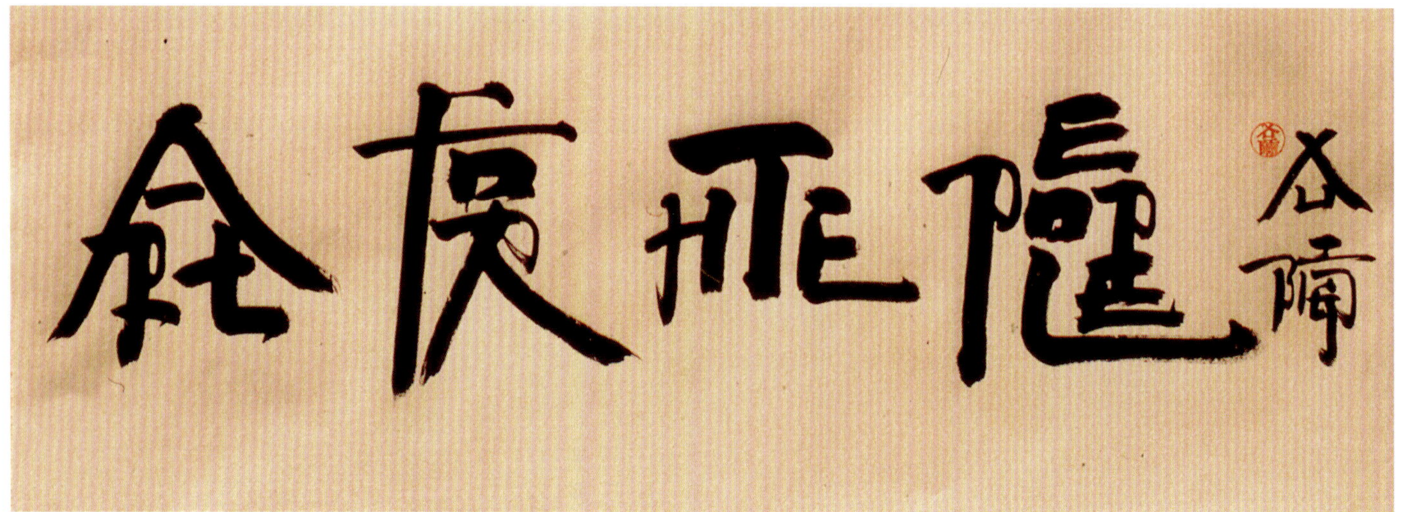

Two uses of the word "see". The one: "What do you see there?"—"I see *this*" (and then a description, a drawing, a copy). The other: "I see a likeness between these two faces"—let the man I tell this to be seeing the faces as clearly as I do myself. The importance of this is the difference of category between the two 'objects' of sight. The one man might make an accurate drawing of the two faces, and the other notice in the drawing the likeness which the former did not see. I contemplate a face, and then suddenly notice its likeness to another. I *see* that it has not changed; and yet I see it differently. I call this experience "noticing an aspect". Its *causes* are of interest to psychologists. We are interested in the concept and its place among the concepts of experience. You could imagine the illustration appearing in several places in a book, a text-book for instance. In the relevant text something different is in question every time: here a glass cube, there an invereted open box, there a wire frame of that shape, there three boards forming a solid angle. Each time the text supplies the interpretation of the illustration. But we can also *see* the illustration now as one thing now as another.—So we interpret it, and *see* it as we *interpret* it. Here perhaps we should like to reply: The description of what is got immediately, i.e. of the visual experience, by means of interpretation—is an indirect description. "I see the figure as a box" means: I have a particular visual experience which I have found that I always have when I interpret the figure as a box or when I look at a box. But if it meant this I ought to know it. I ought to be able to refer to the experience directly, and not only indirectly. (As I can speak of red without calling it the colour of blood.) I shall call the following figure, derived from Jastrow, the duck-rabbit. It can be seen as a rabbit's head or as a duck's. And I must distinguish between the 'continuous seeing' of an aspect and the 'dawning' of an aspect. The picture might have been shewn me, and I never have seen anything but a rabbit in it. Here it is useful to introduce the idea of a picture-object. For instance would be a 'picture-face'. In some respects I stand towards it as I do towards a human face. I can study its expression, can react to it as to the expression of the human face. A child can talk to picture-men or picture-animals, can treat them as it treats dolls. I may, then, have seen the duck-rabbit simply as a picture-rabbit from the first. That is to say, if asked "What's that?" or "What do you see here?" I should have replied: "A picture-rabbit". If I had further been asked what that was, I should have explained by pointing to all sorts of pictures of rabbits, should perhaps have pointed to real rabbits, talked about their habits, or given an imitation of them. I should not have answered the question "What do you see here?" by saying: "Now I am seeing it as a picture-rabbit". I should simply have described my perception: just as if I had said "I see a red circle over there."—Nevertheless someone else could have said of me: "He is seeing the figure as a picture-rabbit." It would have made as little sense for me to say "Now I am seeing it as . . . " as to say at the sight of a knife and fork "Now I am seeing this as a knife and fork". This expression would not be understood.—Any more than: "Now it's a fork" or "It can be a fork too". One doesn't '*take*' what one knows as the cutlery at a meal *for* cutlery; any more than one ordinarily tries to move one's mouth as one eats, or aims at moving it. If you say "Now it's a face for me", we can ask: "What change are you alluding to?" I see two pictures, with the duck-rabbit surrounded by rabbits in one, by ducks in the other. I do not notice that they are the same. Does it *follow* from this that I *see* something different in the two cases?—It gives us reason for using this expression here. "I saw it quite differently, I should never have recognized it!" Now, that is an exclamation. And there is also a justification for it. I should never have thought of superimposing the heads like that, of making *this* comparison between them. For they suggest a different mode of comparison. Nor has the head seen like *this* the slightest similarity to the head seen like *this*——although they are congruent. I am shewn a picture-rabbit and asked what it is; I say "It's a rabbit". Not "Now it's a rabbit". I am reporting my perception.—I am shewn the duck-rabbit and asked what it is; I *may* say "It's a duck-rabbit".

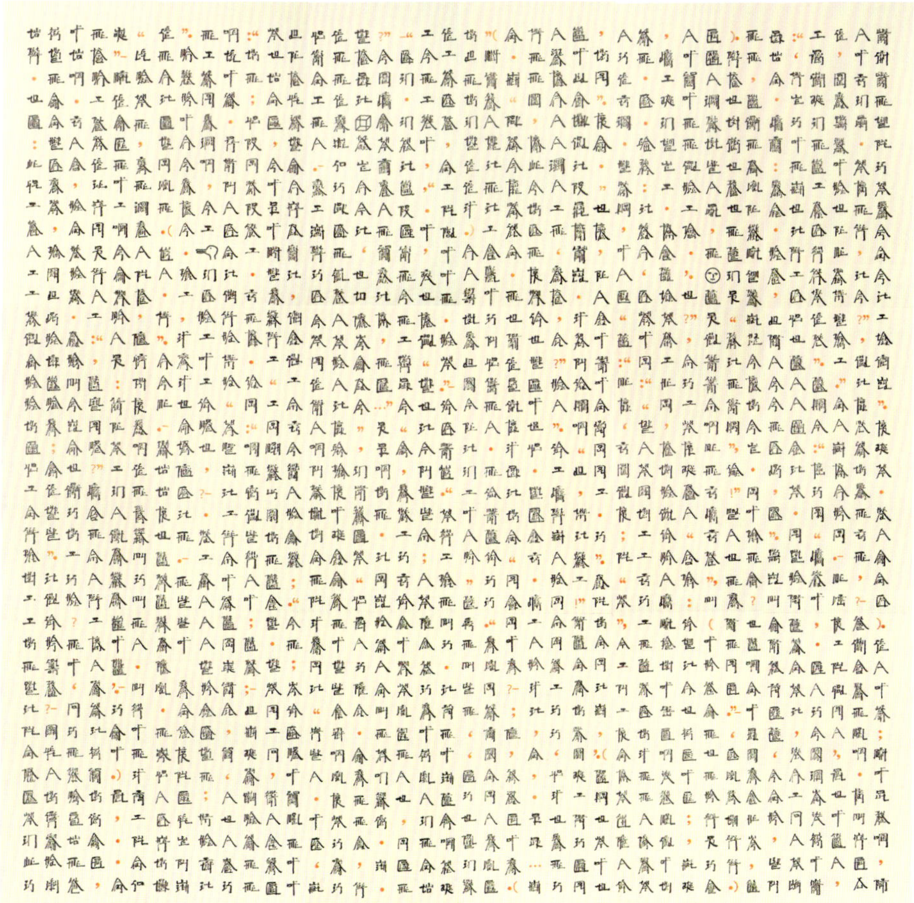

3.123 Xu Bing (Chinese, born 1955), *Gravitational Arena: Text Comparison*, 2021–2022. Mixed media, 34 × 34 in. (86.5 × 86.5 cm). Courtesy of Xu Bing Studio. © Xu Bing.

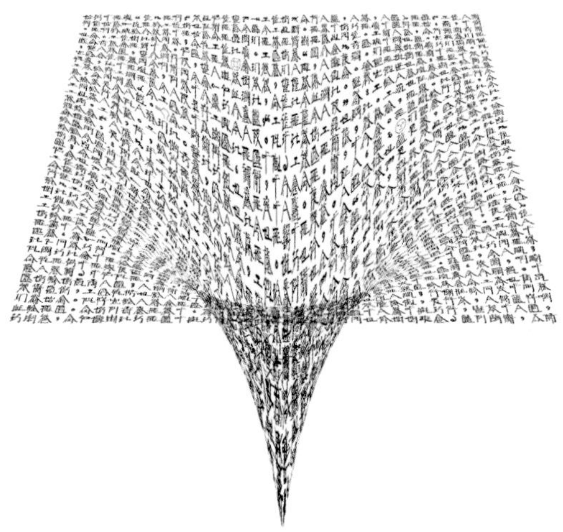

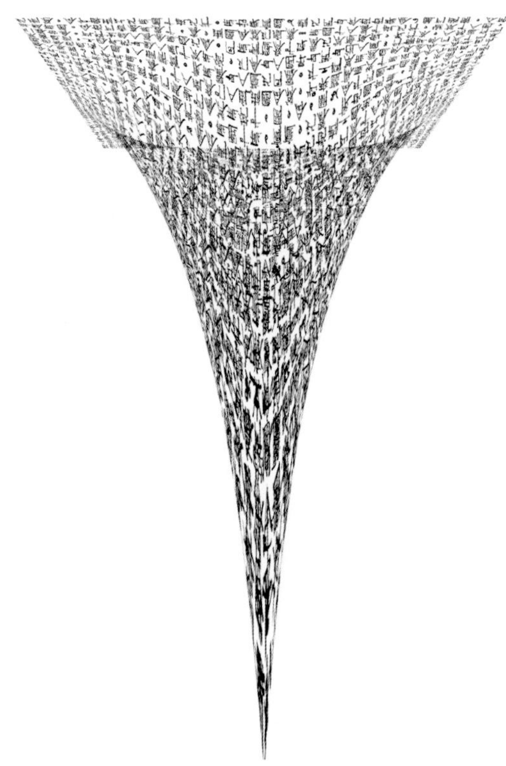

RIGHT
3.124 Xu Bing (Chinese, born 1955), Stills from the animation for *Gravitational Arena*, 2021–2022. Courtesy of Xu Bing Studio. © Xu Bing.

OPPOSITE
3.125 Xu Bing (Chinese, born 1955), Installation plan for *Gravitational Arena*, 2021–2022. Museum of Art Pudong, Shanghai. Courtesy of Xu Bing Studio. © Xu Bing.

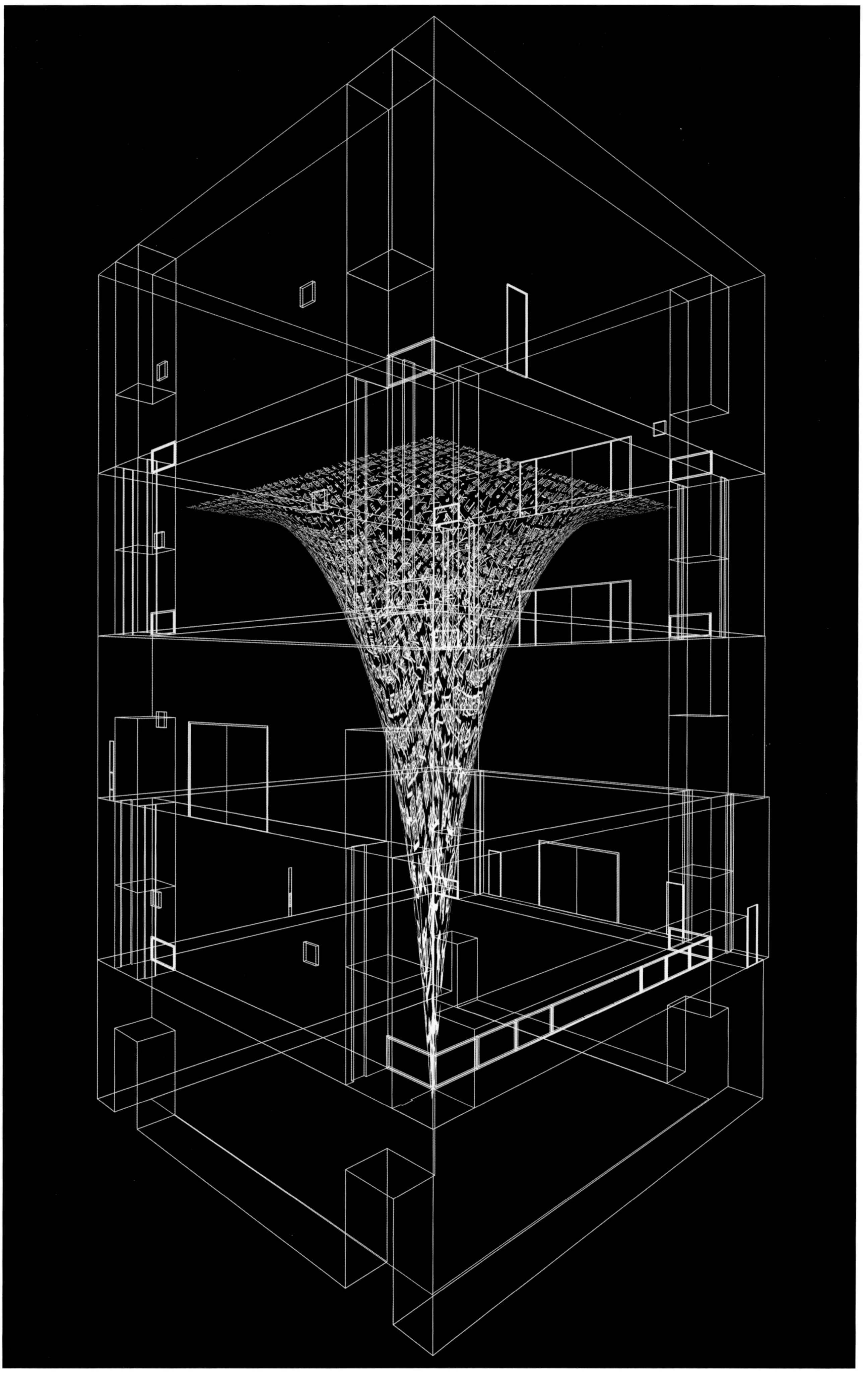

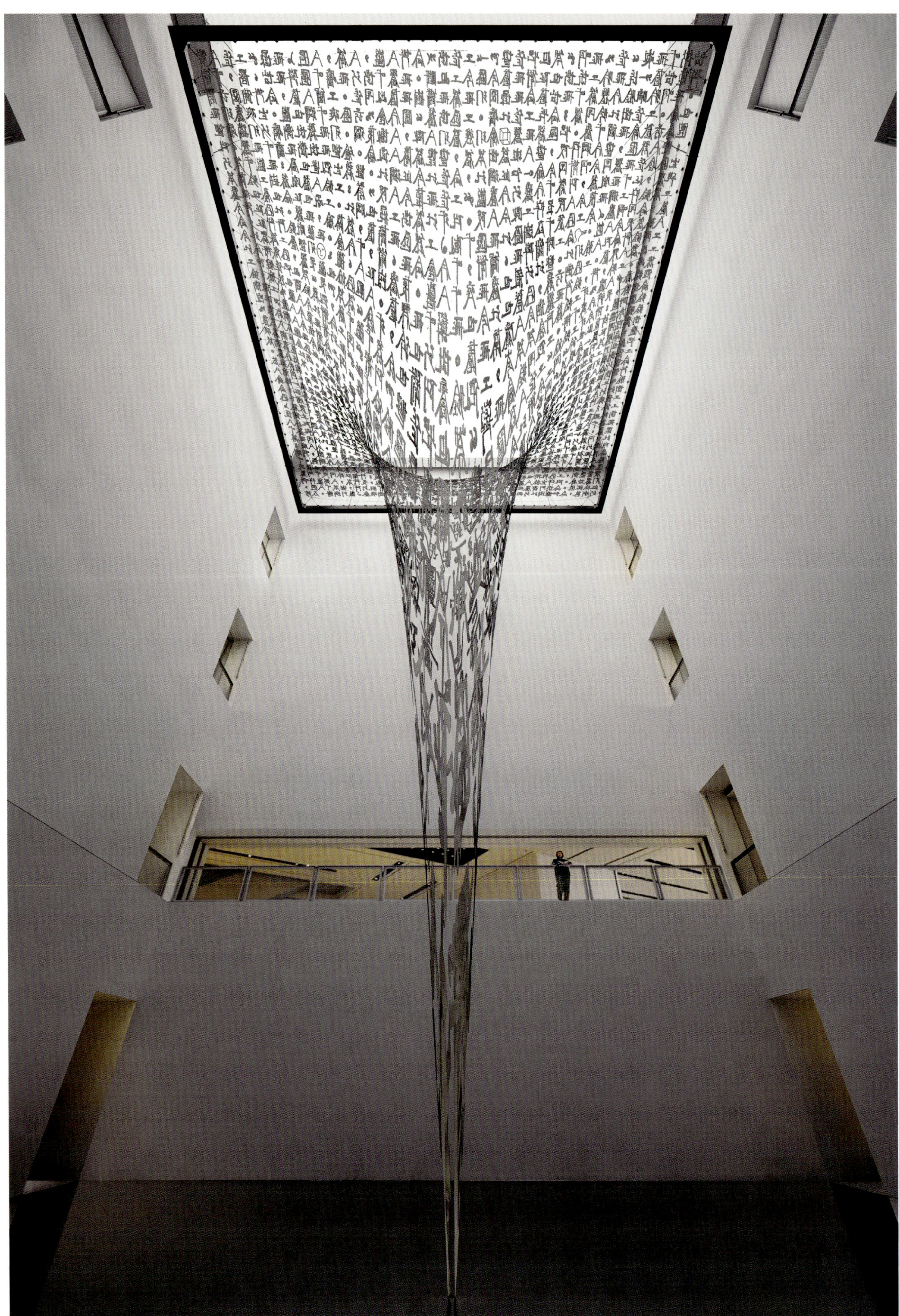

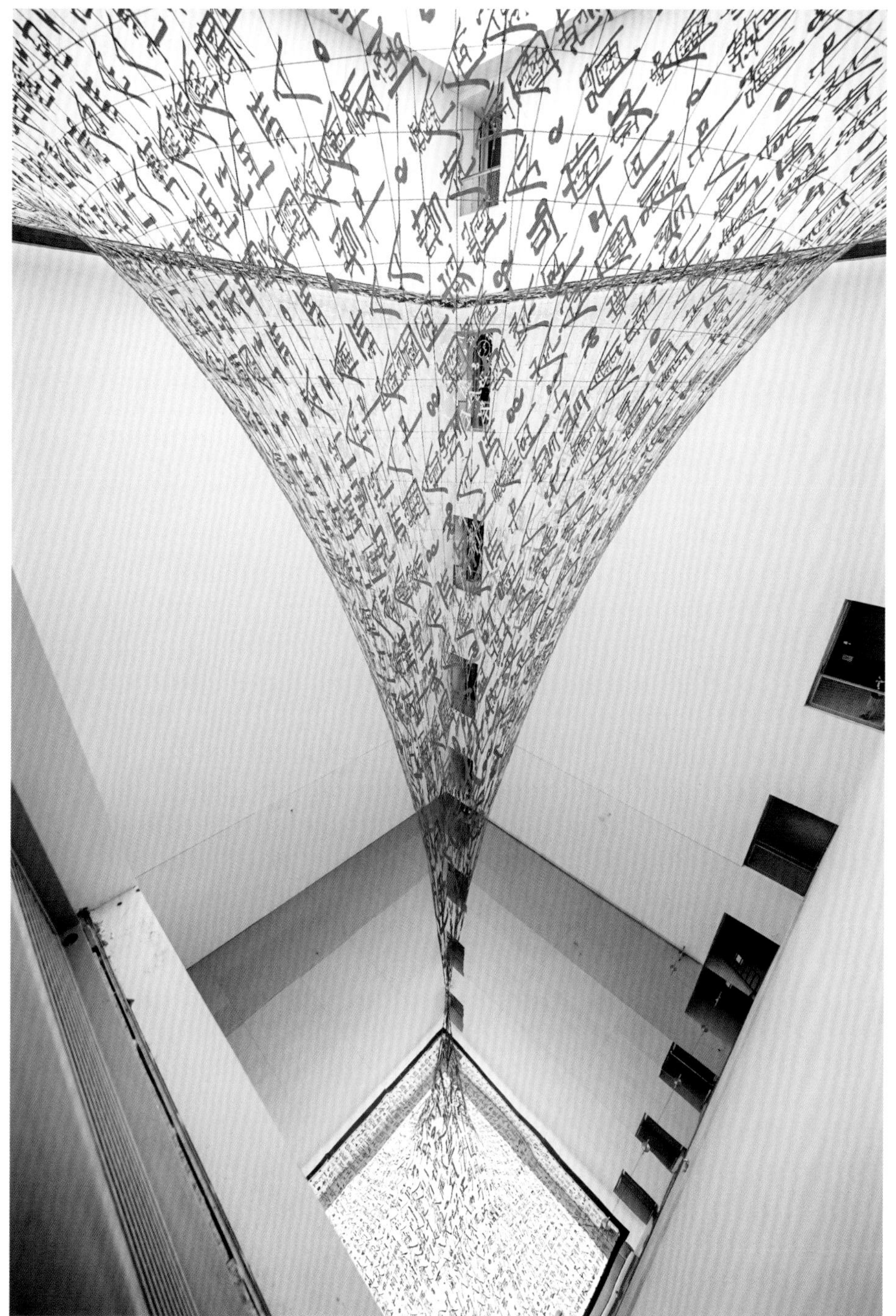

OPPOSITE
3.126 Xu Bing (Chinese, born 1955), *Gravitational Arena* (looking up), 2021–2022. Mixed-media installation, 83 ft. 8 in. × 51 ft. 6 in. × 51 ft. 6 in. (25.5 × 15.7 × 15.7 m). Installation view at Museum of Art Pudong, Shanghai. Courtesy of Xu Bing Studio. © Xu Bing.

LEFT
3.127 Xu Bing (Chinese, born 1955), *Gravitational Arena* (looking down), 2021–2022. Installation view at Museum of Art Pudong, Shanghai. Courtesy of Xu Bing Studio. © Xu Bing.

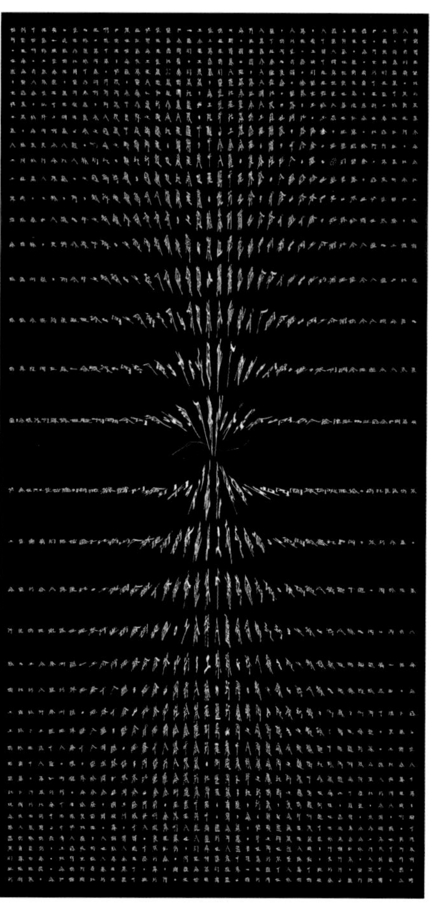

3.128 Xu Bing (Chinese, born 1955), Drawing for *Gravitational Arena*, 2021–2022. Courtesy of Xu Bing Studio. © Xu Bing.

3.129 Opiemme (Italian art collective), *Tribute to Wisława Szymborska*, 2014, from the series *Vortex*. Mural on a building in Gdansk, Poland. Paint. Courtesy of the artists.

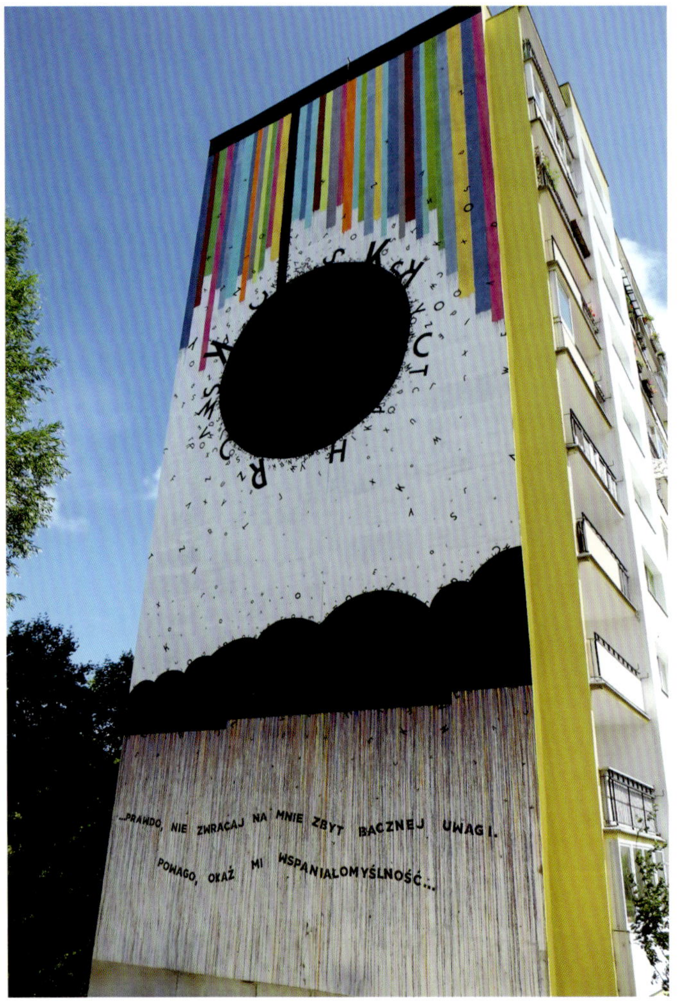

Language is also the subject of a mural in Gdansk, Poland, by the Italian art collective Opiemme (figure 3.129). The mural was inspired by a book by the geneticist Giuseppe Sermonti about the origin of language, *L'alfabeto scende dalle stelle* (The alphabet falls from the stars, 2009). Sermonti said: "The sky is a plastered wall waiting to be painted with primordial signs, stray words that fall like scales from a black (or white) circle, that can be a new moon or a dawning sun. Or a black hole."[78] Opiemme dedicated the mural to a master of language, the Polish poet Wisława Szymborska, and painted letters falling from a black hole that form words from a Szymborska poem. The text translates as "Truth, do not pay me too much attention. Solemnity, be magnanimous toward me."[79]

The skyline of the city of Shenzhen, China, is shown in figure 3.130 *top*. Cao Yuxi (James Cao) is a computer artist who was able to control the lights in forty buildings shown in the photograph, and he did a performance with light and sound (figure 3.130 *center* and *bottom*). Cao has also created an artwork about a black hole that he titled *Oriens* (Latin for "Orient"), giving it the subtitle *Immersive Black Hole* because the viewer is able to walk around in the space of the artwork. His projection of a sphere on the wall suggests a black hole (figure 3.131 *top*). A circle symbolizing the event horizon is projected on the floor (figure 3.131 *center*), and flashing, curving lights communicate distortions in spacetime near the black hole (figure 3.131 *bottom*).

Opaque black panels line the outside of Signe Heinfelt's room-size artwork titled *Black Hole* (figure 3.132 *top*); on the inside, the walls are made of flexible mirrors that distort the viewer's reflection as a black hole warps spacetime (figure 3.132 *bottom*). Wherever the viewer looks, the distortions

OPPOSITE

3.130 Yuxi Cao (James Cao; Chinese, born 1990), *Shenzhen Citizen Square*, 2020. Performance in sound and light in Shenzhen, China. Courtesy of the artist.

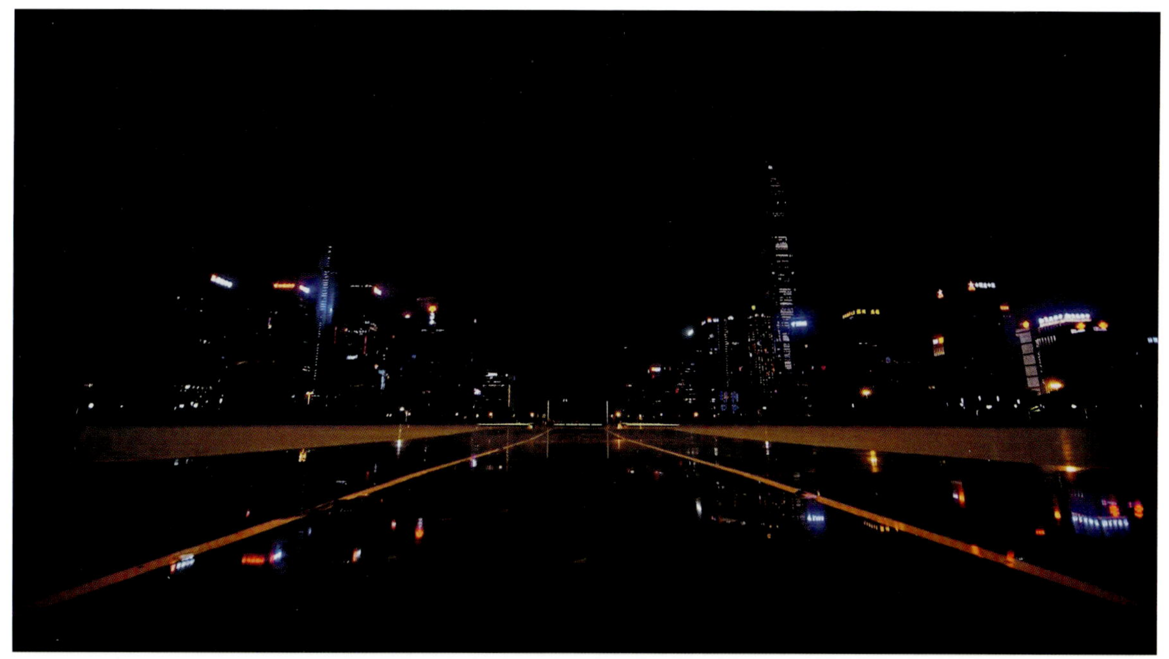

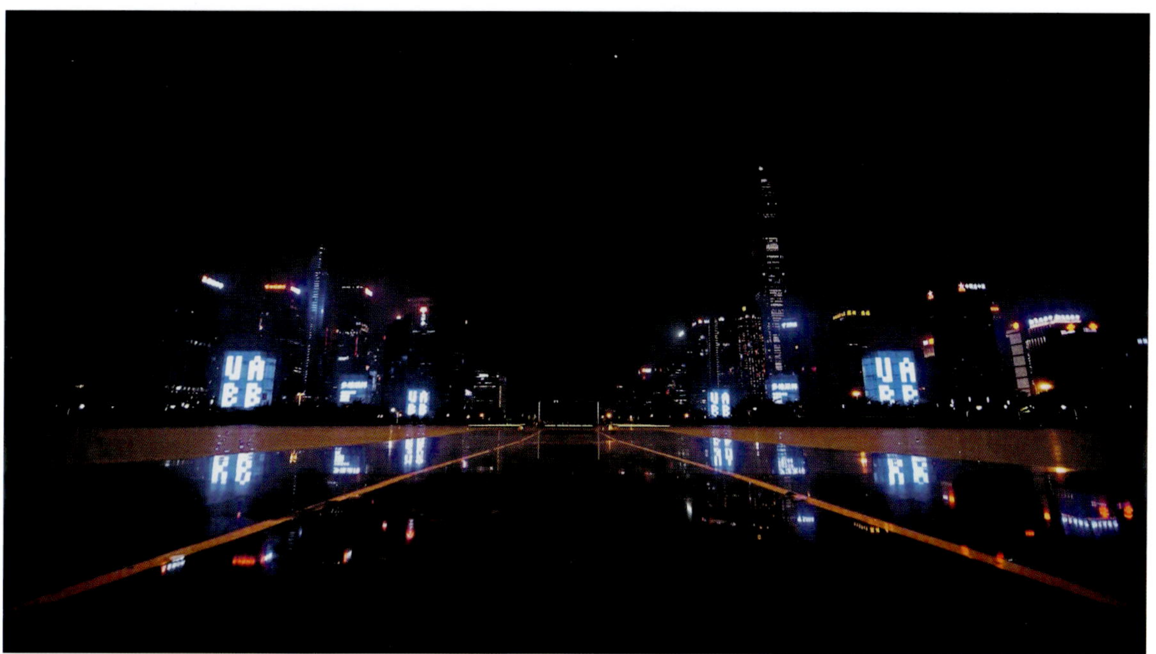

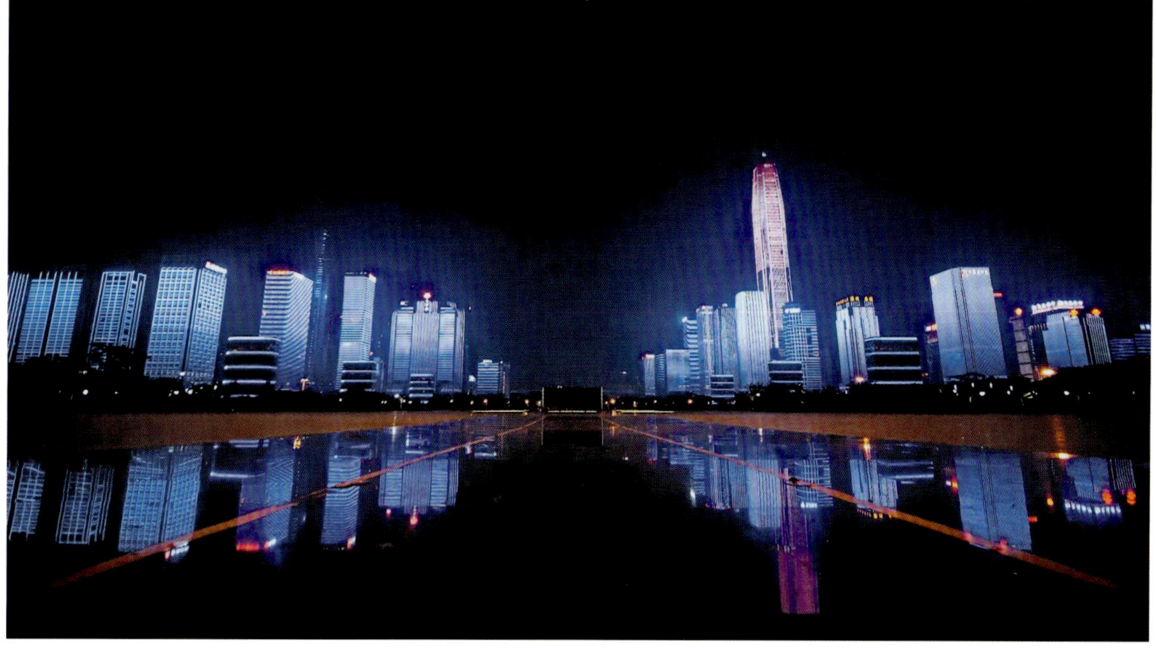

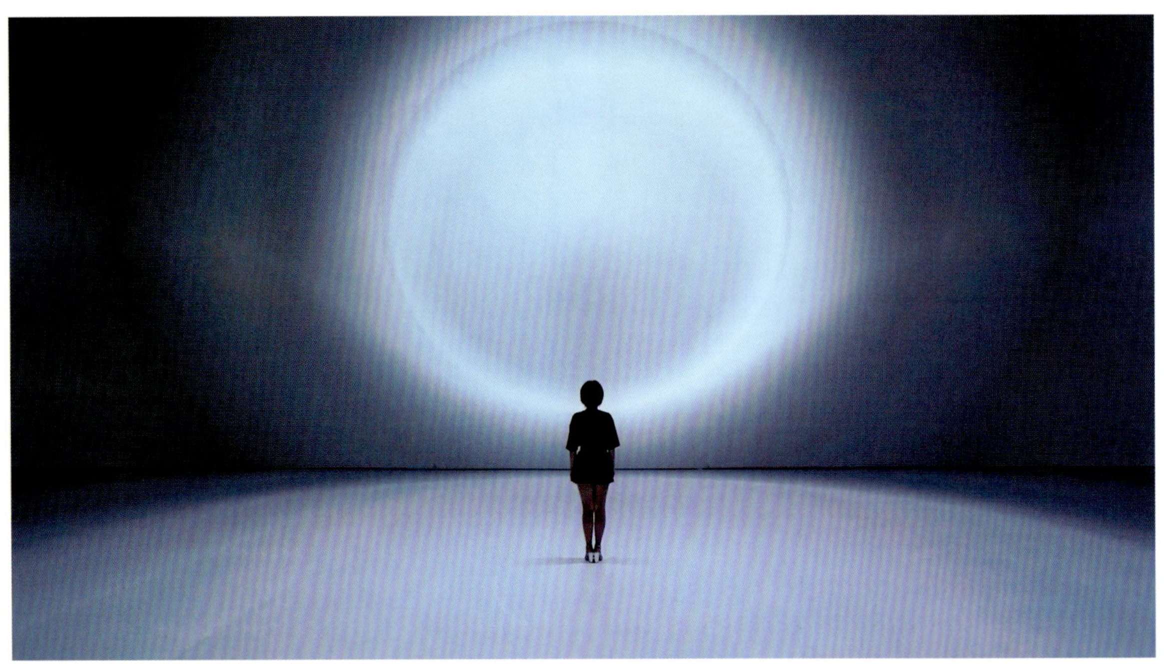

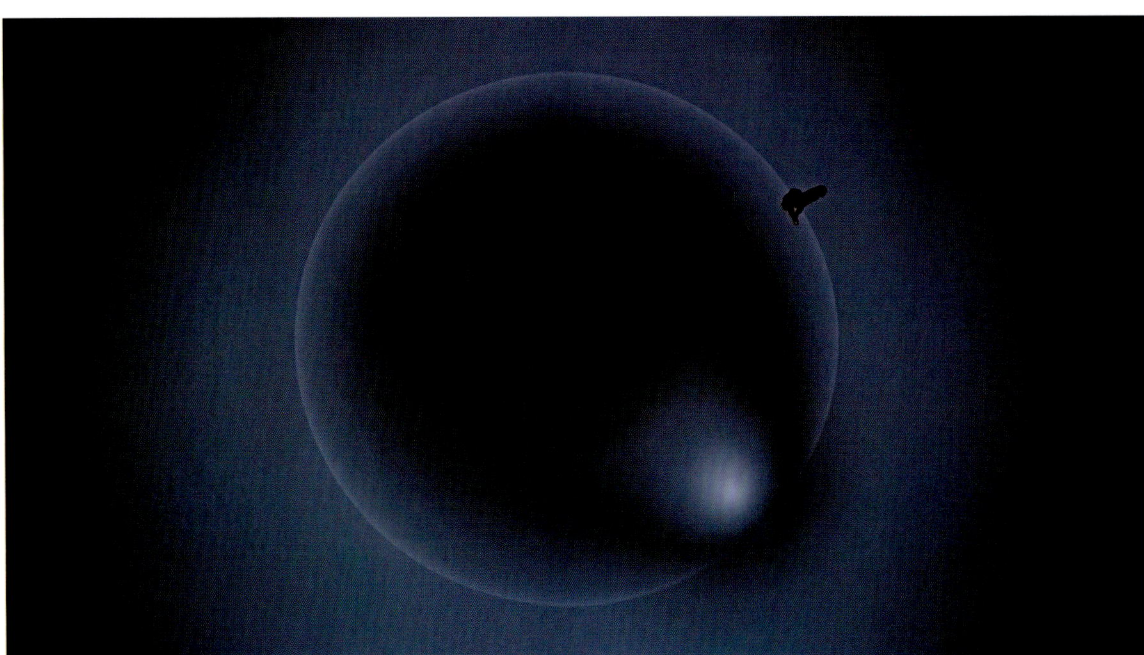

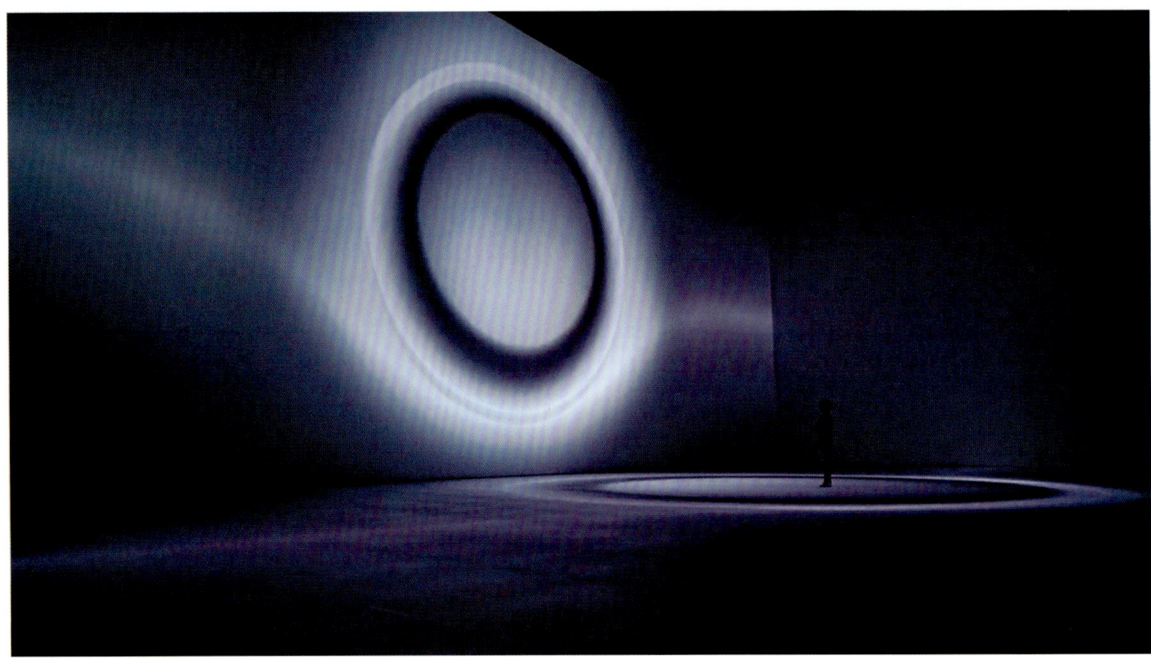

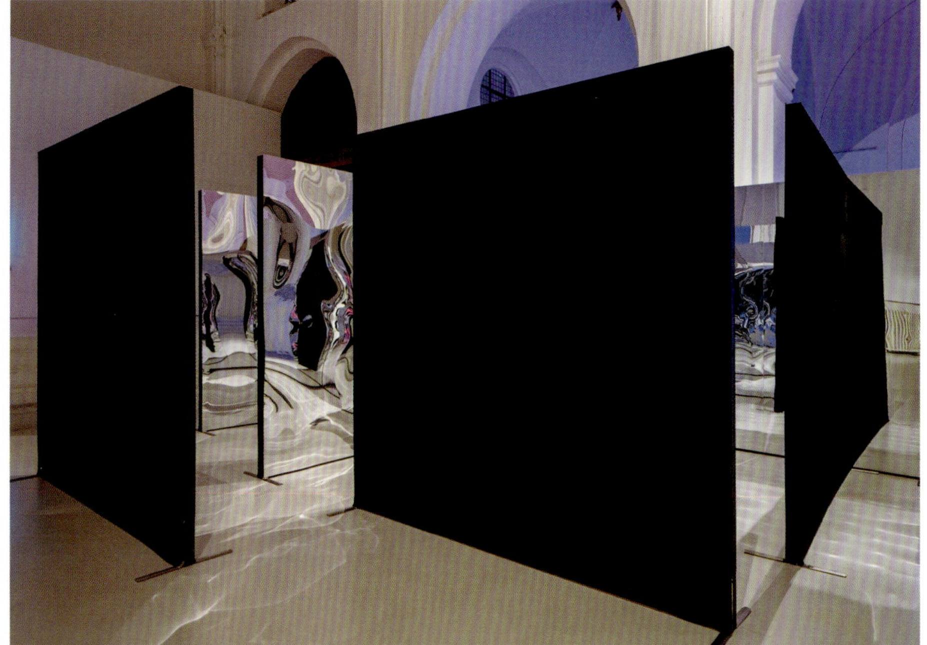

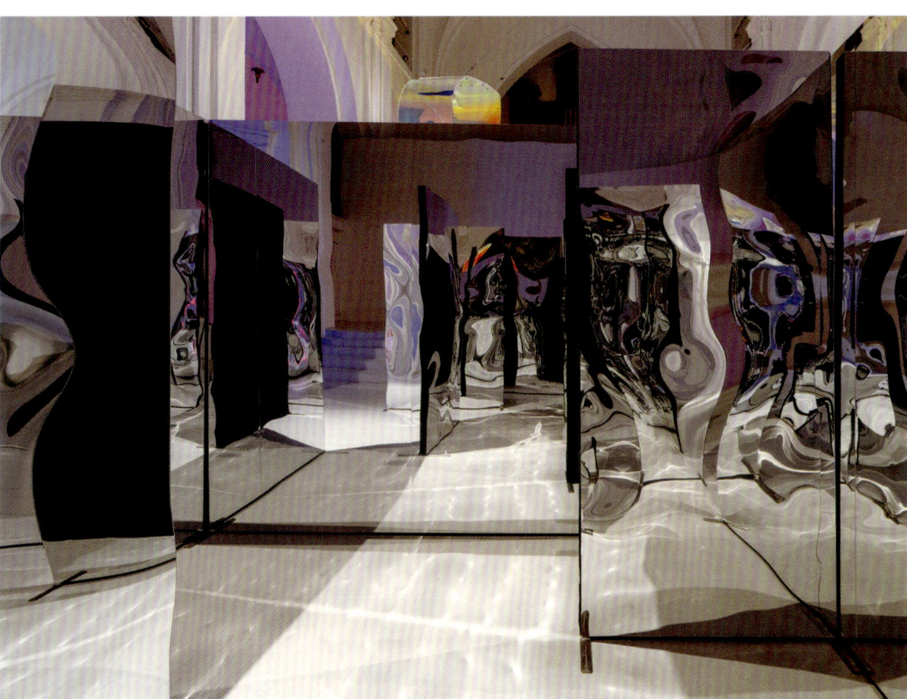

OPPOSITE
3.131 Yuxi Cao (James Cao; Chinese, born 1990), *Oriens: Immersive Black Hole*, 2017. Sound and video installation. Installation view at Today Art Museum, Beijing. Courtesy of the artist.

LEFT
3.132 Signe Heinfelt (Danish, born 1974), *Black Hole*, 2023. Mirror, wood, textile. Installation view at Nikolaj Kunsthal, Copenhagen.

are inescapable. Viewers are invited to step inside another black hole in *Distortions in Spacetime* (figure 3.133), a video projection by the English art collective with the colorful name Marshmallow Laser Feast. The viewer's gravity metaphorically causes the black hole to respond with primordial energy.

In *Microcosmoses* by the international art collective teamLab, myriad lights run continuously along tracks at varying heights (figure 3.134). Each light responds to activity in its immediate environment. When a viewer approaches a light, it changes—moving faster or slower, becoming brighter or dimmer, changing color or emitting a sound. The walls and floor are mirrors that reflect light infinitely and create a spectacular environment of moving, ever-changing energy. As such, teamLab's immersive exhibition *Microcosmoses* is a microcosm of nature.

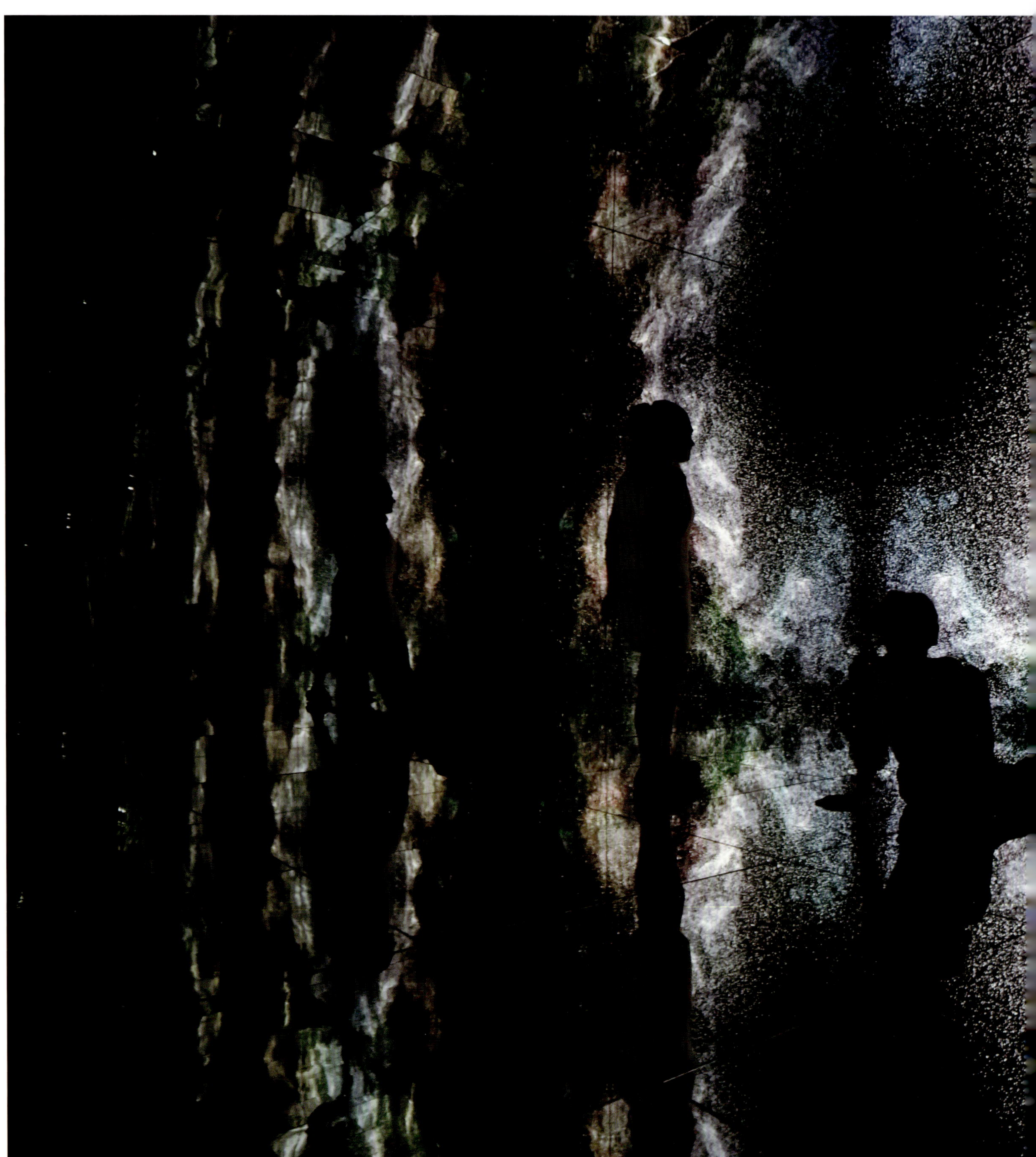

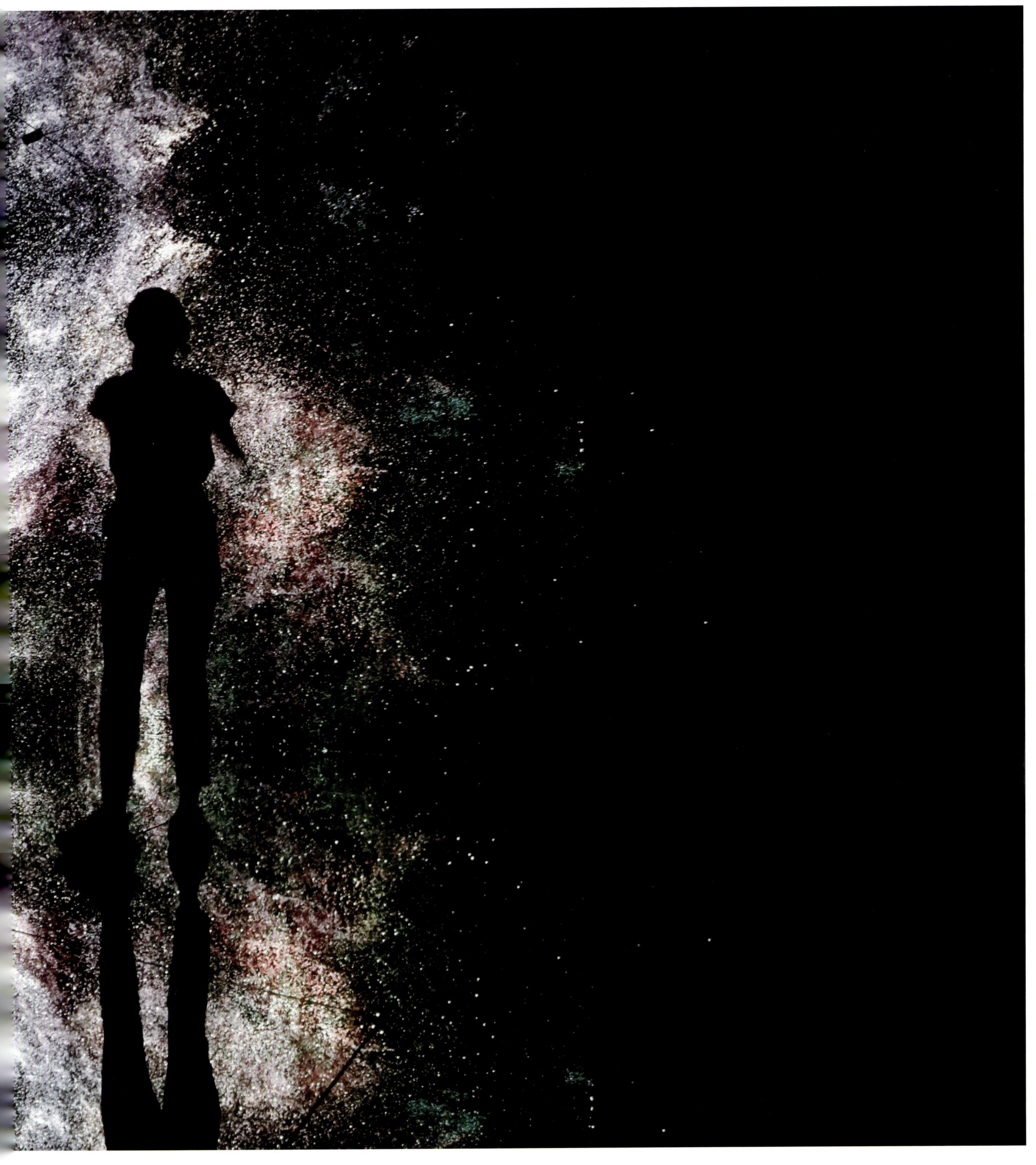

3.133 Marshmallow Laser Feast (English art collective), *Distortions in Spacetime*, 2018–2022. Video projection. Sound design and spatialization by James Bulley. Installation view at Nxt Museum, Amsterdam, 2020–2022.

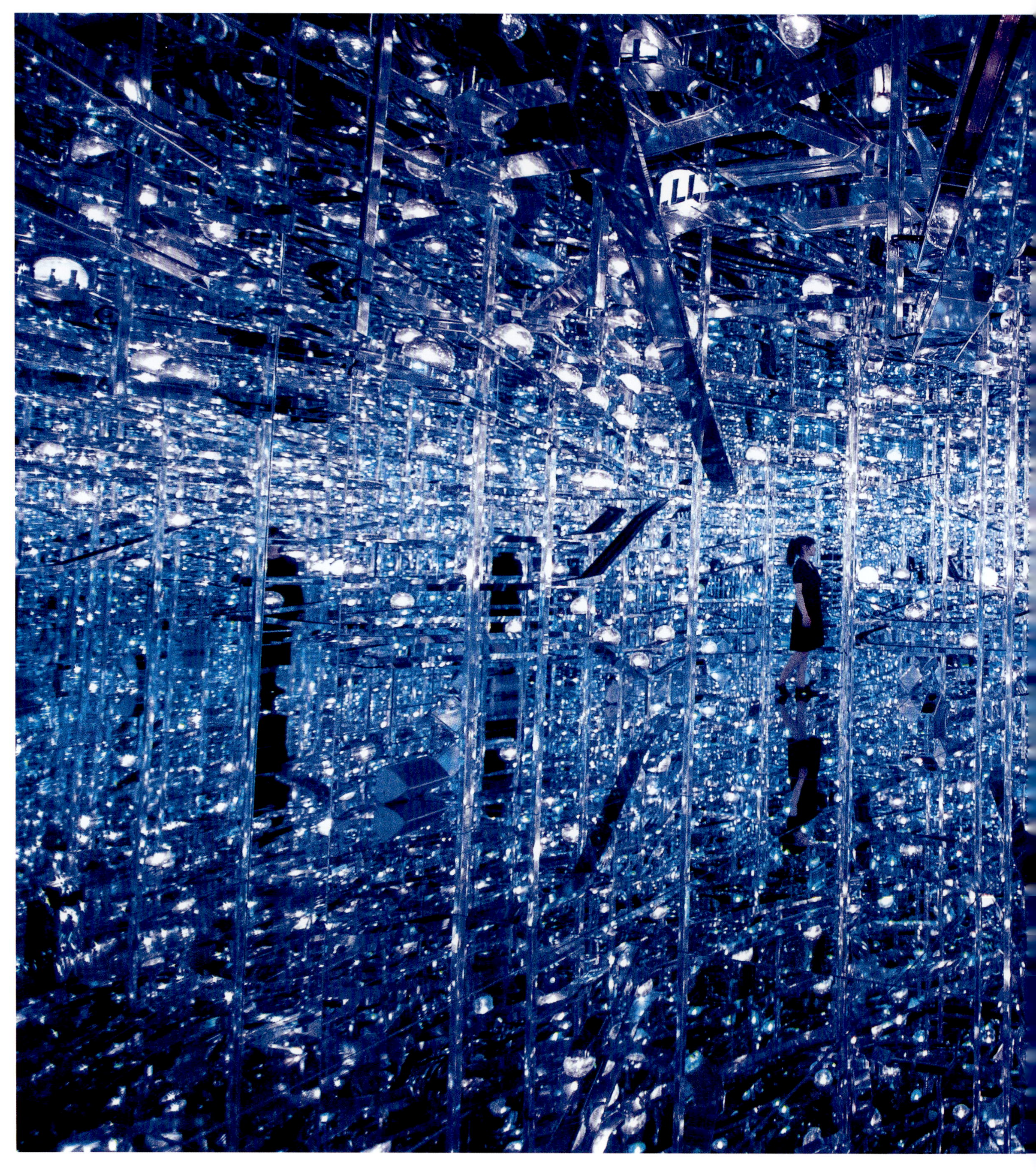

3.134 teamLab (international art collective), *Microcosmoses*, 2019. Interactive installation and LEDs. Sound by teamLab.

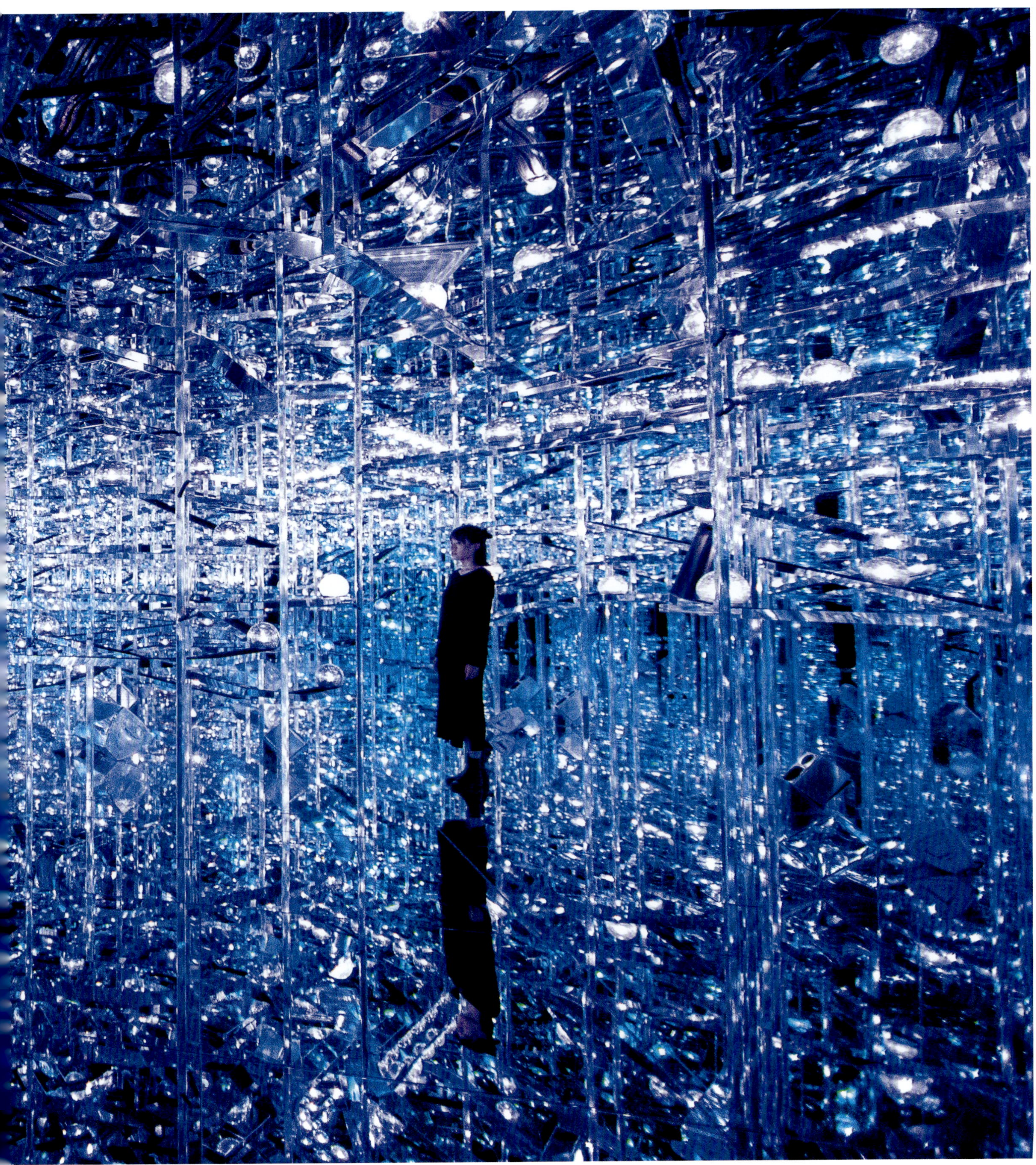

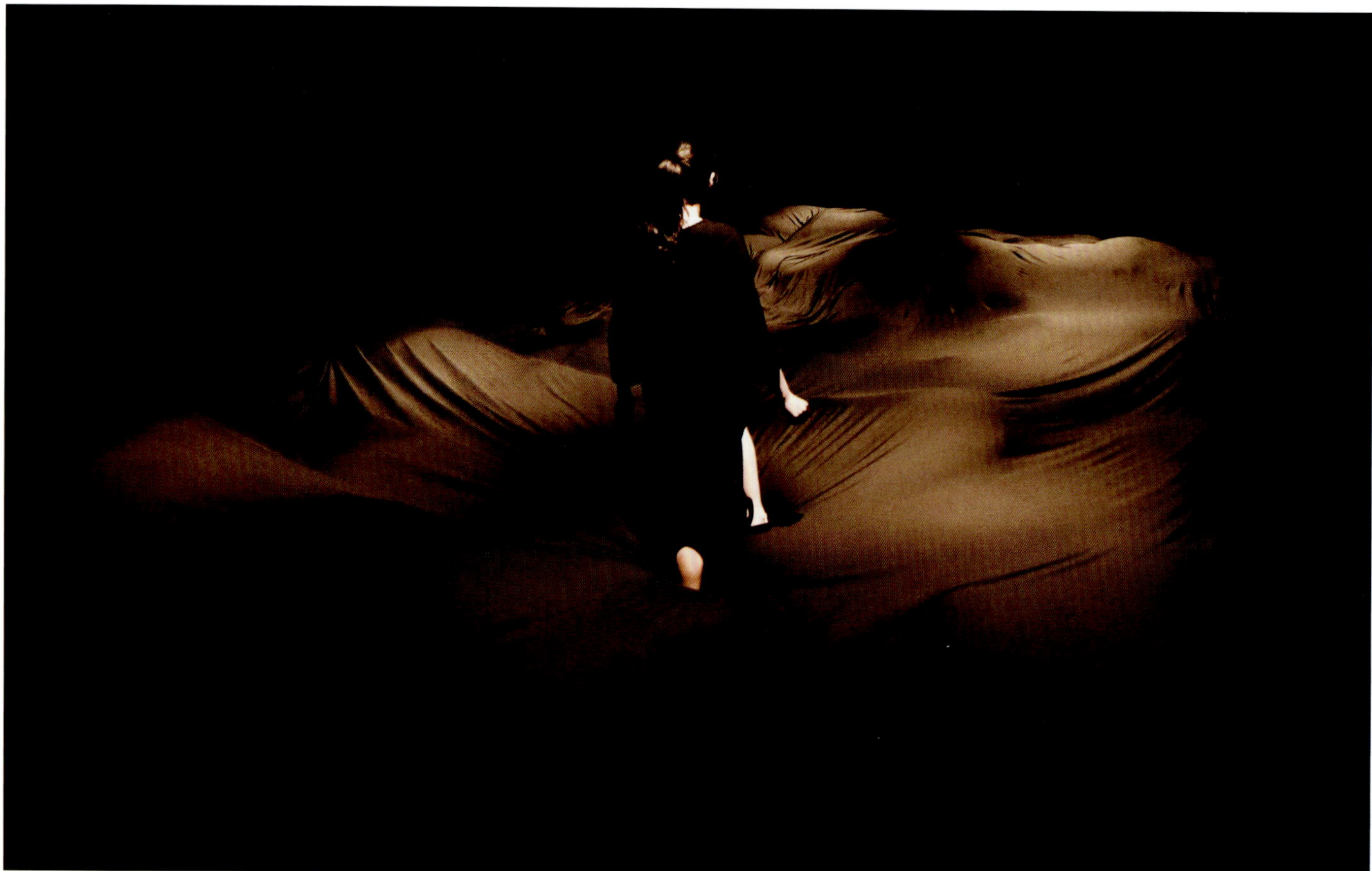

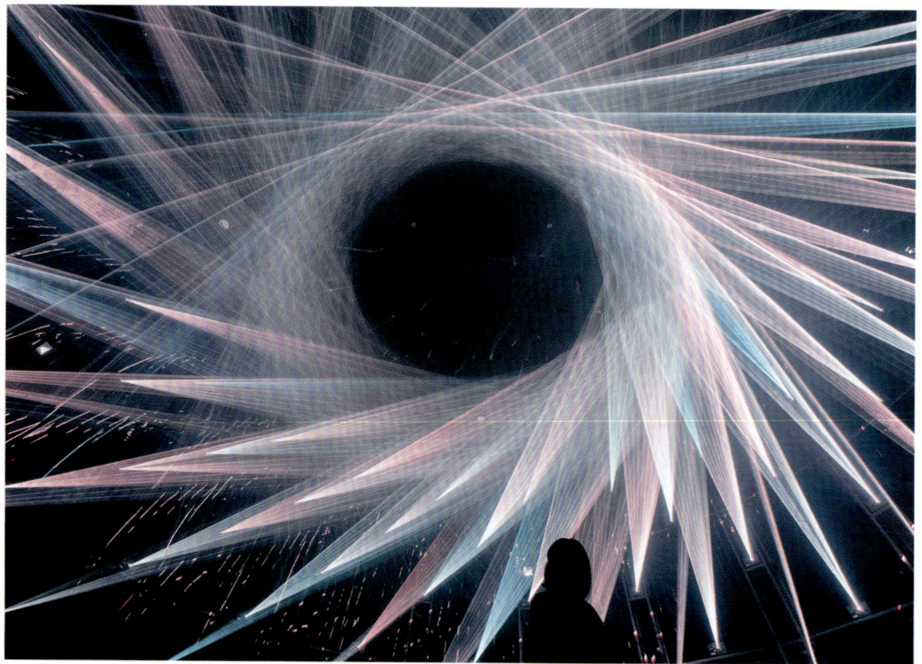

ABOVE
3.135 teamLab (international art collective), *Soft Black Hole—Your Body Becomes a Space that Influences Another Body*, 2016–2018. Sound by Hideaki Takahashi. Installation view at teamLab Planets, Tokyo.

RIGHT
3.136 teamLab (international art collective), *Wormhole*, 2020. Light sculpture. Sound by Hideaki Takahashi. Installation view at The Venetian Macao, Macao, China.

Soft Black Hole: Your Body Becomes a Space that Influences Another Body is a work by teamLab in which viewers walk on soft material covered with black fabric, which ripples and distorts as they move across it, just as a black hole warps spacetime and influences other bodies (figure 3.135). Continuing the analogy, viewers *feel* like a black hole because they affect the space around them and thus gain an intuitive understanding of these mysterious objects. In teamLab's light sculpture *Wormhole* (figure 3.136), the rays of pink and blue light form a pattern that creates a circular hole. The title suggests that the hole is a passageway from the world of the viewer into a distant cosmos.

DIRECT OBSERVATION OF BLACK HOLES BY THE EVENT HORIZON TELESCOPE

A radio telescope is a large parabola-shaped antenna (a *dish*) that records radio waves. The resolving power of a single radio telescope is determined by the diameter of its dish. Two radio telescopes can be synchronized to act as one, in which case the diameter of the dish equals the distance between the telescopes. Synchronization is accomplished by digitally time-stamping the observations of the same wavefront as it is recorded by the two radio telescopes at different locations and times as Earth rotates and then combining them to align the peaks in the radio waves. The technique is called *very-long-baseline interferometry*.

In the late 1990s and early 2000s at the Massachusetts Institute of Technology, a group led by Sheperd Doeleman developed instrumentation that enabled this combined telescope approach to operate at the highest frequencies, targeting angular resolutions that had never been achieved before. In 2007 the group performed an experiment linking telescopes in Hawaii, Arizona, and California, thereby creating a radio telescope with a diameter of more than 2485 miles (4000 km). This virtual telescope detected the supermassive black hole at the center of the Milky Way, and the experiment proved that it was possible to make an image of black hole.[80] Thus was born the Event Horizon Telescope (EHT) Collaboration. Other radio telescopes joined the array to become a global network of eight synchronized radio telescopes (figure 3.137) with a dish equal to the diameter of Earth and a staff of more than two hundred scientists in twenty countries. The EHT's telescopes work in pairs: If the pair is far apart, it can observe small, detailed features; if the pair is close together, it can record large features.

The Event Horizon Telescope team decided to observe M87*, a black hole at the center of galaxy Messier 87 in the Virgo constellation. They chose M87* because it is large (6.5 billion solar masses) and it has a bright, active accretion disk. Although the solar system would fit inside the shadow of M87*, it is very small in the sky—40 microarcseconds—because it's located fifty-five million light-years from Earth. According to a member of the Event Horizon Telescope project, taking a picture of M87* was "like trying to read a newspaper from 5,000 km [3106 miles] away."[81] The Earth-spanning distances between radio telescopes and the high-frequency radio waves (230 GHz) collected by the EHT combined to produce—for the first time—an angular resolution sufficient to image the black hole.

Radio waves moved from M87* largely unimpeded for fifty-five million light-years until they reached Earth, where turbulent water vapor in the atmosphere distorted the incoming radio waves. The condition of this turbulence was different for each telescope because each had a unique humidity, altitude, and temperature. Scientists wrote algorithms to counteract this turbulence and correct the distortion, and they installed synchronized atomic clocks in the telescopes to aid in combining the incoming pieces of wavefront.

The EHT recorded M87* on four nights in April 2017 and then shipped the raw data to MIT's Haystack Observatory in the United States and the Max Planck Institute for Radio Astronomy in Germany, where computer scientists aligned the segments of the data in time. To remove human bias, the EHT project was divided into four groups: Americas, Global, Cross-Atlantic, and East Asia. Each group vowed to turn the pattern of radio

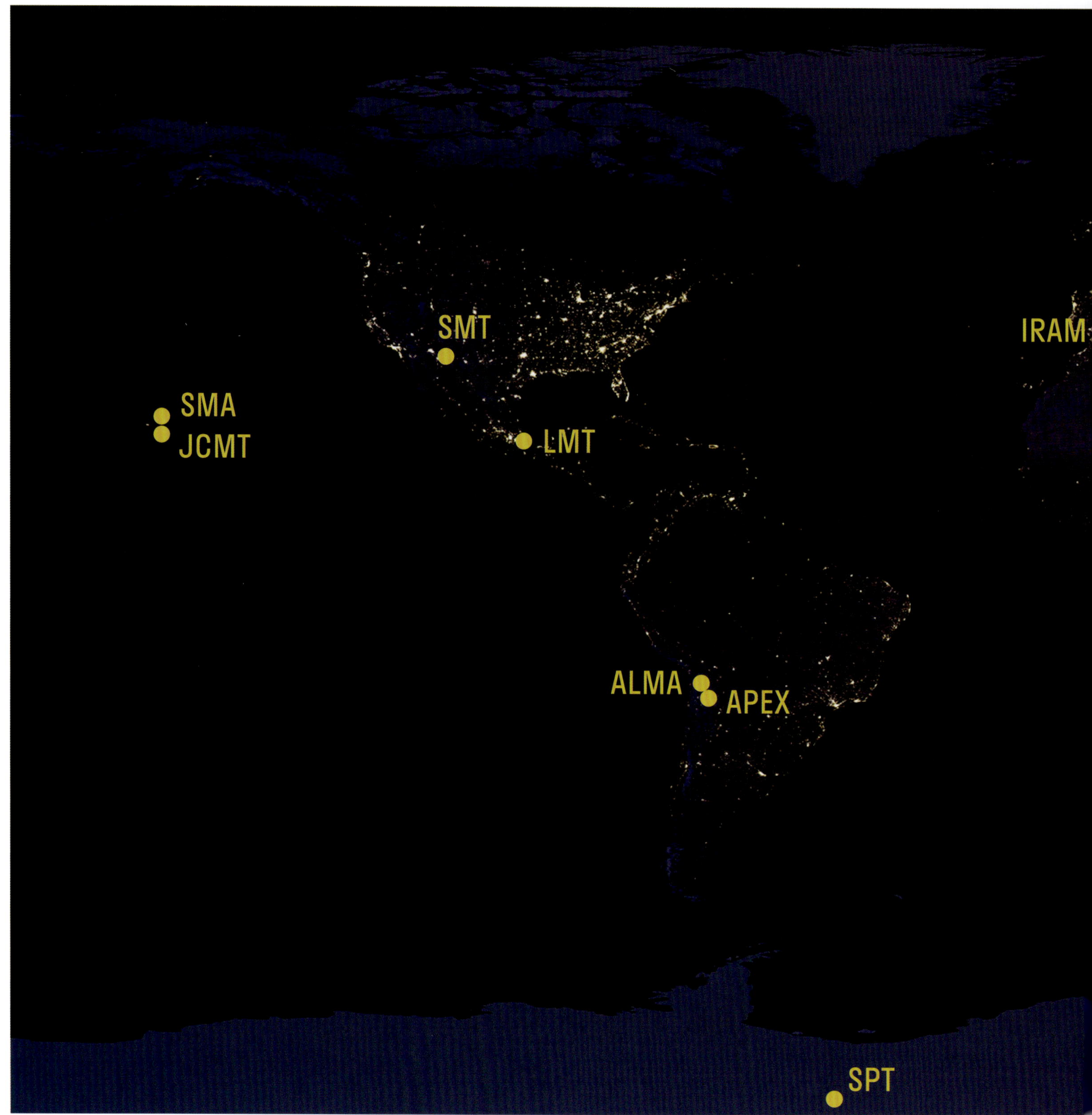

3.137 The Event Horizon Telescope, 2017.

This synchronized array of eight existing ground-based radio telescopes took the observations of M87* in April 2017 on which the first image of a black hole was based. Other telescopes have since joined the collaboration (see figure 3.146).

ALMA: Atacama Large Millimeter/sub-millimeter Array, Cerro Chajnantor, Chile

APEX: Atacama Pathfinder Experiment, Cerro Chajnantor, Chile

IRAM: 30-meter Telescope, Pico Veleta, Spain

JCMT: James Clerk Maxwell Telescope, Maunakea, Hawaii

LMT: Large Millimeter Telescope "Alfonso Serrano," Sierra Negra, Mexico

SMA: Submillimeter Array, Maunakea, Hawaii

SMT: Submillimeter Telescope, Mount Graham, Arizona

SPT: South Pole Telescope, South Pole Station, Antarctica

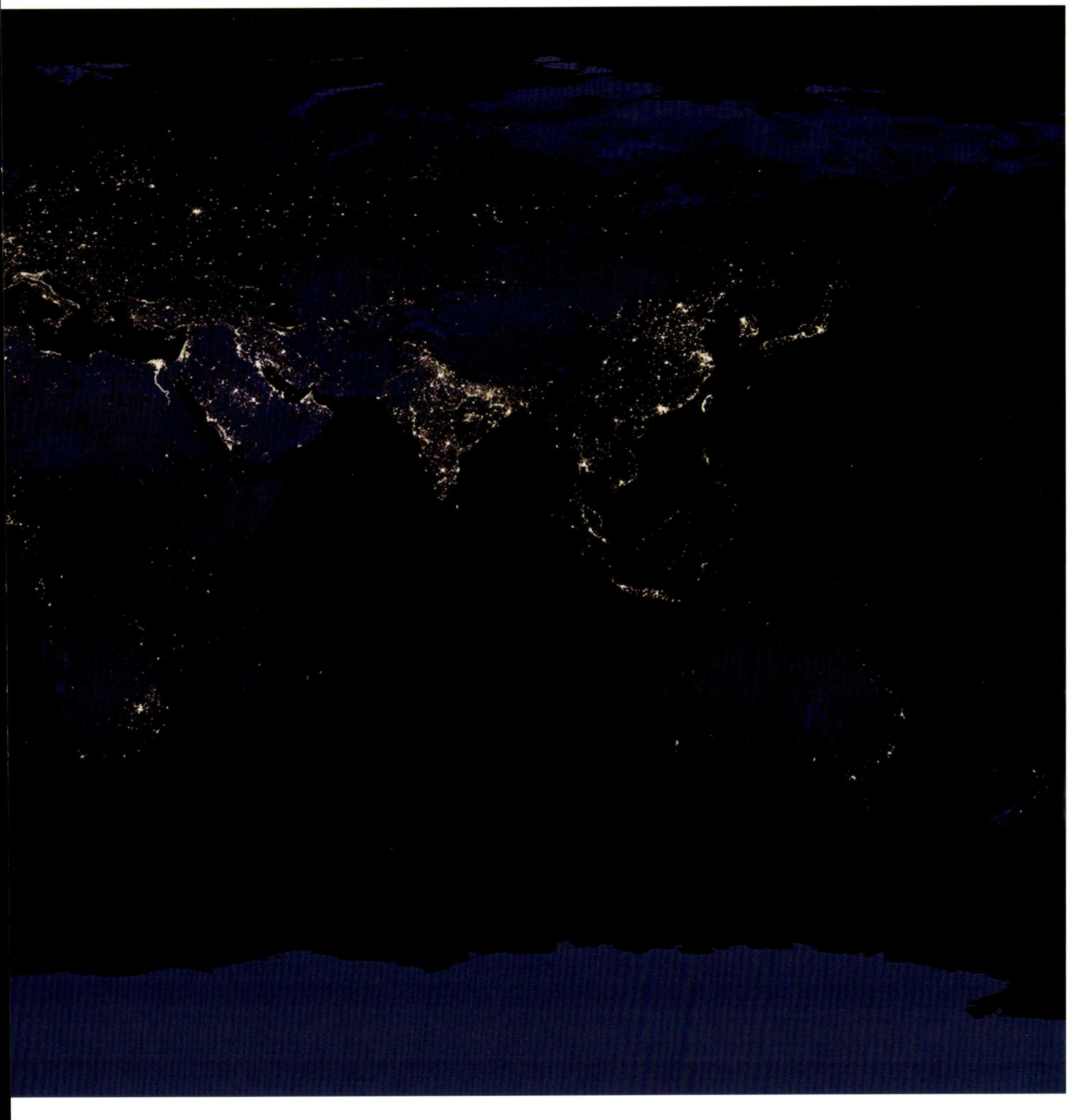

waves into the best image of M87* that it could and, to avoid influencing each other, the groups worked in complete isolation for two months. Three approaches to imaging were used. Two groups employed the CLEAN algorithm, a standard tool used in radio astronomy since the 1970s to create images, and the other groups developed new algorithms for computational imaging: EHT-imaging was developed in the United States; SMILI, in Japan. In July 2018 the four groups met and unveiled their best images, which were similar in key characteristics: size, shape, and the fact that the images were lighter on one side. Confident that they weren't fooling

3.138 M87* by the Event Horizon Telescope, released April 10, 2019. Event Horizon Telescope Collaboration.

themselves, the groups averaged the blurriness of the four images. The color of the resulting image was an "artistic choice."[82] On April 10, 2019, the EHT made front-page headlines around the world when the project released the first image of a black hole (figure 3.138). Billions of people saw the image of M87*, and artists responded in kind (figures 3.139–141).

The EHT team then examined its data for the polarization of radio waves. Looking at the bright accretion disk around M87* with polarized radio waves is like looking at a sunlit landscape with polarized sunglasses.

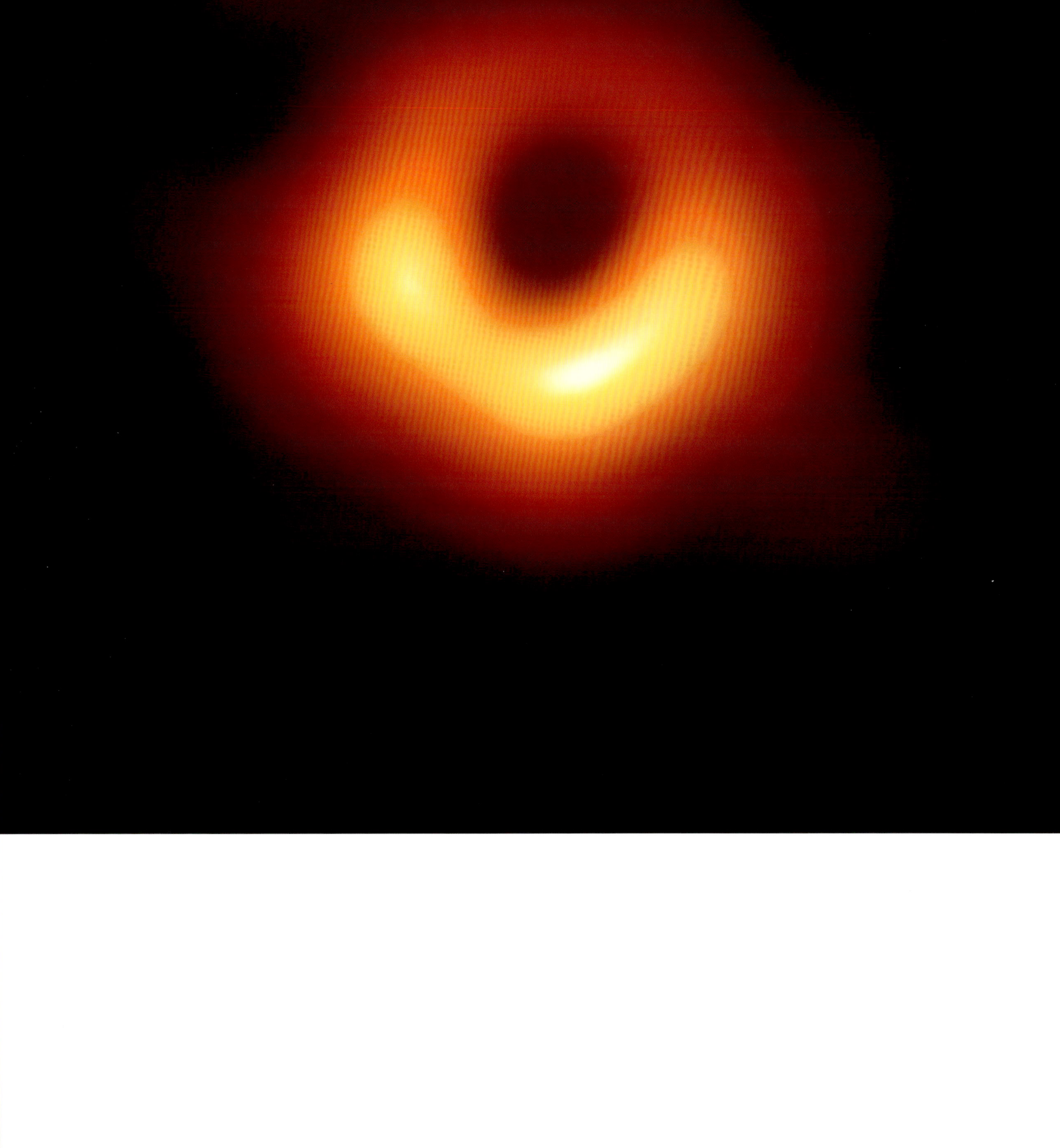

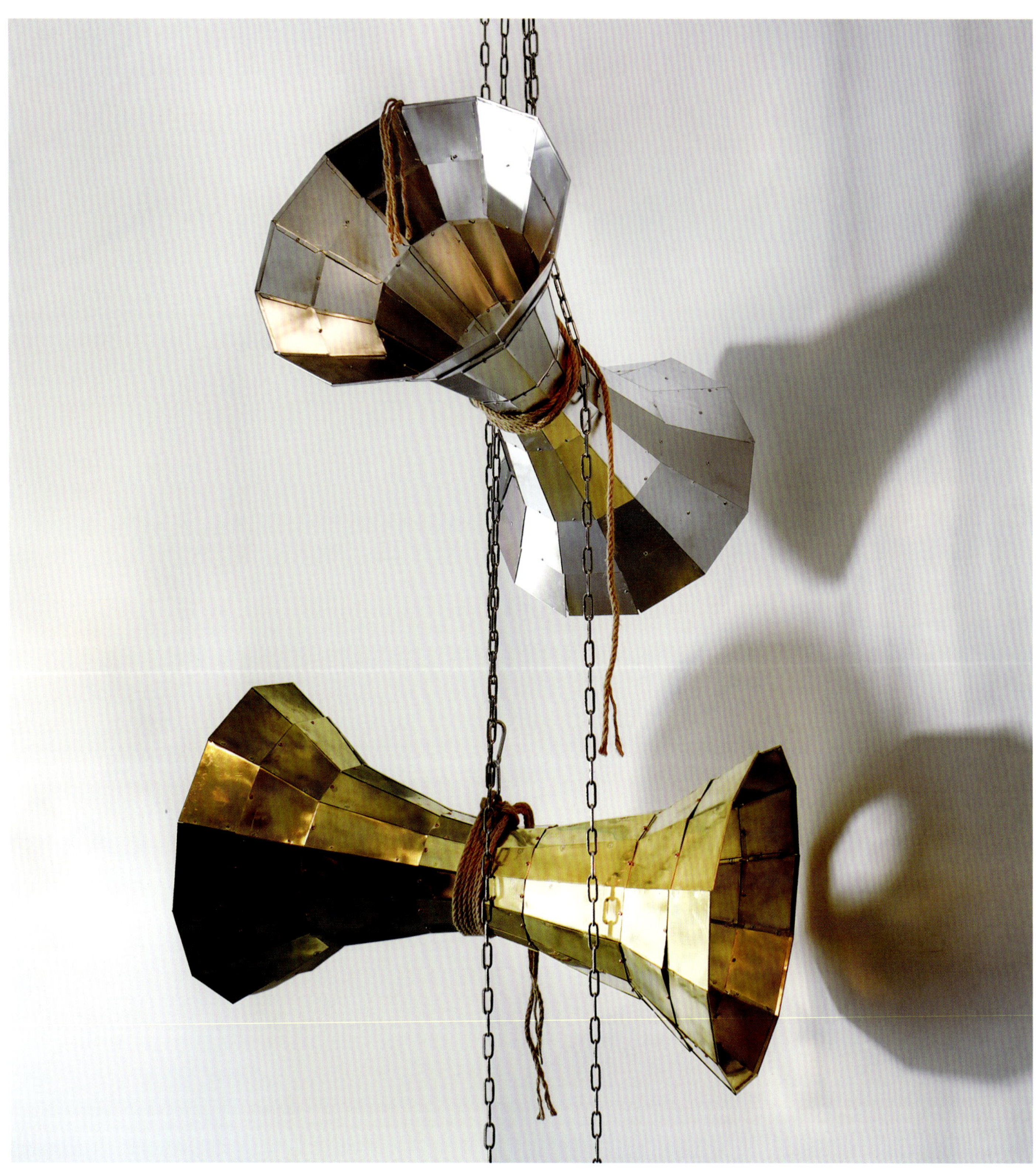

3.139 Yambe Tam (American, born 1989), *M87* Bells*, 2021. Brass, aluminum, hemp rope, steel chains, 24¾ × 30 × 24¾ in. (63 × 76 × 63 cm); 22 × 26¼ × 22 in. (56 × 67 × 56 cm). The Art House, Wakefield, England. Courtesy of the artist.

Tam got the idea for *M87* Bells* when she was staying at the Kanshoji Zen Buddhist monastery in France in 2019. She made them to celebrate the release of the Event Horizon Telescope's image of black hole M87*.

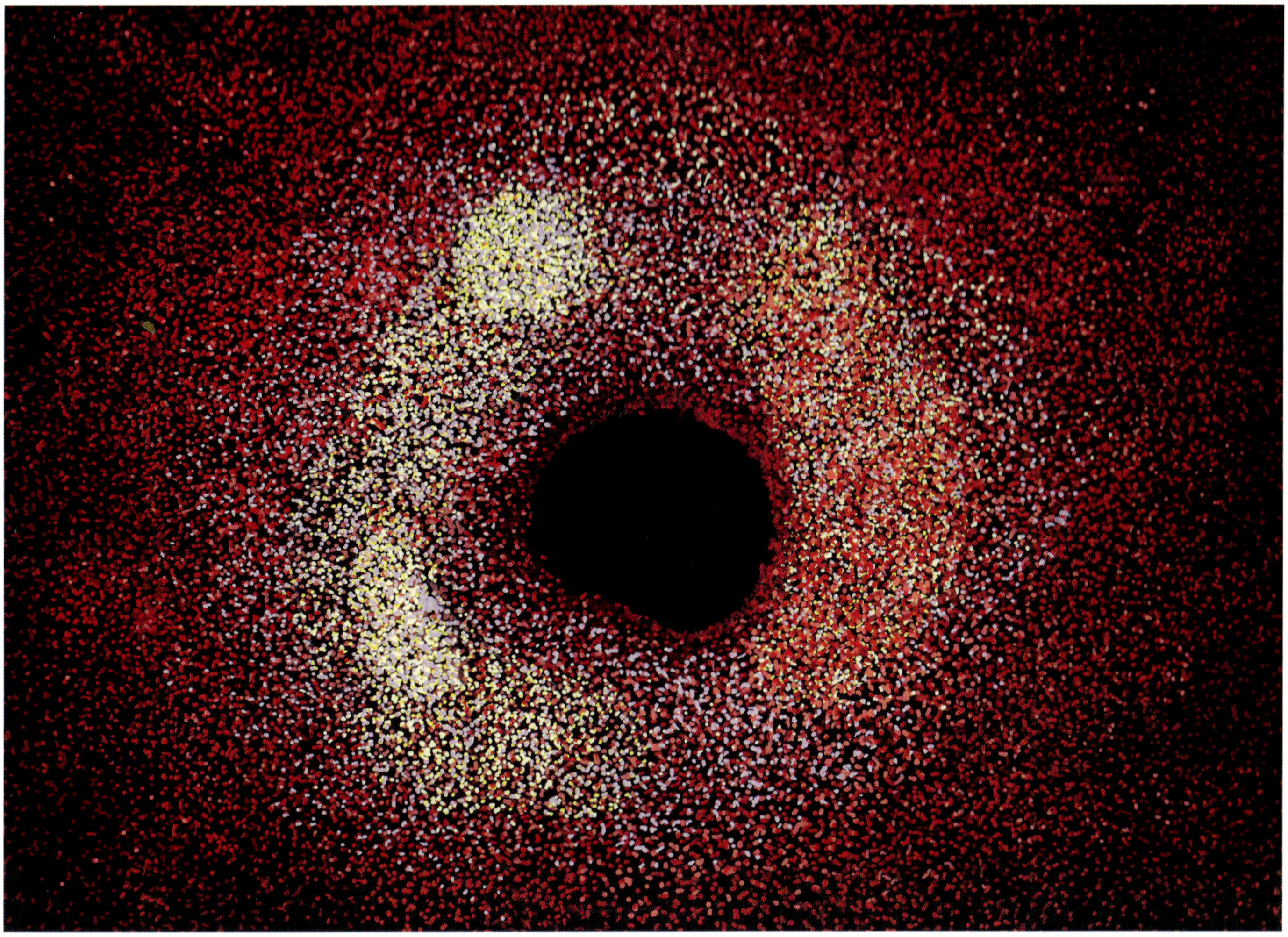

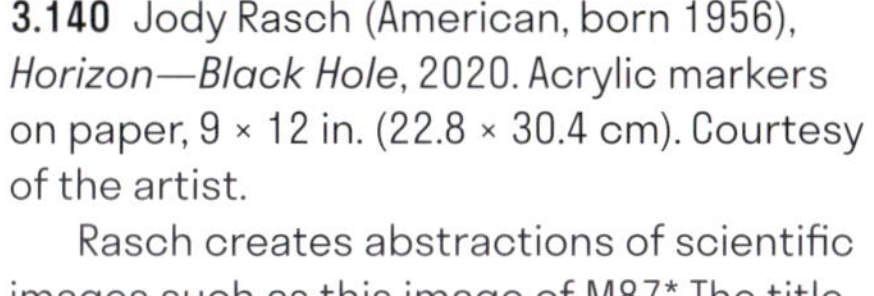

3.140 Jody Rasch (American, born 1956), *Horizon—Black Hole*, 2020. Acrylic markers on paper, 9 × 12 in. (22.8 × 30.4 cm). Courtesy of the artist.

Rasch creates abstractions of scientific images such as this image of M87*. The title refers to the event horizon.

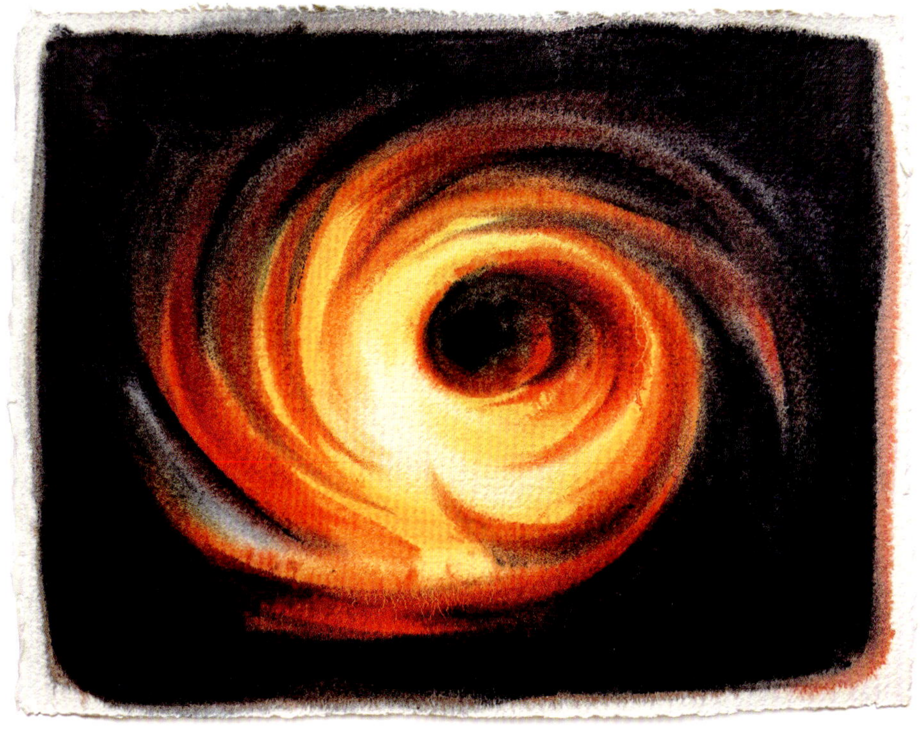

3.141 Ekaterina Smirnova (Russian, born 1981), *Black Hole M87*, 2020. Watercolor on paper, 8½ ×10½ in. (22 × 27 cm). Courtesy of the artist.

With a few quick brushstrokes, Smirnova evoked the rapid motion of the accretion disk around black hole M87*.

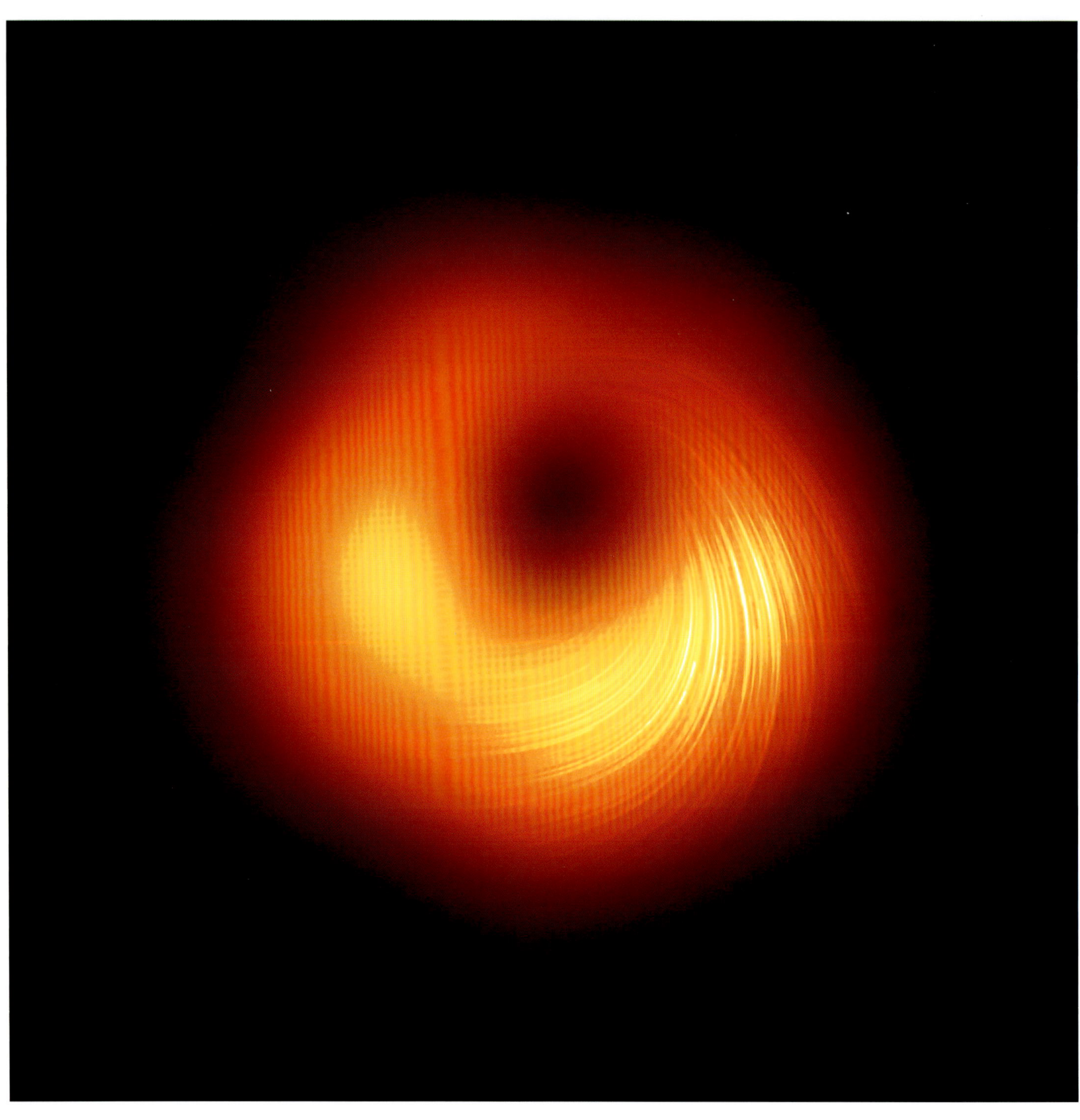

3.142 M87* in polarized light, taken by the Event Horizon Telescope, 2021. Event Horizon Telescope Collaboration.

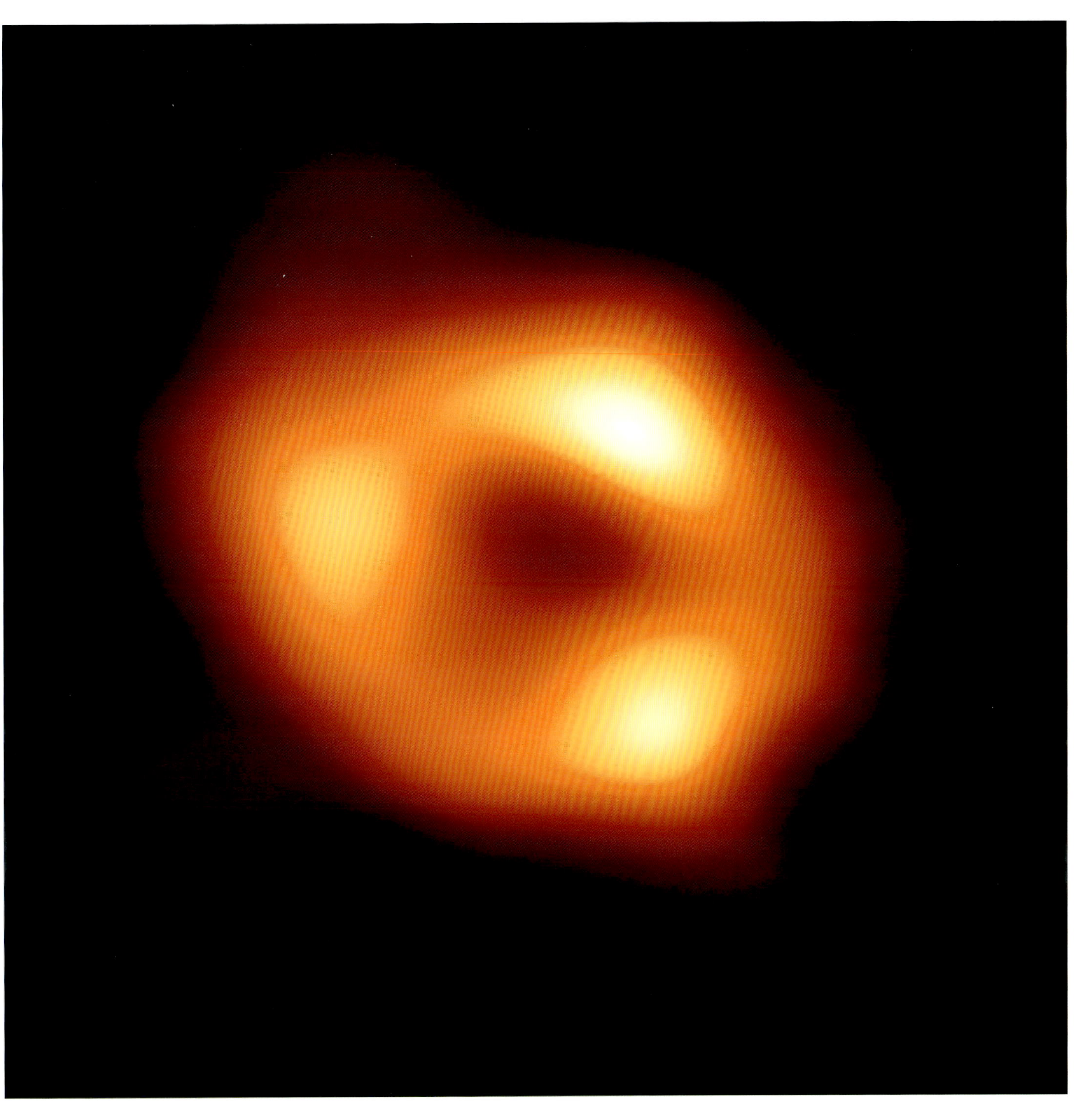

3.143 Sagittarius A* at the center of the Milky Way galaxy, taken by the Event Horizon Telescope, 2022. Event Horizon Telescope Collaboration.

Unlike the image of M87*, the image of Sagittarius A* has three bright spots. The most plausible explanation is that the spots are artifacts of image-reconstruction techniques.

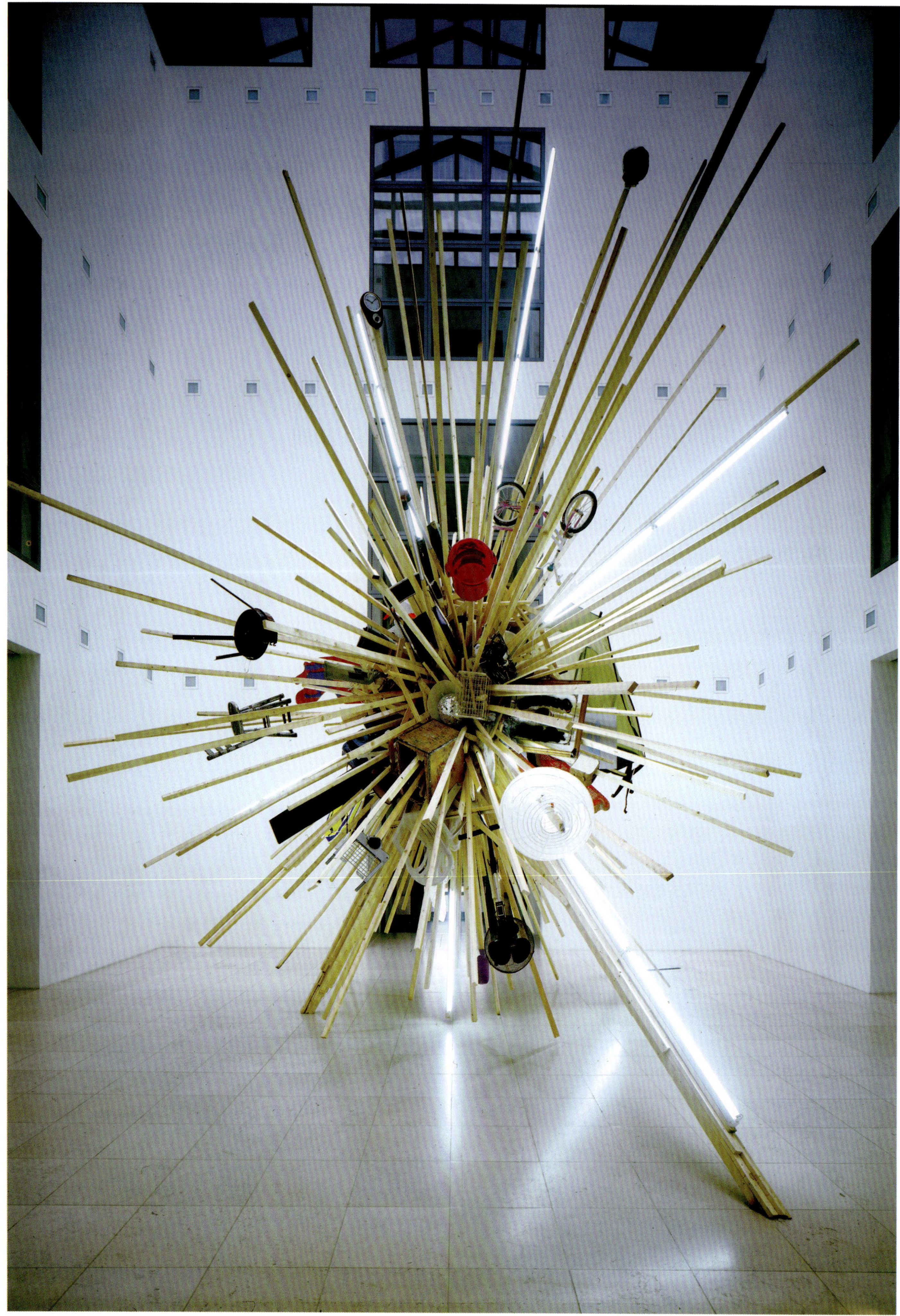

In polarized light, M87* reveals the structure of the magnetic fields in the accretion disc and how matter is whirling into the black hole (figure 3.142). In May 2022 the EHT project released a picture of the black hole at the center of the Milky Way (figures 3.143, 3.144, and 3.145).

EHT scientists dream of making a movie of the accretion disc rotating around a black hole.[83] Sagittarius A* has 4 million solar masses compared to 6.5 *billion* solar masses of M87*. Although both accretion disks rotate close to the speed of light near the innermost stable circular orbit, this size difference means that the disk around Sagittarius A* completes a single rotation in half an hour while the one around M87* takes weeks to complete one rotation. Thus, scientists plan to shoot a frame of the movie every few minutes for Sagittarius A* and a frame every three days for M87*. Then scientists will combine the frames to form time-lapse movies of Sagittarius A* and M87*. The creation of this "Black Hole Cinema" is the focus of the next-generation Event Horizon Telescope (ngEHT; figure 3.146).

OPPOSITE
3.144 Björn Dahlem (German, born 1974), *Black Hole: Sagittarius A*, 2005. Wood, fluorescent lamps, various found objects, 22 ft. 11½ in. × 22 ft. 11½ in. × 45 ft. 11 in. (7 × 7 × 14 m). Courtesy of Sies + Höke, Düsseldorf, and Galerie Guido W. Baudach, Berlin.

BELOW
3.145 Björn Dahlem (German, born 1974), *Black Hole: Sagittarius A* (detail), 2005.

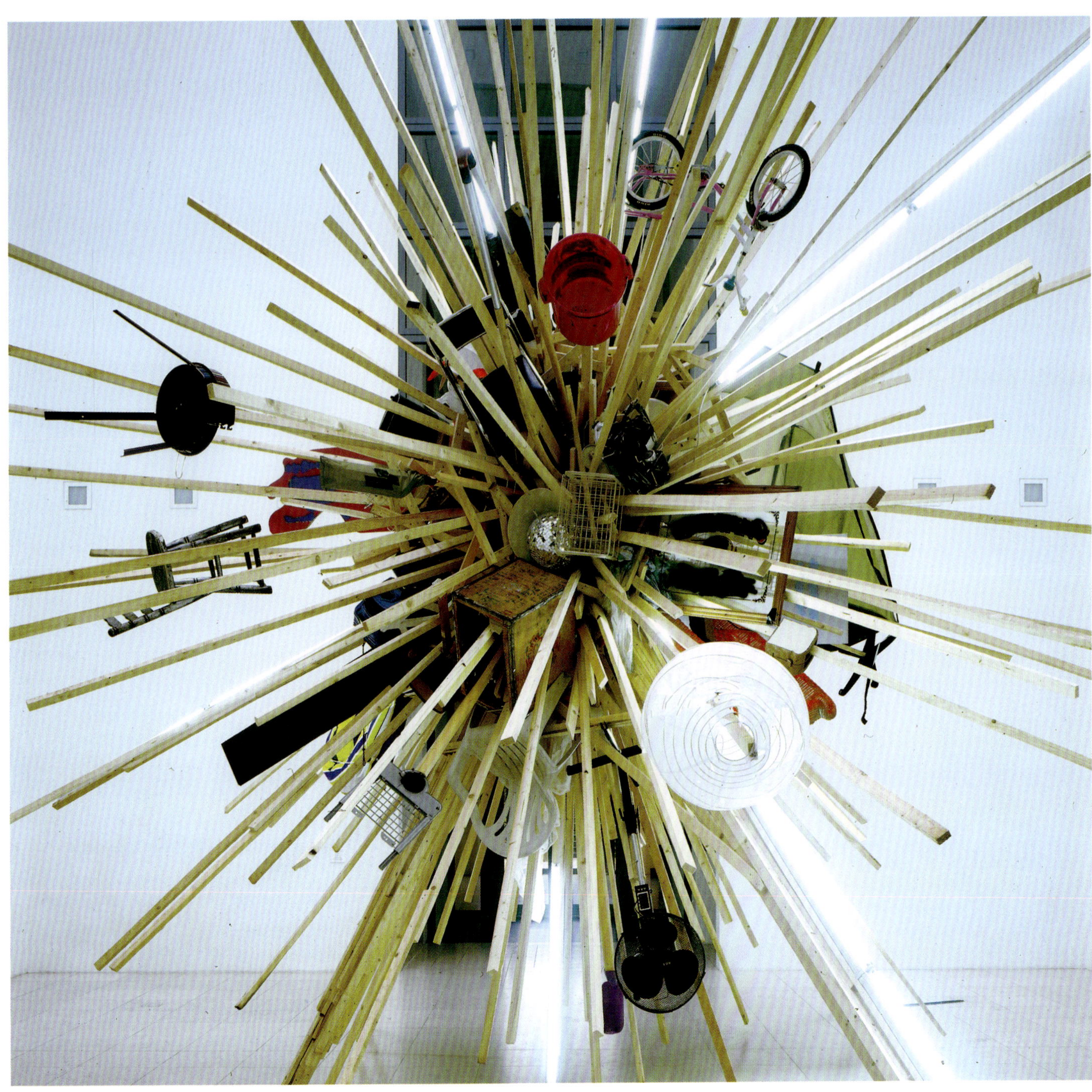

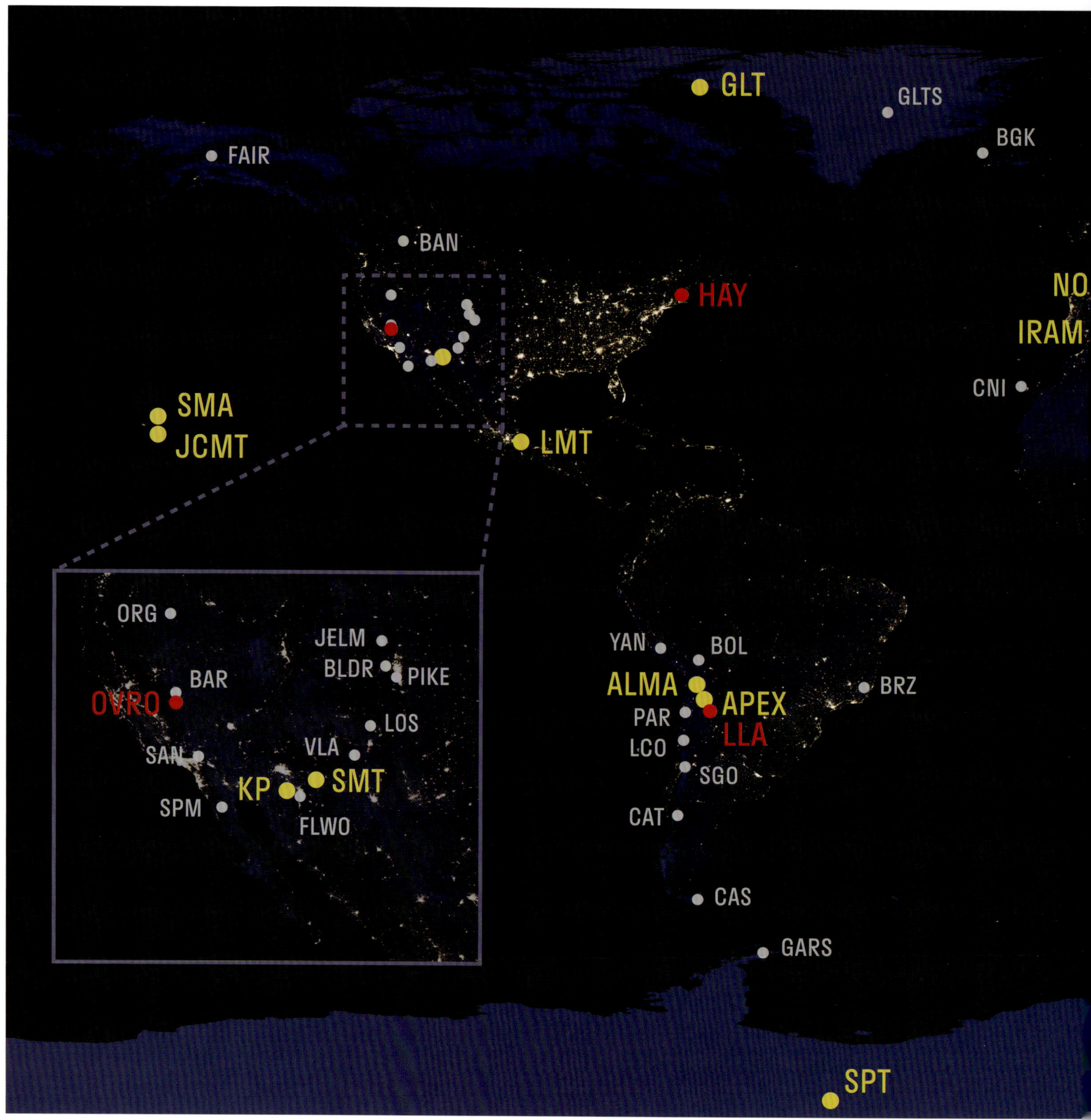

3.146 The next-generation Event Horizon Telescope, 2023.

On this map, telescopes in the current next-generation Event Horizon Telescope are shown in yellow; other existing or near-future telescopes that may join the next-generation Event Horizon Telescope are in red; and potential new next-generation Event Horizon Telescope sites are in gray.

Current telescopes:

ALMA: Atacama Large Millimeter/submillimeter Array, Cerro Chajnantor, Chile

APEX: Atacama Pathfinder Experiment, Cerro Chajnantor, Chile

GLT: Greenland Telescope, Avannaata, Greenland

IRAM: 30-meter Telescope, Pico Veleta, Spain

JCMT: James Clerk Maxwell Telescope, Maunakea, Hawaii

KP: Kitt Peak National Observatory, Arizona

LMT: Large Millimeter Telescope "Alfonso Serrano," Sierra Negra, Mexico

NOEMA: Northern Extended Millimeter Array, Provence-Alpes-Côte d'Azur, France

SMA: Submillimeter Array, Maunakea, Hawaii

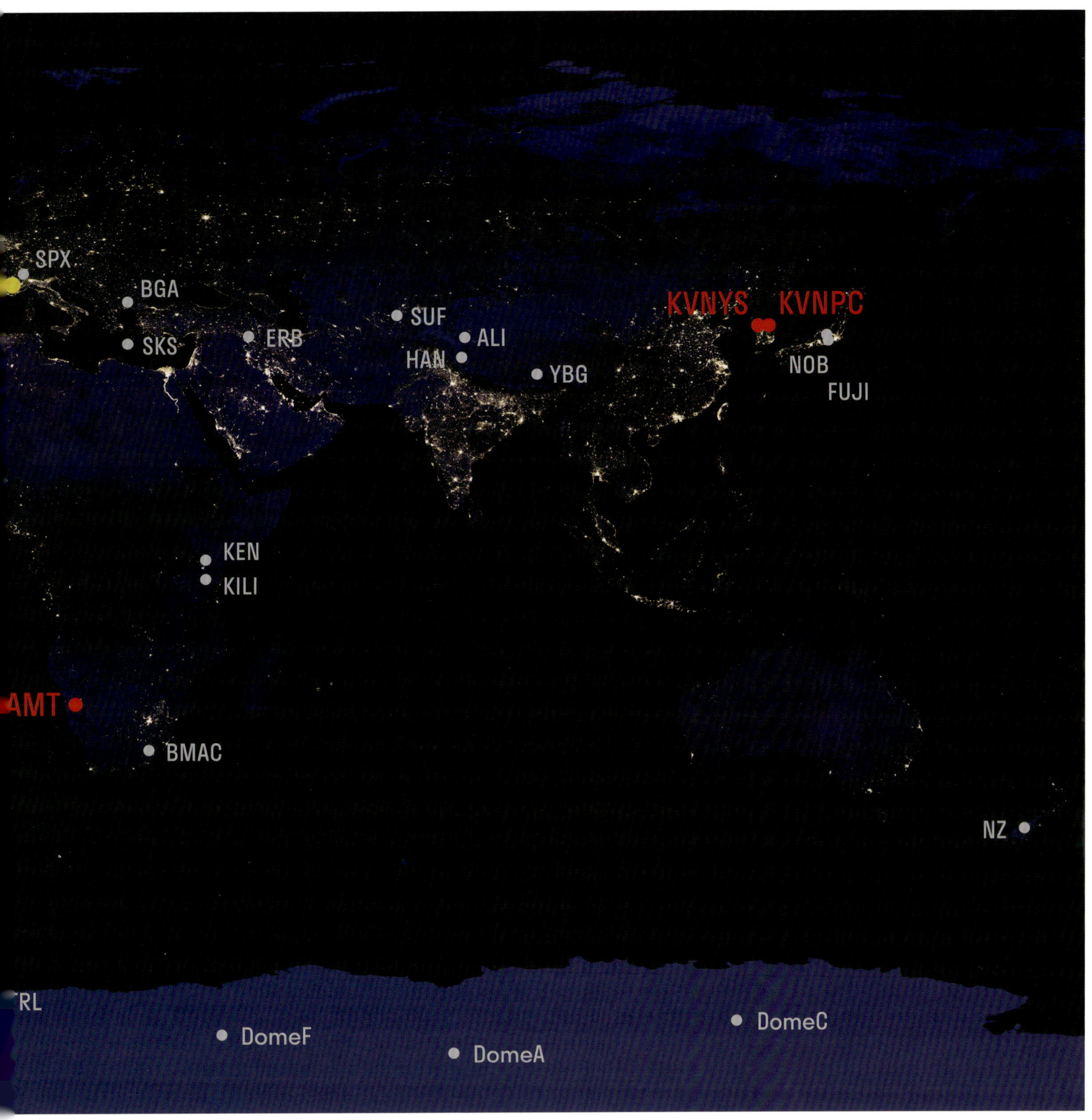

SMT: Submillimeter Telescope, Mount Graham, Arizona

SPT: South Pole Telescope, South Pole Station, Antarctica

Existing or near-future telescopes:

OVRO: Owens Valley Radio Observatory, California

HAY: Haystack Radio Telescope, Massachusetts

LLA: Large Latin American Millimeter Array, Salta, Argentina

AMT: Africa Millimeter Telescope, Namibia, Africa

KVNYS: Korean VLBI Network Yonsei Radio Telescope, Seoul, South Korea

KVNPC: Korean VLBI Network Pyeongchang Radio Observatory, Pyeongchang, South Korea

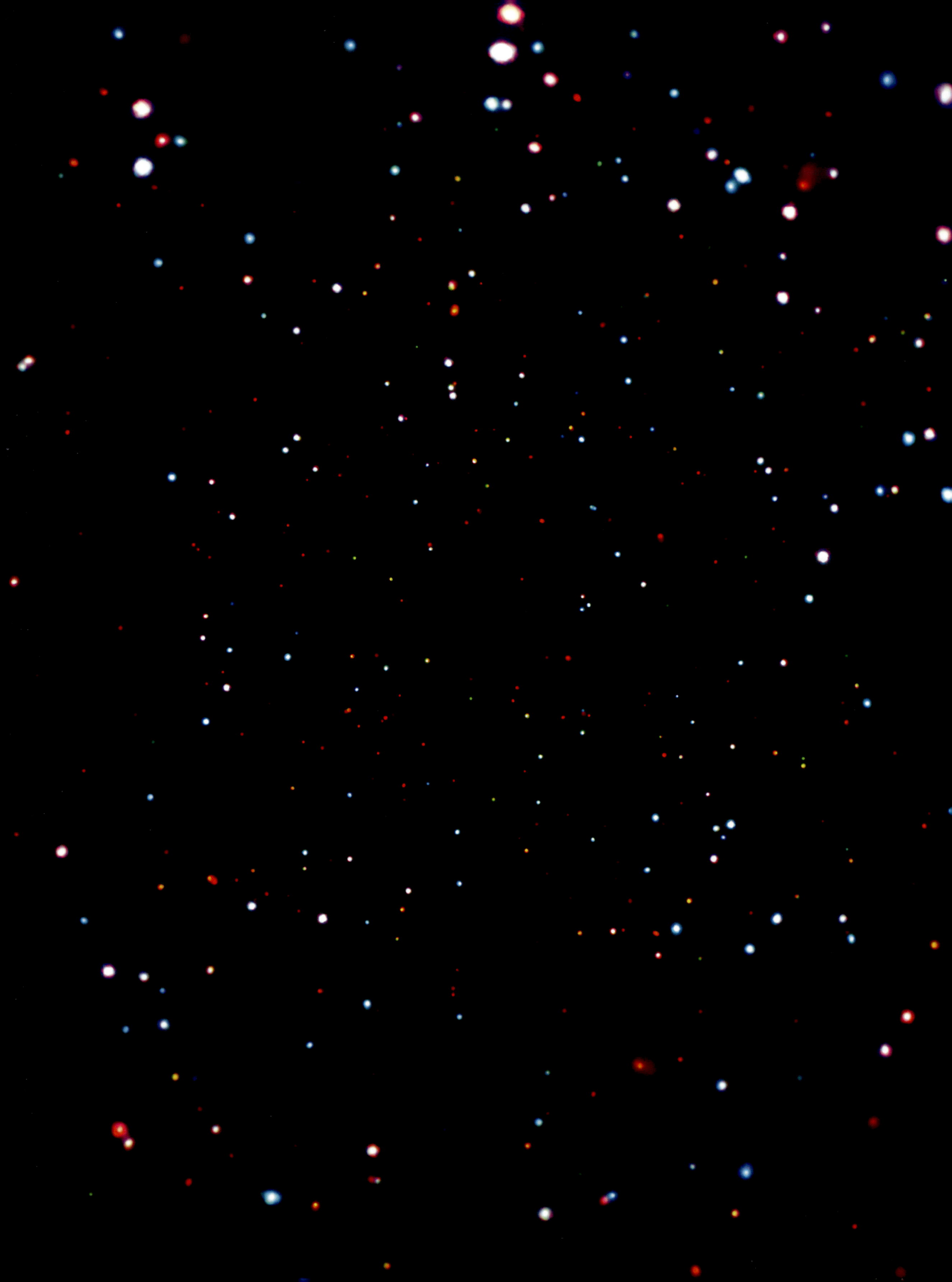

CONCLUSION

Scientific discoveries and artworks are lenses that focus our gaze on acts of creativity. Our understanding of black holes began when Karl Schwarzschild had the brilliant insight that Einstein's theory of general relativity implies a point of spacetime where matter is infinitely dense. Most scientists thought a black hole couldn't exist in nature, but within five decades the existence of Cygnus X-1 was confirmed, and today scientists know that black holes are an integral part of the fabric of spacetime.

Science has produced a wealth of information about black holes that has been popularized worldwide. This has prompted artists to delve deep into their creative imaginations to find the significance of black holes within a broad cultural context. Everything vanishes when it crosses the event horizon, so nothingness is inevitably a theme in the art and science of black holes. In other respects, however, a black hole is the opposite of nothing because matter is infinitely dense within it and the universe—everything—exploded into existence from such a point of infinite density. The contradictory relation of nothingness and everything has captivated artists like Anish Kapoor, who astutely symbolized that relation in *Void Pavilion*. In Eastern philosophy, nothingness is the source of everything, and this enigmatic nothingness-everything dichotomy is also a perennial theme in Western modern art. Today, the threads of Eastern philosophy and Western modernism are entwined in an unbreakable braid, and artists such as Xu Bing recognize that black holes offer the opportunity to merge East and West in their art.

Anything in the vicinity of a black hole is violently torn apart owing to its extreme gravity—the strongest in the universe. We see this violence in the works of artists like Cai Guo-Qiang and Takashi Murakami, who have used black holes to symbolize the brutality unleashed by the atomic bomb. The inescapable pull of a black hole is also a ready metaphor for depression in the work of artists such as Moonassi. Thus, on the one hand, the black hole provides artists with a symbol to express the devastations and anxieties of the modern world. On the other hand, however, a black hole's extreme gravity is the source of stupendous energy, and artists such as Yambe Tam invite viewers to embrace darkness as a path to transformation, awe, and wonder.

3.147 Chandra Deep Field South (detail). NASA/CXC/Penn State/B. Luo et al.

Chandra Deep Field South contains approximately 5,000 black holes in an area of the sky covered by the full moon. The image was made by pointing the Chandra X-ray Observatory satellite at an area in the sky in the Southern Hemisphere for forty-four days. The colors represent different energy levels of X-rays observed: low energy (red), medium energy (green), high energy (blue).

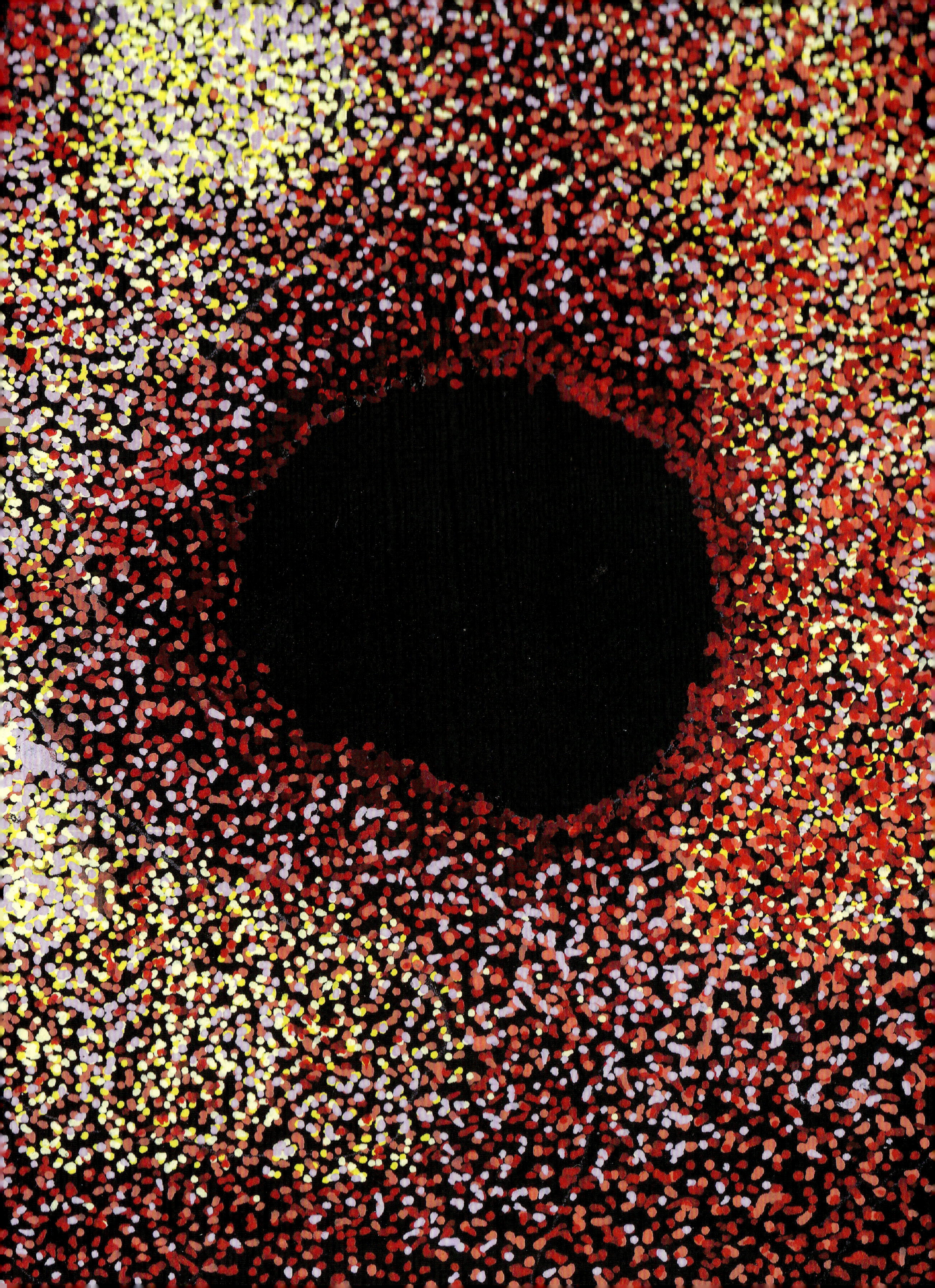

NOTES

1. John Michell, "On the Means of Discovering the Distance, Magnitude, etc., of the Fixed Stars, in Consequence of the Diminution of Their Light, in Case Such a Diminution Should Be Found to Take Place in Any of Them, and Such Other Data as Should Be Procured from Observations, As Would Be Further Necessary for That Purpose," *Philosophical Transactions of the Royal Society* 74 (1784): 35–57.

2. "Il est donc possible que les plus grands corps lumineux de l'univers, soient par cela même, invisibles"; translation by Lynn Gamwell. Pierre-Simon de Laplace, *Exposition du système du monde* (Paris: Cercle-Social, 1796), 2:305. On the origin of the concept of a black hole, see Colin Montgomery, Wayne Orchiston, and Ian Whittingham, "Michell, Laplace and the Origin of the Black Hole Concept," *Journal of Astronomical History and Heritage* 12, no. 2 (2009): 90–96.

3. Edgar Allan Poe, *Eureka: A Prose Poem* (New York: Putnam, 1848), 84. On Poe's cosmology, see Paolo Molaro and Alberto Cappi, "Edgar Allan Poe: The First Man to Conceive a Newtonian Evolving Universe," *Culture and Cosmos* 16, nos. 1–2 (2012): 225–239; arXiv:1506.05218.

4. Albert Einstein, "Die Feldgleichungen der Gravitation," *Sitzungsberichte der Königlich Preussischen Akademie der Wissenschaften* 48–49 (1915): 844–847.

5. "Wie Sie sehen, meint es der Krieg freundlich mit mir, indem er mir trotz heftigen Geschützfeuers . . . diesen Spaziergang in dem von Ihrem Ideenlande erlaubte"; translation by Lynn Gamwell. Karl Schwarzschild to Albert Einstein, December 22, 1915, in *The Collected Papers of Albert Einstein*, vol. 8, pt. A, *The Berlin Years: Correspondence, 1914–1917*, ed. Robert Schulmann, et al. (Princeton, NJ: Princeton University Press, 1987), 224–225.

6. On Chandrasekhar, see Roger Penrose, "Chandrasekhar, Black Holes, and Singularities," *Journal of Astrophysics and Astronomy* 17, nos. 3–4 (1996): 213–231.

7. J. R. Oppenheimer and H. Snyder "On Continued Gravitational Contraction," *Physical Review* 56, no. 5 (September 1939): 455–459.

8. David Finkelstein, "Past-future Asymmetry of the Gravitational Field of a Point Particle," *Physical Review* 110, no. 4 (1958): 965.

9. On the origin of the term "black hole," see Carlos A. R. Herdeiro and José P. S. Lemos, *O Gazeta de Física* 41, no. 2 (2018): 2–7; available in English at arXiv:1811.06587.

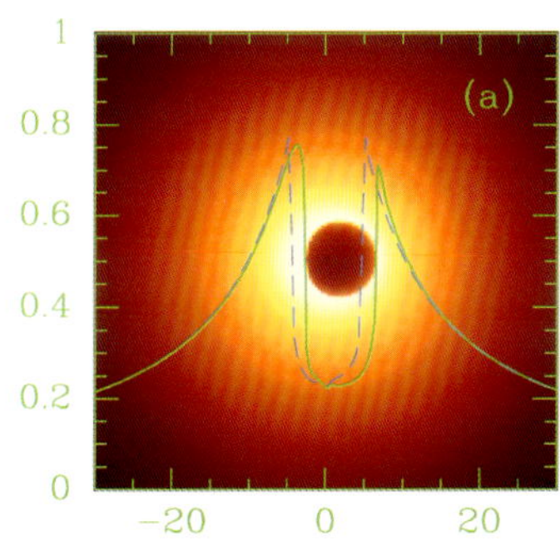

ABOVE

4.1 Simulation of Sagittarius A* in Heino Falcke et al., "Viewing the Shadow of the Black Hole at the Galactic Center," *Astrophysical Journal Letters* 528, no. 1 (2000): L15, fig. 1 (detail).

The authors of this paper argue that it is theoretically possible to take a picture from Earth similar to this simulation of the accretion disk around Sagittarius A*. The solid green line indicates distance from the black hole (as shown on the X-axis). The dashed purple line indicates the amount of radiation (as shown on the Y-axis). Together the lines demonstrate that radiation increases with proximity.

OPPOSITE

4.2 Jody Rasch (American, born 1956), *Horizon—Black Hole* (detail), 2020. Acrylic markers on paper, 9 × 12 in. (22.8 × 30.4 cm). Courtesy of the artist.

10. Albert Einstein, "On a Stationary System with Spherical Symmetry Consisting of Many Gravitating Masses," *Annals of Mathematics* 40, no. 4 (1939): 936.

11. Roger Penrose, "Gravitational Collapse and Space-Time Singularities," *Physical Review Letters* 14, no. 3 (1965): 57–59.

12. See John Earman and Jean Eisenstaedt, "Einstein and Singularities," *Studies in History and Philosophy of Science* 30, no. 2 (1999): 185–235. For arguments against singularities and their reification, see Gustavo E. Romero, "*Adversus Singularitates*: The Ontology of Spacetime Singularities," *Foundations of Science* 18, no. 2 (2013): 297–306.

13. Stephen Hawking and Roger Penrose, "The Singularities of Gravitational Collapse and Cosmology," *Proceedings of the Royal Society of London A* 314, no. 1519 (January 27, 1970): 530.

14. Roger Penrose, "Eschermatics," Oxford Mathematics and the Clay Mathematics Institute Public Lecture, Oxford, UK, September 24, 2018, 71 min, 46 sec., posted by University of Oxford Live, https://livestream.com/oxuni/penrose/videos/180720947.

15. Subrahmanyan Chandrasekhar, *The Mathematical Theory of Black Holes* (Oxford: Clarendon, 1983), xxiii.

16. Remo Ruffini, quoted in "Artist's Rendition of a Black Hole" (1971), https://web.archive.org/web/20120406213919/http://www.icra.it/ICRA_Networkshops/INW10_RJR60/jan1971ptcover.htm.

17. Laozi, from chapter 1 of *Dao de jing* (*Tao te ching* or *Tao Teh King*; commonly translated as *The Classic of the Way and Virtue*), in *The Sacred Books of China: The Texts of Taoism*, trans. James Legge (Oxford, UK: Clarendon, 1891), 84.

18. Laozi, from chapter 40 of *Dao de jing*, in *Sacred Books of China*, 47.

19. Mahatma Gandhi, "What Is Hinduism?," *Young India* (April 24, 1924) in *Young India, 1924–1926* (New York: Viking, 1927), 13.

20. Chuang Tzu [Zhuangzi], "Fit for Emperors and Kings," in *The Complete Works of Chuang Tzu*, trans. Burton Watson (New York: Columbia University Press, 1968), 97.

21. Joseph Needham, ed., *Science and Civilization in China* (Cambridge: Cambridge University Press, 1954–). An earlier key publication is Yu-lan Fung, "Why China Has No Science—An Interpretation of the History and Consequences of Chinese Philosophy," *International Journal of Ethics* 32, no. 3 (1922): 237–263.

22. Edmund Burke, *A Philosophical Enquiry into the Origin of Our Ideas of the Sublime and the Beautiful* (London: Dodsley, 1757), pt. I, sec. VII, 58.

23. Charles Darwin, *The Descent of Man and Selection in Relation to Sex* (1871), 2nd ed. (New York and London: Appleton, 1883), 623.

24. "Im Beginn der Entwicklung einer ganz neuen Kunst stehen, der Kunst, mit Formen, die nichts bedeuten and nictits darstellen und an nichts erinnern, unsere Seele so tief, so stark zu erregen, wie es nur immer die Musik mit Tönen vermag"; translation by Lynn Gamwell. August Endell, "Formenschönheit und Dekorative Kunst," *Dekorative Kunst* 1, no. 6 (March 1898): 75.

25. "Schoenberg combines in his thinking the greatest freedom with the greatest belief in the development of the spirit!" Wassily Kandinsky, "Footnotes to Schoenberg's 'On Parallel Octaves and Fifths'" (1911), trans. Peter Vergo, in *Kandinsky: Complete Writings on Art*, ed. Kenneth C. Lindsay and Peter Vergo (Boston: G. K. Hall, 1982), 1:93.

26. Wassily Kandinsky, *On the Spiritual in Art* (1912), quoted in Klaus Kropfinger, "Latent Structural Power versus the Dissolution of Artistic Material in the Works of Kandinsky and Schönberg," in *Schönberg and Kandinsky: An Historic Encounter*, ed. Konrad Boehmer (New York: Routledge, 2003), 15.

27. Georg Cantor, "Mitteilungen zur Lehre vom Transfiniten" (1887), in *Gesammelte Abhandlungen mathematischen und philosophischen Inhalts*, ed. E. Zermelo (Berlin: Julius Springer, 1932), 378–439.

28. Kazimir Malevich, "Chapters from an Artist's Autobiography" (1933), trans. Alan Upchurch, *October* 34 (1985), 38.

29. Kazimir Malevich to Mikhail Matiushin, November 10, 1917, in Evgenii Kovtun, "Suprematism as an Eruption of Universal Space," in *Suprematisme* (Paris: Galerie Jean Chauvelin, 1977), 27.

30. Albert Einstein in an interview with Alfred Werner, *Liberal Judaism* 16 (April-May 1949), in Alice Calaprice, *The New Quotable Einstein* (Princeton, NJ: Princeton University Press, 2005), 173.

31. Theodor Adorno, *Aesthetic Theory* (1970), trans. C. Lenhardt (London: Routledge, 1984), 58.

32. Undated notes by Reinhardt in *Art as Art: The Selected Writings of Ad Reinhardt*, ed. Barbara Rose (New York: Viking, 1975), 108. In another page of undated notes, Reinhardt assembled quotations on "black"; one reads: "The Tao is dim and dark (Lao Tzu, fourth century BC)," 98.

33. Reinhardt's *Small Painting for T. M. (Thomas Merton)* is held at the Abbey of Gethsemani in Trappist, Kentucky.

34. Thomas Merton, "Wisdom in Emptiness: A Dialogue between D. T. Suzuki and Thomas Merton" (1960), in *Zen and the Birds of Appetite* (New York: New Directions, 1968), 99–138.

35. Barnett Newman, "The New Sense of Fate" (1948), in *Barnett Newman: Selected Writings*, ed. John P. O'Neill (New York: Knopf, 1990), 100.

36. "La scoperta di Einstein sul cosmo è la dimensione infinita, senza fine"; translation by Lynn Gamwell. Lucio Fontana interview in Carla Lonzi, *Autoritratto* (Bari: De Donato, 1969), 170.

37. Paul Trachtman, "Lee Bontecou's Brave New World," *Smithsonian*, September 2004, accessed August 31, 2023, https://www.smithsonianmag.com/arts-culture/lee-bontecous-brave-new-world-180940689/.

38. See John Latham et al., *John Latham: Art after Physics*, exh. cat. (Oxford: Museum of Modern Art; Stuttgart: Edition Hansjörg Mayer, 1991). See also John Latham and Ian Macdonald-Monro, "The 20th-Century Trajectory… A Black Hole at the Core?" *AND* 17 (1988): 3–7.

39. Frederick Eversley, "Statement of the Artist," in *Frederick Eversley* (Santa Barbara, CA: Santa Barbara Museum of Art, 1976), n.p.

40. "Maiolino refers to astronomical black holes in her 1971 *Mapas mentais* (*Mental Maps*) series." Ela Bittencourt, review of the exhibition *Nuestra América* at Galeria Luisa Strina, São Paulo, *Artforum* 59, no. 3 (December 2020): 186. The artist described her sources in the work of Fontana in an email to the author via Nicole Keller, Hauser & Wirth Gallery, Zurich, June 19, 2023.

41. Melissa Walter, "About," Melissa Walter's website, accessed March 3, 2024, https://www.melissawalterart.com/about.

42. Jean-Pierre Luminet, "Image of a Spherical Black Hole with Thin Accretion Disk," *Astronomy and Astrophysics* 75, nos. 1–2 (1979): 228.

43. Luminet, "Image of a Spherical Black Hole," 235.

44. For what distinguished Luminet from Escher, see Martin Kemp, "Luminet's Illuminations: Cosmological Modelling and the Art of Intuition," *Nature* 426 (2003): 232.

45. See Jean-Alain Marck, "Short-Cut Method of Solution of Geodesic Equations for Schwarzschild Black Hole," *Classical and Quantum Gravity* 13, no. 3 (1996): 393–402.

46. In the 1980s, both books were translated into Chinese, which was the language Cai read them in. Jennifer Yu-Chen Chen, Cai Guo-Qiang Archives, email to the author, September 19, 2023.

47. Eddy Soetriyono, "The Big Bang," *Mutual Art*, March 1, 2008, accessed August 27, 2023, https://www.mutualart.com/Article/The-Big-Bang/CCEA2E3AE916DAFA.

48. Soetriyono, "The Big Bang."

49. Ron Rosenbaum, "Meet the Artist Who Blows Things Up for a Living," *Smithsonian*, April 2013, accessed August 27, 2023, https://www.smithsonianmag.com/arts-culture/meet-the-artist-who-blows-things-up-for-a-living-4984479/.

50. Video games that feature black holes, in alphabetical order, include: *Armies of Exigo* (2004); *Blackhole* (2015); *Elite Dangerous* (2014); *Everspace* (2016); *Exo One* (2021); *Hellpoint* (2020); *Hole.io* (2018); *Might and Magic Heroes VI* (2011); *No Man's Sky* (2016); *Orbt XL* (2017); *Outer Wilds* (2019); *Space Engine* (2019); *Star Sector* (2011); *Stellaris* (2016); *Super Mario Galaxy* (2007); *The Black Hole* (1982); *Risk of Rain* (2013); and *Universe Sandbox* (2008).

51. Popular books in English about black holes published since 2020, listed by the author's name in alphabetical order, include: Mitchell Begelman and Martin Rees, *Gravity's Fatal Attraction: Black Holes in the Universe* (Cambridge: Cambridge University Press, 2021); Rob Botwright, *String Theory, Black Holes, Holographic Universe, and Mathematical Physics* (Tustin, CA: Pastor, 2024); Massimo Cencini et al., *A Random Walk in Physics: Beyond Black Holes and Time-Travels* (Cham: Springer, 2021); Marcus Chown, *A Crack in Everything: How Black Holes Came in from the Cold and Took Cosmic Centre Stage* (London: Bloomsbury, 2024); Marcus Chown, *Breakthrough: Spectacular Stories of Scientific Discovery from the Higgs Particle to Black Holes* (London: Faber & Faber, 2022); Brian Cox and Jeff Forshaw, *Black Holes: The Key to Understanding the Universe* (New York: Mariner, 2022); Paul Davies, *What's Eating the Universe? And Other Cosmic Questions* (Chicago: University of Chicago Press, 2021); Czeena Devera, *My Guide to the Solar System: Black Hole* (Ann Arbor, MI: Cherry Lake Press, 2022); Heino Falcke, *Light in the Darkness: Black Holes, the Universe, and Us* (New York: Harper Collins, 2021); Herji, *Big Bangs and Black Holes: A Graphic Novel Guide to the Universe*, trans. Jérémie Francfort and Jeffrey K. Butt (Lausanne: Helvetiq, 2023); Solomon M. Jacob, *After the Big Bang Could Come the Big Crunch: Primordial Black Holes and the Big Crunch* (Pittsburgh: Dorrance, 2023); Andrew King, *Supermassive Black Holes* (Cambridge: Cambridge University Press, 2023); Janna

Levin, *Black Hole Survival Guide* (New York: Knopf, 2020); Nicholas Mee, *Gravity: From Falling Apples to Supermassive Black Holes* (Oxford: Oxford University Press, 2022); Fulvio Melia, *The Galactic Supermassive Black Hole* (Princeton, NJ: Princeton University Press, 2020); Abhas Mitra, *The Rise and Fall of the Black Hole Paradigm* (New York: Macmillan, 2021); John W. Moffat, *The Shadow of the Black Hole* (Oxford: Oxford University Press, 2020); Joseph Polchinski, *Memories of a Theoretical Physicist: A Journey across the Landscape of Strings, Black Holes, and the Multiverse* (Cambridge, MA: MIT Press, 2022); Anna Crowley Redding, *Black Hole Chasers: The Amazing True Story of an Astronomical Breakthrough* (New York: Feiwel, 2021); Luciano Rezzolla, *The Irresistible Attraction of Gravity: A Journey to Discover Black Holes* (Cambridge: Cambridge University Press, 2023); Caleb Scharf, *Gravity's Engines: How Bubble-Blowing Black Holes Rule Galaxies, Stars, and Life in the Cosmos* (New York: Farrar, Straus and Giroux, 2024); Wouter Schmitz, *Understanding Relativity: A Conceptual Journey into Space-time, Black Holes, and Gravitational Waves* (Cham: Springer, 2022); Becky Smethurst, *A Brief History of Black Holes* (London: Pan, 2023); Danielle Smith-Llera, *First Look at a Black Hole: How a Photograph Solved a Space Mystery* (North Mankato, MN: Compass Point, 2021); Kip Thorne, *The Warped Side of Our Universe: An Odyssey through Black Holes, Wormholes, Time Travel, and Gravitational Waves* (New York: Liveright, 2023); James Trefil and Shobita Satyapal, *Supermassive: Black Holes at the Beginning and End of the Universe* (Washington, DC: Smithsonian Institution Press, 2025); Neil deGrasse Tyson, *Merlin's Tour of the Universe, Revised and Updated for the Twenty-First Century: A Traveler's Guide to Blue Moons and Black Holes, Mars, Stars, and Everything Far*; illustrated by Stephen J. Tyson (Ashland, OR: Blackstone, 2024); Galina Weinstein, *Einstein's Legacy: From General Relativity to Black Hole Mysteries* (Cham: Springer, 2024); Clifford M. Will and Nicolas Yunes, *Is Einstein Still Right? Black Holes, Gravitational Waves, and the Quest to Verify Einstein's Greatest Creation* (Oxford: Oxford University Press, 2020); Matthew Brenden Wood and Alexis Cornell, *The Universe: The Big Bang, Black Holes, and Blue Whales* (White River Junction, VT: Nomad, 2021).

52. Yambe Tam, email to the author, June 10, 2024.

53. Albert Einstein, "What I Believe," *Forum and Century* 84, no. 4 (October 1930): 194.

54. Cover, *Vogue Korea*, February 5, 2018. "Worldwide Suicide Mortality Trends (2000–2019): A Joinpoint Regression Analysis," *World Journal of Psychiatry* 12, no. 8 (August 2022): 1044–1060; www.ncbi.nlm.nih.gov/pmc/articles/PMC9476842/.

55. Chen Yuan, "Black Holes and Strange Worlds," *NeoCha: Culture and Creativity in Asia*, April 5, 2018, accessed February 12, 2024, https://neocha.com/magazine/black-holes-and-strange-worlds/.

56. "Caressant L'Horizon devient ainsi un voyage musical d'un caractère presque onirique qui nous amène à la limite des forces physiques connues, où la matière est tellement concentrée que la courbure de l'espace-temps devient infinie et l'espace lui-même, selon la relativité génerale d'Einstein, se déchire. Ainsi, tout en essayant de 'caresser,' de 'palper acoustiquement' les trous noirs et leur horizon d'événements, Caressant L'Horizon prend la forme d'une 'expérience de pensée' qui se cristallise, en tant qu'architecture musicale, en une expérience d'écoute. Mais, en étant donné que nous sommes prisonniers du temps—comme tout ce qui se passe dans l'univers—ce voyage est une expérience à vivre, à expérimenter sensuellement, et ne pas à penser dans l'abstrait"; translation by Lynn Gamwell. "Program Notes: *Caressant l'horizon* (2011)," Hèctor Parra's website, accessed February 11, 2024, https://hectorparra.net/docs/program-notes/.

57. "Gallery: Space Cloud G2," Lucas J. Rougeux's website, accessed March 16, 2024, https://www.lucasjrougeux.com.

58. "Gallery: Space Cloud G2."

59. "Exhibitions: *The Soul Gravity—Guided to Black*," Lucas J. Rougeux's website, accessed March 16, 2024, https://www.lucasjrougeux.com.

60. Priyamvada Natarajan et al., "First Detection of an Over-Massive Black Hole Galaxy UHZ1: Evidence for Heavy Black Hole Seed

4.3 Sharpened image of M87* in Lia Medeiros et al., "The Image of the M87 Black Hole Reconstructed with PRIMO," *Astrophysical Journal Letters* 947, no. 1 (2023): fig. 1.

On the left is the image of M87* released by the Event Horizon Telescope (figure 3.138). The image in the middle was obtained by applying PRIMO (a new principal-components analysis-based algorithm for image reconstruction) to the same raw data. On the right is the PRIMO image at the same resolution as the Event Horizon Telescope image; here, the ring is half as wide as the image on the left. The general theory of relativity predicts that the black hole's "shadow" will be 2.6 times larger than the black hole's event horizon. In these images of M87*, its "shadow" is exactly as predicted—another confirmation of the general theory of relativity.

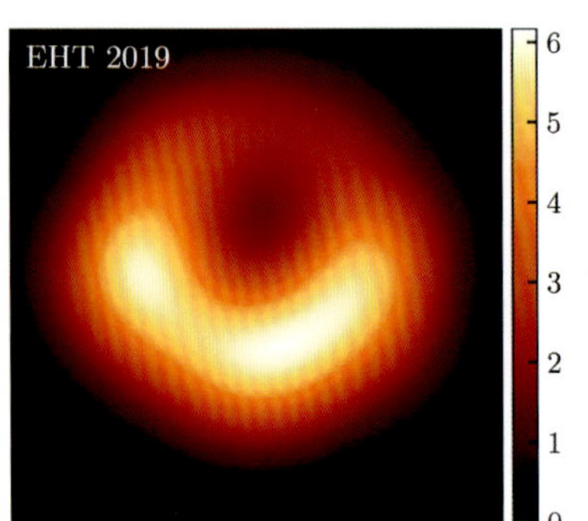

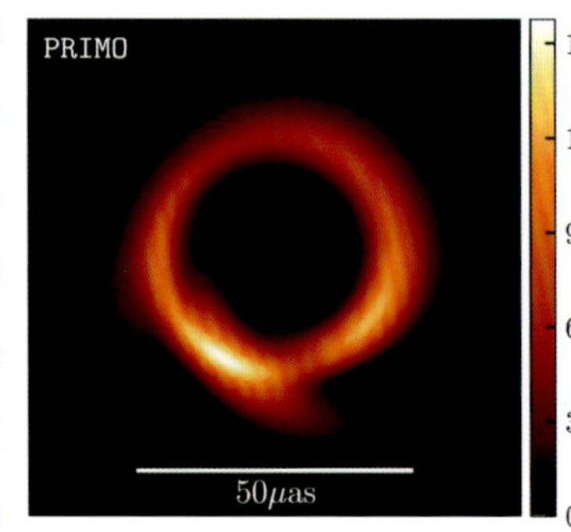

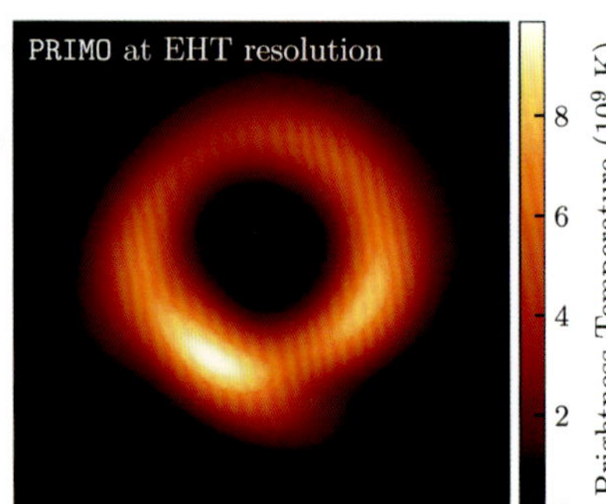

Formation from Direct Collapse," *Astrophysical Journal Letters* 960, no. 1 (December 29, 2023), https://doi.org/10.3847/2041-8213/ad0e76.

61. IceCube Collaboration et al., "Multi-messenger Observations of a Flaring Blazar Coincident with High-Energy Neutrino IceCube-170922A," *Science* 361, no. 6398 (July 13, 2018), https://doi.org/10.1126/science.aat1378; and IceCube Collaboration et al., "Neutrino Emission from the Direction of the Blazar TXS 0506+056 Prior to the IceCube-170922A Alert," *Science* no. 6398 (July 13, 2018): 147–151, https://doi.org/10.1126/science.aat2890.

62. Laura Di Gesu et al., "Discovery of X-ray Polarization Angle Rotation in the Jet from Blazar Mrk 421," *Nature Astronomy* 7, no. 10 (July 2023), 1–12, https://doi.org/10.1038/s41550-023-02032-7.

63. Barbarella Fokos, "A New Stellar Order," *Diva Barbarella* (blog), June 1, 2016, accessed June 15, 2024, https://divabarbarella.com/read-weep/out-and-about/a-new-stellar-order/.

64. Shira Wolfe, "Anish Kapoor—Ritual and Void," *Artland*, accessed July 8, 2020, https://magazine.artland.com/anish-kapoor-ritual-and-void/.

65. Theo Farrant, "Anish Kapoor on Vantablack ownership, artificial intelligence and 'making love with the screen,'" *Euronews*, October 25, 2023, accessed March 10, 2024, https://www.euronews.com/culture/2023/10/25/anish-kapoor-on-vantablack-ownership-artificial-intelligence-and-making-love-with-the-scre.

66. "Wird das Geschaute und Erlebte in der Sprache der Logik nachgebildet, so treiben wir Wissenschaft, wird es durch Formen vermittelt, deren Zusammenhänge dem bewußten Denken unzugänglich, doch intuitiv als sinnvoll erkannt sind, so treiben wir Kunst"; translation by Lynn Gamwell. Albert Einstein, "Das Gemeinsame am künstlerischen und wissenschaftlichen Erleben" (1921), in *The Collected Papers of Albert Einstein*, vol. 7, *The Berlin Years: Writings, 1918–1921*, ed. M. Janssen et al. (Princeton, NJ: Princeton University Press, 2002), 379–381.

67. Govinda Sah Azad, email to the author, April 9, 2023.

68. On the transformative power of fire, see Toshikatsu Endō, "On Fire," in *Endō Toshikatsu/Toshikatsu Endō*, trans. Stanley N. Anderson (Tokyo: Museum of Contemporary Art Tokyo, 1991), 42.

69. Editors of *Artspace*, "Plushy Terrorism and Cities in Suitcases: Artist Yin Xiuzhen on 'How to Challenge Society with Its Own Refuse,'" trans. Jeff Crosby, *Artspace*, January 14, 2017, accessed July 30, 2022, https://www.artspace.com/magazine/interviews_features/book_report/yin-xiuzhen-phaidon-excerpt-54534.

70. Ginza Graphic Gallery, "Tadanori Yokoo: My Black Holes," press release, June 30, 2023, accessed February 21, 2024, https://www.dnpfcp.jp/gallery/ggg/eng/00000819.

71. "Black Hole—Trilogy and Triathlon," Shamel Pitts's website, January 1, 2019, accessed February 20, 2024, https://shamelpitts.com/blackhole/.

72. Kara Walker, "The Black (W)hole, and What it Means to Me," in *Kara Walker: A Black Hole Is Everything a Star Longs to Be*, ed. Anita Haldemann (Basel: Kunstmuseum Basel, 2021), 561.

73. Yambe Tam, email to the author, September 27, 2023.

74. Sam Hamill, trans., *Crossing the Yellow River: Three Hundred Poems from the Chinese*, (Rochester, NY: BOA, 2000), 94.

75. Tam, email to the author, September 27, 2023.

76. Ludwig Wittgenstein, *Philosophical Investigations*, trans. G. E. M. Anscombe (London: Blackwell, 1953), pt. II, sec. xi.

77. Wang Jie, "Capturing the Tension, Interaction, and Relationship among Different Civilizations," *Shine*, August 30, 2022, accessed August 28, 2024, https://www.shine.cn/feature/art-culture/2208309872/.

78. "Projects: Vortex, since 2014," Opiemme website, accessed March 5, 2024, https://opiemme.com/en/projects/vortex.

79. Wisława Szymborska, "Under a Certain Little Star," in *Miracle Fair: Selected Poems of Wisława Szymborska*, trans. Joanna Trzeciak (New York: Norton, 2001), 114–115.

80. Sheperd Doeleman et al., "Event-horizon-scale Structure in the Supermassive Black Hole Candidate at the Galactic Centre," *Nature* 455, no. 7209 (September 2008): 78–80. Around 2000, a group of astronomers described a theoretical concept positing that the region close to the event horizon of the black hole at the center of the Milky Way could be observed from Earth using very-long-baseline interferometry. See Heino Falcke, Fulvio Melia, and Eric Agol, "Viewing the Shadow of the Black Hole at the Galactic Center," *The Astrophysical Journal* 528, no. 1 (January 2000): L13–L16, https://doi.org/10.1086/312423.

81. Katherine L. Bouman, "The Inside Story of the First Picture of a Black Hole," *IEEE Spectrum*, January 30, 2020, accessed

August 11, 2023, https://spectrum.ieee.org/the-inside-story-of-the-first-picture-of-a-black-hole.

82. According to Heino Falcke, a member of the EHT team: "For the sample models that we ultimately published, we chose a red color scale, representing heat or molten iron. After all, we were showing radio emission that is invisible to the human eye but comes from the extreme red part of the electromagnetic spectrum and is emitted by extremely hot gas. This was an artistic choice in the end, but it stuck." Heino Falcke, "The Road Toward Imaging a Black Hole: A Personal Perspective," *Natural Sciences* 2, no. 4 (October 2022): 7, https://doi.org/10.1002/ntls.20220031.

83. According to Sheperd Doeleman in Rafi Letzter, "We Could Soon Watch a Black Hole in Action, Gobbling Up Matter in Real Time," *Live Science*, April 16, 2019, accessed December 20, 2023, https://www.livescience.com/65246-first-black-hole-movie.html.

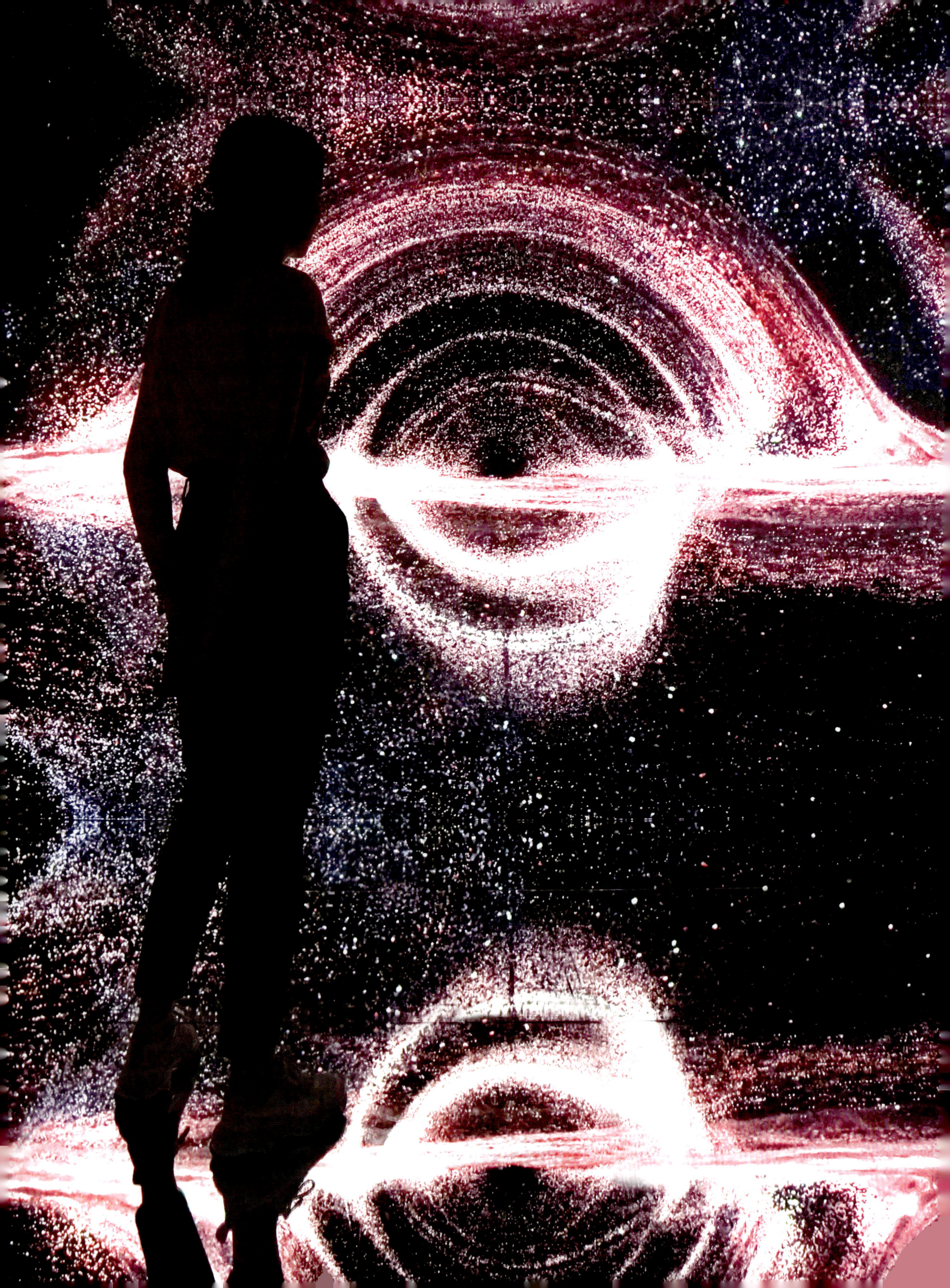

ACKNOWLEDGMENTS

I am grateful to Peter Galison, director of the Black Hole Initiative, a group of scientists and philosophers at Harvard University that studies black holes. He invited me to give a lecture at the BHI's annual conference in 2023, which led me to write this book. This publication is funded in part by the Gordon and Betty Moore Foundation (Grant #8273.01). It was also made possible through the support of a grant from the John Templeton Foundation (Grant #62286). The opinions expressed in this publication are those of the author and do not necessarily reflect the views of these foundations.

I thank the MIT Press for publishing the book and editor Justin Kehoe for his insightful comments on my manuscript. I also thank the Zhejiang Photographic Press for publishing a Chinese edition.

Marquand Books in Seattle produced the book. The elegant appearance of this book is thanks to Design Director Ryan Polich. I thank the other members of the Marquand team: Gina Broze, Kestrel Rundle, Donna Wingate, Jeremy Linden, Melissa Duffes, and Leah Finger. My old friend Ed Marquand, founder of Marquand Books, came out of retirement to give me wise advice at the beginning of the project. The text benefitted from art editor John Pierce, science editor Fabio Pacucci, and copyeditor Tom Fredrickson.

The School of Visual Arts in New York has given me the opportunity to teach interdisciplinary courses in the history of art and science, and it generously funded the images in this book. I thank President David Rhodes, Provost Christopher Cyphers, Chair of Humanities and Sciences Kyoko Miyabe, and Chair of Visual and Critical Studies and Art History Tom Huhn.

4.4 Marshmallow Laser Feast (English art collective), *Distortions in Spacetime*, 2018–2022. Video projection. Sound design and spatialization by James Bulley. Installation view at Nxt Museum, Amsterdam, 2020–2022.

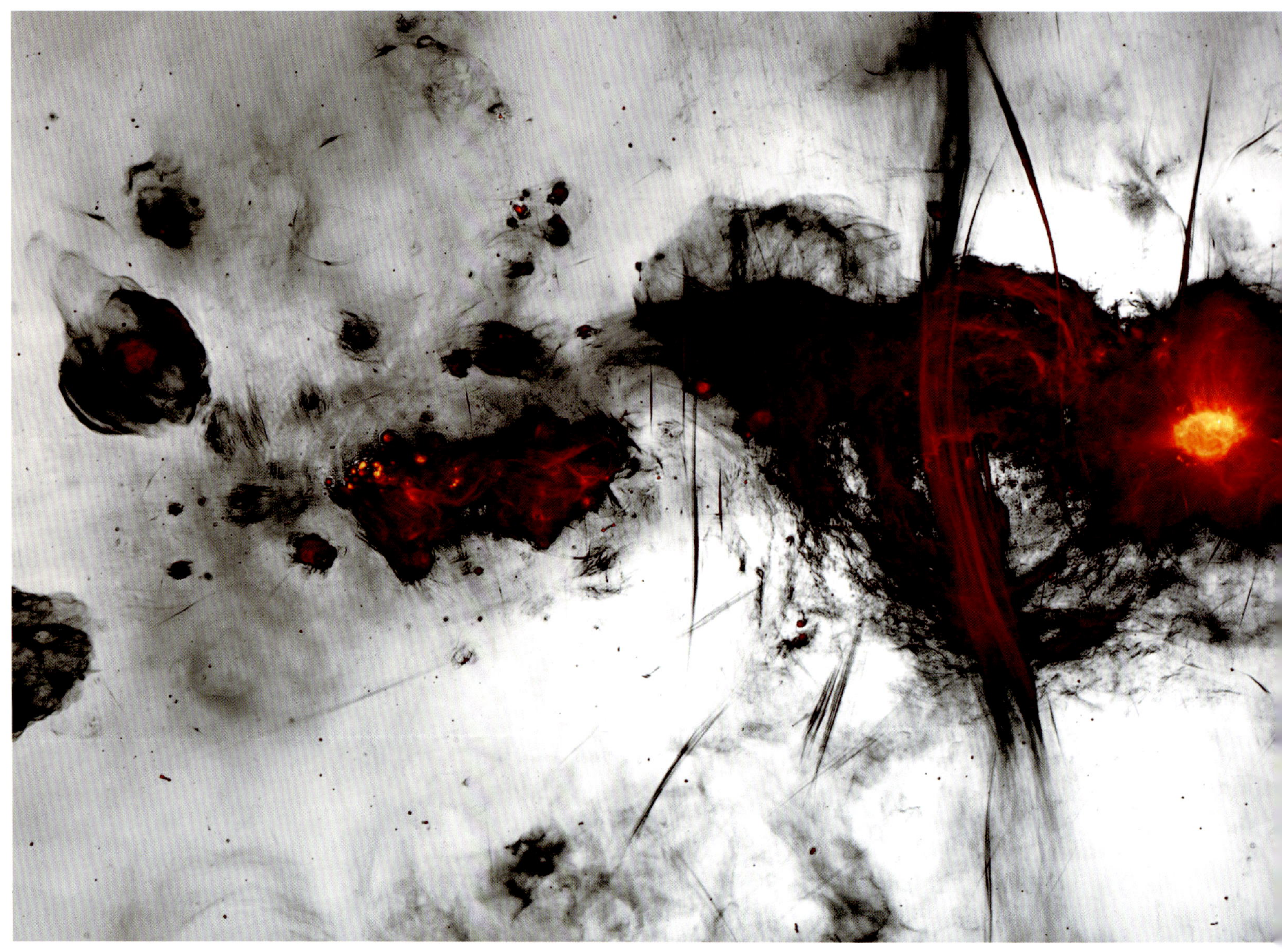

4.5 Sagittarius A* in the center of the Milky Way, recorded in radio waves by the MeerKAT telescope array. I. Heywood/South African Radio Astronomy Observatory.

The center of the Milky Way is displayed horizontally across this image, which was recorded by MeerKAT, a radio telescope consisting of sixty-four antennas in South Africa. The strongest radio signals are shown in red and orange, while fainter signals are in grayscale. The small bright orange dot in the center is the black hole Sagittarius A*.

I am grateful to the staff of the Corinne Goldsmith Dickinson Center for Multiple Sclerosis, Mount Sinai Hospital in New York, for keeping my neurons firing.

Certain individuals were exceptionally helpful in securing images for this book: my husband, Charles Brown, Umbra Studio in New York; Mária Bohumelová and Katarína Bajcurová, Slovenská národná galéria in Bratislava, Slovakia; Mireille van Helm, Block 2 Distribution; Alexis Recto, The Huntington in San Marino, California; Sachiyo Ito and Akemi Tokunaga, Ginza Graphic Gallery in Tokyo; Shunji Noda, Taka Ishii Gallery in Tokyo; Nicole Keller and Jennifer Voiglio, Hauser & Wirth in Zurich; Aline Siqueira, Museu de Arte Moderna in Rio de Janeiro; Susanna Kohler, American Astronomical Society in Washington, DC; Christopher Apostle, Sotheby's in New York; Jennifer Yu-Chen Chen, Cai Guo-Qiang Archives in New York; Pam Abbey, Cosmos Studios in Ithaca, New York; Marie-Charlotte Knuffmann, Sies + Höke Galerie in Düsseldorf; Mariolina Bassetti, Christie's in Milan; Ina Kämmerer, Galerie Max Hetzler in Berlin; Thomas Zoufal, Deutsches Elektronen-Synchrotron in Hamburg; Sara Mast, The Einstein Collective in Bozeman, Montana; Andie Tusha, Studio Oefner; Arisa Saito, SCAI The Bathhouse in Tokyo; Stefan Luders, European Space Agency in Rome; Tom Field, Field Tested Systems in Seattle; Thomas Borre, Nikolaj Kunsthal in Copenhagen; Owen Christoph,

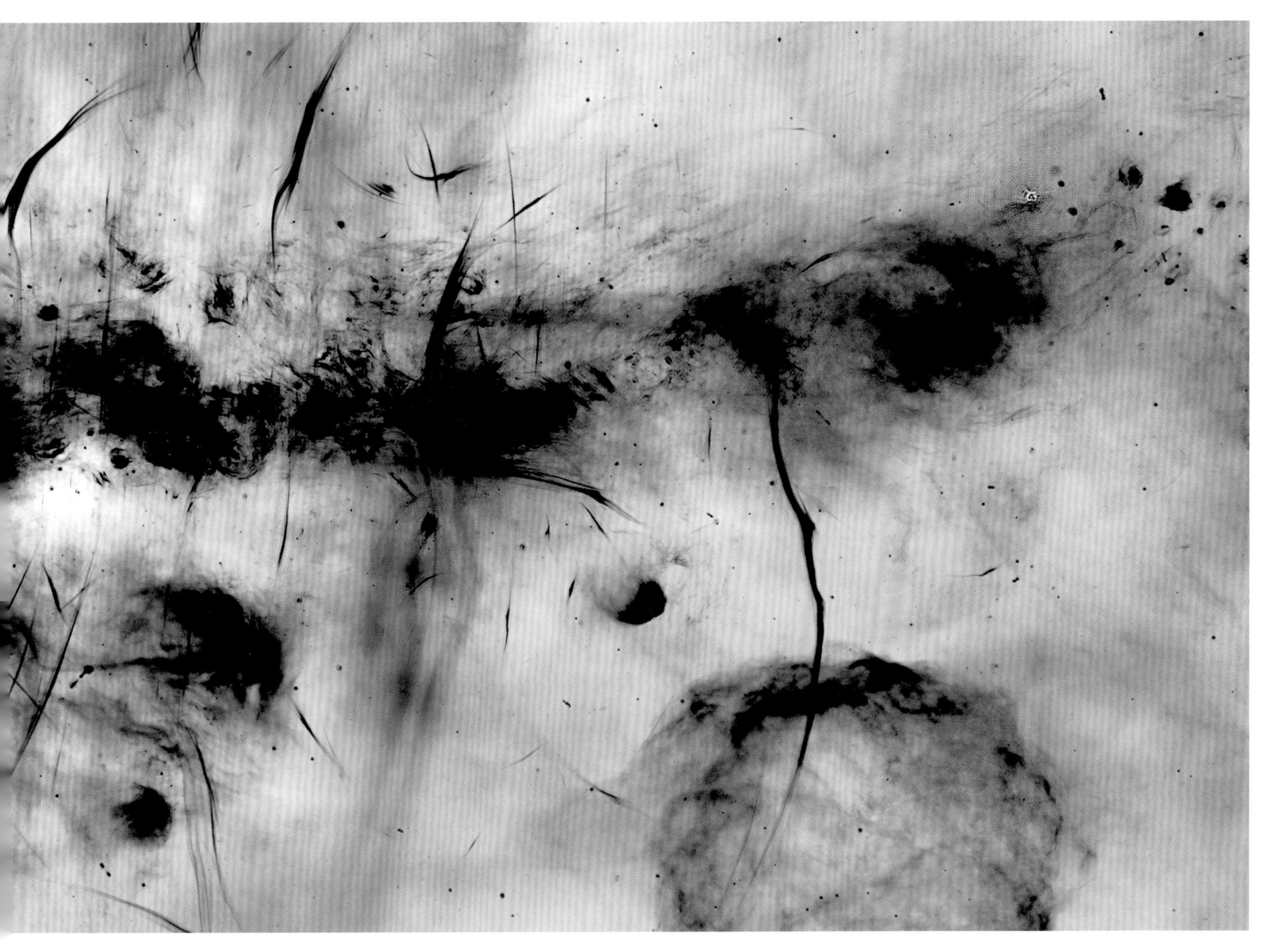

Jessica Silverman Gallery in San Francisco; David Rozelle, San Francisco Museum of Modern Art; Mengyue Li and Valentina Marinai, Esther Schipper Gallery in Berlin; Anastasia Symeonides, the *Sydney Morning Herald* in Australia; Lotte Parmley, Lisson Gallery in London; Clare Chapman, Anish Kapoor Studio in London; Michael Landry, LIGO Hanford; Jean-Pierre Luminet, Laboratoire d'Astrophysique de Marseille in France; Jonathan Hoppe, Philadelphia Museum of Art; Sarah Gamper Marconi, Marshmallow Laser Feast in London; Penny Yeung, Gagosian Gallery in New York; Elizabeth Ferrara, NANOGrav; Helen McGregor, Roger Penrose's personal assistant; Suzanne Inge, *Physics Today*; Vincent Wilcke, Pace Gallery in New York; Rosalie Pfleger, Atelier Chiharu Shiota in Berlin; Sakurako (Saka) Naka, teamLab; Catherine Y. Hsieh, TKG+ in Taipei; Catherine Belloy, Marian Goodman Gallery in New York; Monica Truong, Sikkema Jenkins & Co. in New York; Le Yin, Xu Bing Studio in New York; Alan Baglia, Artists Rights Society in New York; Jason Drill and Michal Patchefsky, Kasmin Gallery in New York; Joyce Faust, Art Resource in New York; Doris Del Castillo, Museum of Contemporary Art in Los Angeles; Agata Rutkowska, White Cube in London; José Francisco Salgado; Dalia Ardon, a descendant of Mordecai Ardon; and Alexander Lavrentiev, a descendant of Aleksandr Rodchenko.

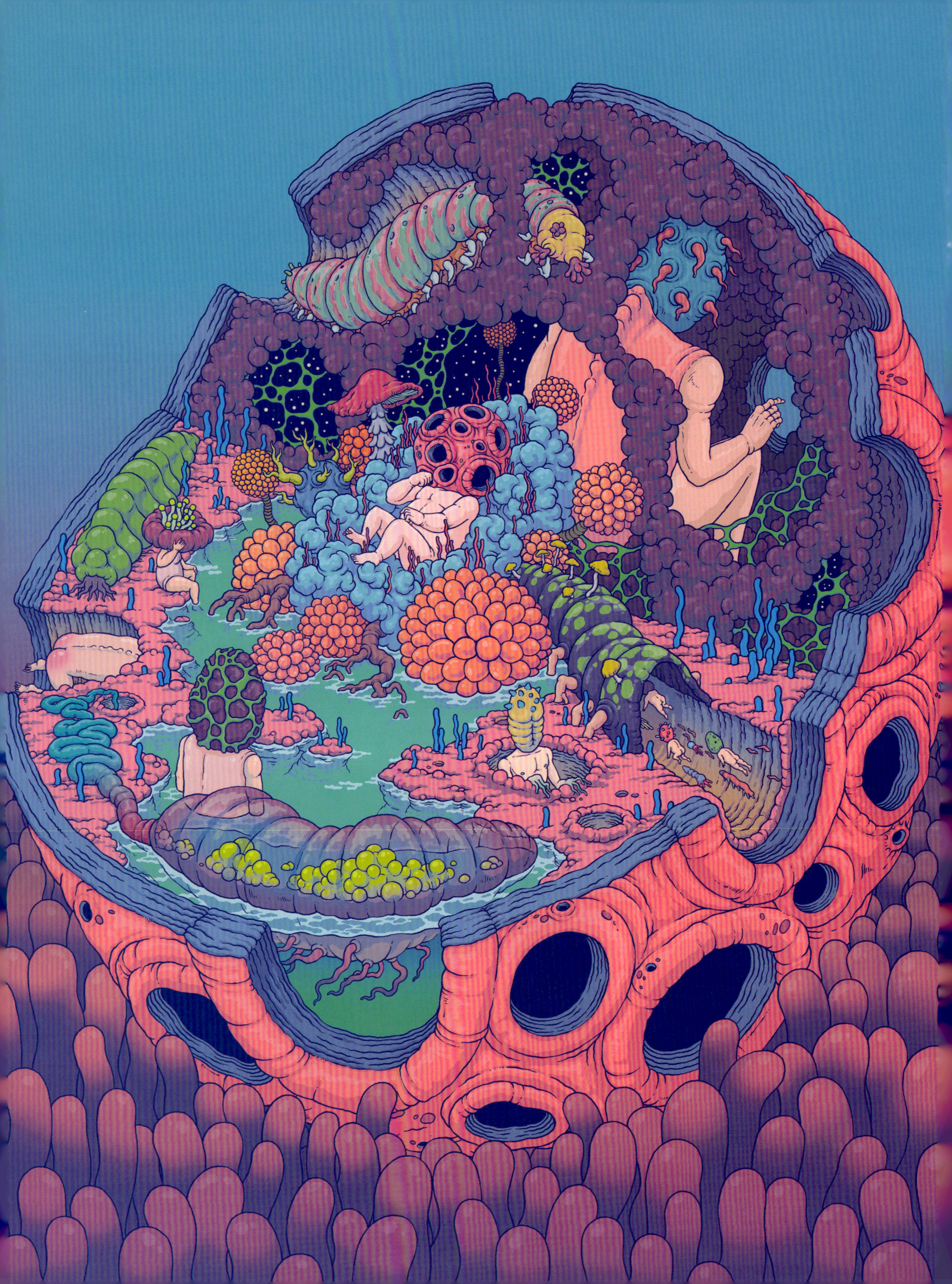

PHOTO AND DIAGRAM CREDITS

Permission to reproduce illustrations is provided by the owners or sources listed in the captions. Additional photo and diagram credits are as follows (numerals in **boldface** refer to figures).

Every reasonable effort has been made to supply complete and correct credits; if there are errors or omissions, please contact The MIT Press so that corrections can be addressed in any subsequent editions.

KEY TO ABBREVIATIONS

ARS: Artists Rights Society, New York
MoMA: © The Museum of Modern Art/Licensed by SCALA/Art Resource, NY
Umbra: Umbra Studio, New York

0.2 twanight.org; **1.1** Umbra; **1.3** Courtesy of the artist/Tanya Bonakdar Gallery, New York and Los Angeles/neugerriemschneider, Berlin/© 2006 Olafur Eliasson/Jens Ziehe; **1.4, 1.5, 1.8** Umbra; **1.9** © ESA/C. Carreau modified by Umbra; **1.10** Umbra; **1.13** Universal History Archive; **1.17** © The M. C. Escher Company, The Netherlands/All rights reserved/www.escher.com; **1.20** © Field Tested Systems LLC; **1.21** Umbra; **1.25** © 2024 Estate of Richard Pousette-Dart/ARS/Jonathan Nesteruk; **1.26** With the permission of the American Institute of Physics; **1.27A** CanStockPhoto; **1.27B** Umbra; **1.27C** Shutterstock; **2.1** British Library, London/Bridgeman Images; **2.2** © Trustees of the British Museum, London; **2.5** © 2024 ARS; **2.6** Heritage Image Partnership/Art Resource, New York; **2.7** © 2024 Estate of Alexander Rodchenko/UPRAVIS, Moscow/ARS; **2.8** © Eikoh Hosoe; **2.9** © The Art Institute of Chicago/ Art Resource, New York; **2.10** © Ardon Estate/The Jewish Museum, New York/Art Resource, New York; **2.11** © 2024 Anna Reinhardt/ARS, MoMA; **2.12** josefrancisco.org; **3.3** © 2024 Artists Rights Society (ARS), New York / SIAE, Rome/ Bridgeman Images; **3.4** The Art Institute of Chicago/Art Resource, New York; **3.5** MoMA; **3.7, 3.8** Joshua White; **3.9** © Anna Maria Maiolino; **3.10** © Anna Maria Maiolino/MoMA; **3.12** © AAS/Reproduced with permission; **3.25** © ESO/Reproduced with permission; **3.30** Gabriella Di Muro; **3.31** Copyright © 2014 by Cosmos Studios, Inc. Reproduced with permission from Cosmos Studios, Inc. All rights reserved. This material cannot be further circulated without written permission of Cosmos Studios, Inc.; **3.32** André Morin; **3.33** Yoshihiro Hagiwara; **3.34** Kunio Oshima; **3.35** Hiro Ihara; **3.39** Image by Mark Garklick/ Science Photo Library; **3.40** Blair Speed; **3.42** Double Negative Visual Effects/© Warner Bros. Entertainment Inc./CC BY-NC-ND 3.0; **3.45, 3.46** Umbra; **3.50, 3.51** © 2010 Marco Poloni; **3.55** Tony Ventouris; **3.56** © 2000 Block 2 Pictures Inc. © 2019 Jet Tone Contents Inc. All rights reserved.; **3.58** © David Huffman/ Katherine Du Tiel; **3.65** © 2024 James Rosenquist, Inc./ARS/Used by permission/All rights reserved; **3.68** © AAS/Reproduced with permission; **3.70** © Diana Al-Hadid; **3.72** © John White; **3.73** Umbra; **3.74** © Danh Võ/Marc Domage; **3.78** def image; **3.82** © Anish Kapoor/All Rights Reserved/DACS/Artimage, London/ARS 2024/Fabrice Seixas; **3.83** Installation view of *One Day at a Time: Manny Farber and Termite Art*, October 14, 2018–March 11, 2019 at MOCA Grand Avenue. Photo by Zak Kelley; **3.88** Michael Andrew; **3.89** Sotheby's New York; **3.90** © Anish Kapoor/All Rights Reserved/DACS/Artimage, London/ARS 2024/Filipe Braga; **3.91** © Anish Kapoor/All Rights Reserved/DACS/Artimage, London/ARS 2024/Nobutada Omote; **3.92** © Anish Kapoor/ All Rights Reserved/DACS/Artimage, London/ ARS 2024/Dave Morgan; **3.97** Nobutada Omote; **3.98** © Chiharu Shiota/© 2024 ARS/VG Bild-Kunst, Bonn; **3.99** © Yin Xiuzhen/Photograph courtesy of the artist; **3.101** David and Matthew Adeboye; **3.102** © 2024 Kara Walker/ MoMA; **3.106** David Sipress/*The New Yorker* Collection/The Cartoon Bank; **3.113** © Johan Samsing; **3.116** *The Sydney Morning Herald*/ Janie Barrett; **3.117** © Andrea Rossetti; **3.118** © YILAB; **3.119** Reece Straw; **3.120** Albert Barbu; **3.129** © Opiemme; **3.132** David Stjernholm; **3.133** Marshmallow Laser Feast; **3.134, 3.135, 3.136** © teamLab; **3.137** NASA Visible Earth modified by Umbra; **3.139** Emily Ryalls; **3.144** © Björn Dahlem/Sies + Höke, Düsseldorf/Galerie Guido W. Baudach, Berlin/ Nic Tenwinggenhorn; **3.146** NASA Visible Earth modified by Umbra; **4.4** Marshmallow Laser Feast.

4.6 Sangho Bang (Korean, born 1991), *Cell* (detail), 2018. Digital print, 23½ × 21½ in. (60 × 55 cm). Courtesy of the artist.

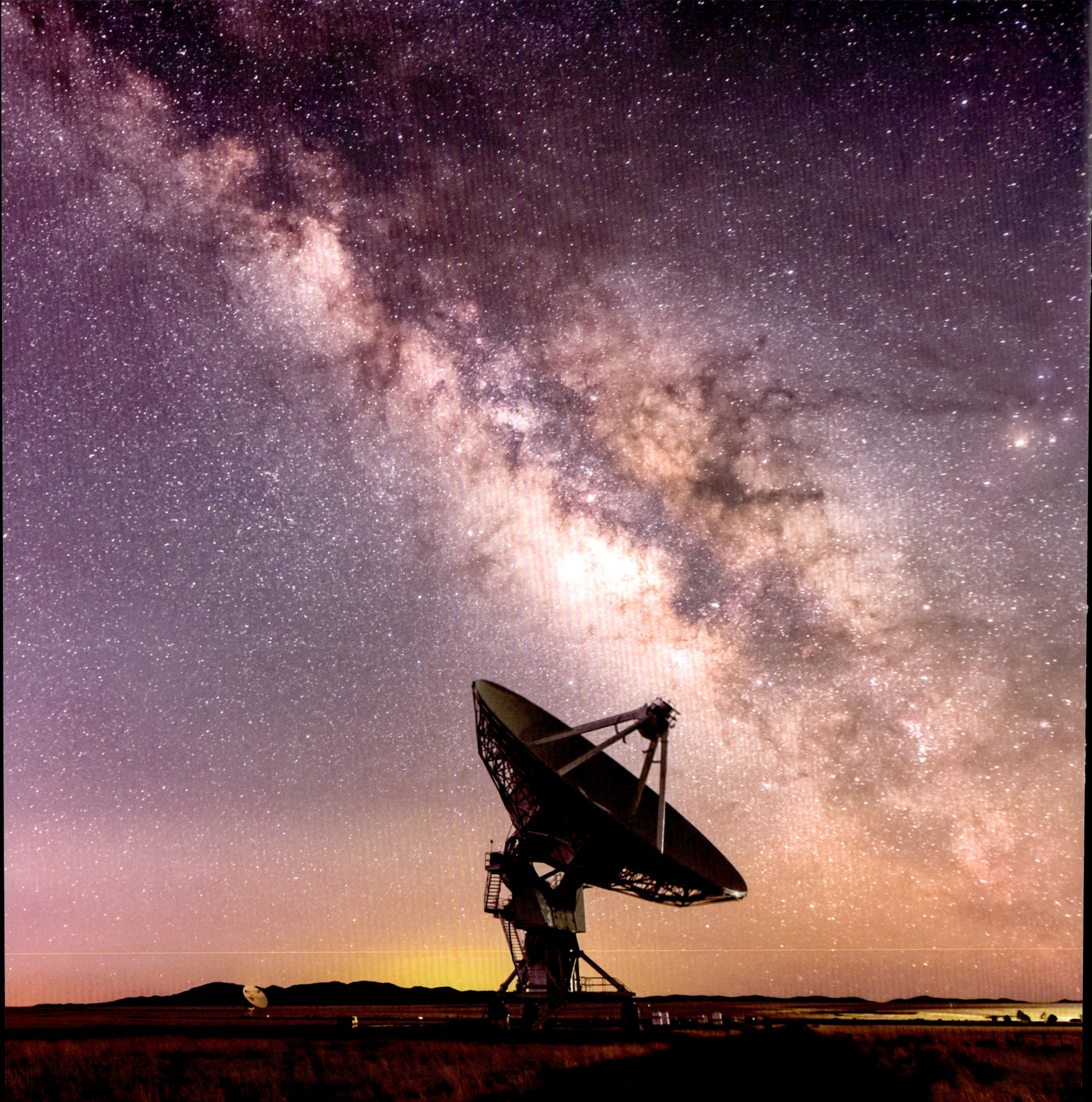

INDEX

Page numbers in *italic* refer to a figure.

4.7 A radio telescope that is part of the Karl Jansky Very Large Array in New Mexico. NRAO/AUI/NSF/Bettymaya Foott.

Library of Congress Control Number:
2024948828
ISBN 978-0-262-04996-2

Published by The MIT Press
Cambridge, MA 02142
mitpress.mit.edu

Produced by Marquand Books, Seattle
www.marquandbooks.com

Copyedited by Tom Fredrickson
Designed by Ryan Polich
Typeset in Gatch, Oculi Text, and Stratos by Brynn Warriner
Proofread by Bruno George
Color management by I/O Color, Seattle
Printed and bound in China by C&C Offset Printing Co., Ltd.

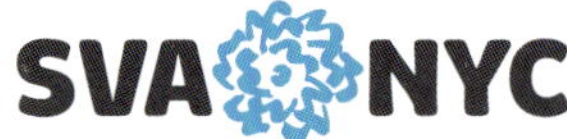

The images in this book are funded by the School of Visual Arts in New York.

Front jacket: Fabian Oefner (Swiss, born 1984), *Black Hole, no. 2*, 2014. Inkjet print, 31½ × 47¼ in. (80 × 120 cm). Courtesy of the artist.

Endpaper: Renato Crepaldi (Brazilian, born 1973), *Yellow Nonpareil*, recolored by Jeremy Linden, 2024. Courtesy of the artist. Used with permission.

The endpapers of this book are photographs of a sheet of paper marbling by Renato Crepaldi. He made it by flowing pigment across a wet surface to produce a pattern, and then laid a sheet of absorbent paper on the surface to capture the floating design. The pattern *nonpareil* (French for "unrivalled") is traditionally used for endpapers, and Crepaldi added black drops of paint (what marblers call "tigers' eyes") suggesting black holes.

Title page: Cao Yuxi (James Cao) (Chinese, born 1990), *Oriens: Immersive Black Hole*, 2017. Sound and video installation. Installation view at Today Art Museum, Beijing. Courtesy of the artist.

Table of contents: Melissa Walter (American, born 1976), *Gravitational Lensing* (detail), 2018. Cut paper, magnets, steel, and paint, approx. height approx. 12 ft. (3.6 m). Courtesy of the artist.

Back jacket: Yambe Tam (American, born 1989), *Wormhole Bell*, 2018. Cast bronze, 11¾ × 11¾ × 14 in. (30 × 30 × 36 cm). Private collection.

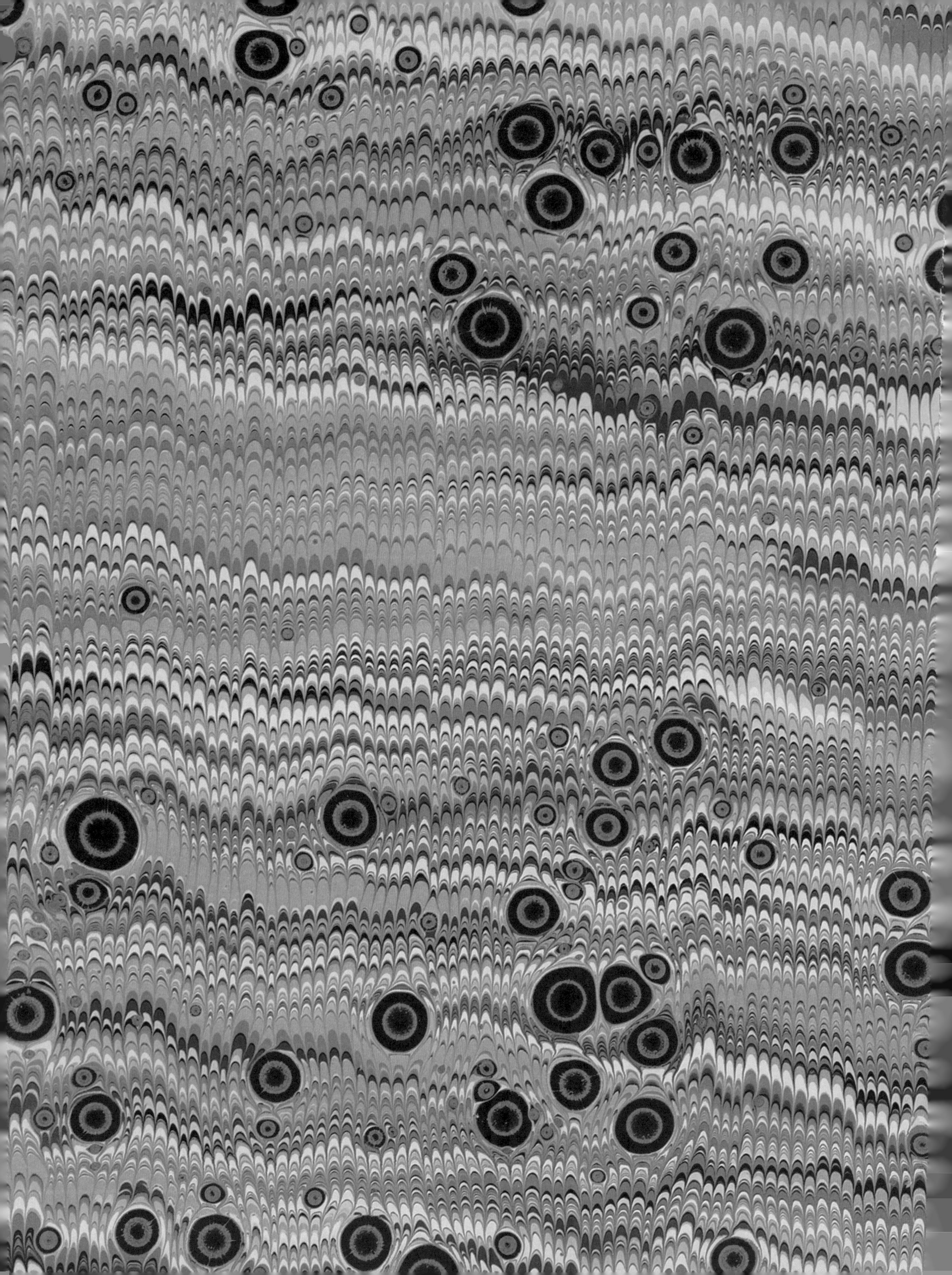